STERLING
Test Prep

College Chemistry

Practice Questions

8th edition

Customer Satisfaction Guarantee

Your feedback is important because we strive to provide the highest quality prep materials. Email us comments or suggestions.

info@sterling–prep.com

We reply to emails – check your spam folder

8 7 6 5 4 3 2 1

ISBN-13: 978-1-9547253-4-8

Sterling Test Prep materials are available at quantity discounts.

Contact info@sterling–prep.com

Sterling Test Prep
6 Liberty Square #11
Boston, MA 02109

© 2022 Sterling Test Prep

Published by Sterling Test Prep

 Printed in the U.S.A.

Thousands of students use our study aids to achieve higher grades!

To achieve a high grade in college chemistry, you need to do well on your tests and final exam. This book helps you develop and apply knowledge to quickly choose the correct answers to questions typically tested in college chemistry courses. Solving targeted practice questions builds your understanding of fundamental concepts and is a more effective strategy than merely memorizing terms.

This book has 1,140 practice questions covering inorganic chemistry topics. Chemistry instructors with years of teaching experience prepared this practice material to build your knowledge and skills crucial for success in a college chemistry course. Our editorial team reviewed and systematized the content for targeted preparation.

The detailed explanations describe why an answer is correct and – more important for your learning – why another attractive choice is wrong. They provide step-by-step solutions for quantitative questions and teach the scientific foundations and details of essential chemistry topics needed to answer conceptual questions. Read the explanations carefully to understand how they apply to the question and learn important inorganic chemistry principles and the relationships between them. With the practice material contained in this book, you will significantly improve your understanding, test scores, and your grade.

We wish you great success in your academic achievements and look forward to being an important part of your preparation!

220315gdx

College Chemistry Review provides comprehensive coverage of inorganic chemistry topics taught in college chemistry. The content covers foundational principles and theories necessary to understand the material and answer test questions.

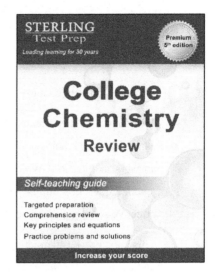

Visit our Amazon store

College study aids by Sterling Test Prep

Cell and Molecular Biology Review

Organismal Biology Review

Cell and Molecular Biology Practice Questions

Organismal Biology Practice Questions

Physics Review

Physics Practice Questions

Organic Chemistry Practice Questions

United States History 101

American Government and Politics 101

Environmental Science 101

Table of Contents

College Level Examination Program (CLEP)

Biology Review

Biology Practice Questions

Chemistry Review

Chemistry Practice Questions

Introductory Business Law Review

College Algebra Practice Questions

College Mathematics Practice Questions

History of the United States I Review

History of the United States II Review

Western Civilization I Review

Western Civilization II Review

Social Sciences and History Review

American Government Review

Introductory Psychology Review

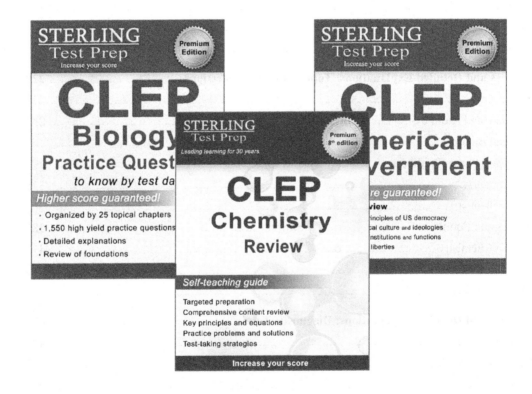

Visit our Amazon store

Topical Practice Questions

Electronic and Atomic Structure; Periodic Table

==

Practice Set 1: Questions 1–20

==

1. The property defined as the energy required to remove one electron from an atom in the gaseous state is:

A. electronegativity

B. ionization energy

C. electron affinity

D. hyperconjugation

E. none of the above

2. Which of the following elements most easily accepts an extra electron?

A. He   B. Ca   C. Cl   D. Fr   E. Na

3. Which of the following pairs has one metalloid element and one nonmetal element?

A. ^{82}Pb and ^{83}Bi

B. ^{19}K and ^{9}F

C. ^{51}Sb and ^{20}Ca

D. ^{33}As and ^{14}Si

E. ^{32}Ge and ^{9}F

4. Periods on the periodic table represent elements:

A. in the same group

B. with consecutive atomic numbers

C. known as isotopes

D. with similar chemical properties

E. known as ions

5. How many electrons can occupy the $n = 2$ shell?

A. 2   B. 6   C. 8   D. 18   E. 32

6. Ignoring hydrogen and helium, which area(s) of the periodic table contain(s) both metals and nonmetals?

I. *s* area   II. *p* area   III. *d* area

A. I only

B. II only

C. III only

D. I and II only

E. I, II and III

7. What is the number of known nonmetals relative to the number of metals?

A. About two times greater

B. About fifty percent

C. About five times less

D. About twenty-five percent greater

E. About three times greater

8. Based on experimental evidence, Dalton postulated that:

    A. atoms of different elements have the same mass

    B. not all atoms of a given element are identical

    C. atoms can be created and destroyed in chemical reactions

    D. each element consists of indivisible minute particles called atoms

    E. none of the above

9. Which statement is true regarding the average mass of a naturally occurring isotope of iron with an atomic mass equal to 55.91 amu?

    A. 55.91 / 1.0078 times greater than a ^{1}H atom   **C.** 55.91 times greater than a ^{12}C atom

    B. 55.91 / 12.000 times greater than a ^{12}C atom   **D.** 55.91 times greater than a ^{1}H atom

    E. none of the above

10. Which is the principal quantum number?

    A. n **B.** m **C.** l **D.** s **E.** $+\frac{1}{2}$

11. How many electrons can occupy the $4s$ subshell?

    A. 1   **B.** 2   **C.** 6   **D.** 8   **E.** 10

12. An excited hydrogen atom emits a light spectrum of specific, characteristic wavelengths. The light spectrum is a result of:

    A. energy released as H atoms form H_2 molecules

    B. the light wavelengths not absorbed by valence electrons when white light passes through the sample

    C. particles being emitted as the hydrogen nuclei decay

    D. excited electrons being promoted to higher energy levels

    E. excited electrons dropping to lower energy levels

13. Which has the largest radius?

    A. Br^- **B.** K^+ **C.** Ar   **D.** Ca^{2+} **E.** Cl^-

14. In its ground state, how many unpaired electrons does a sulfur atom have?

    A. 0   **B.** 1   **C.** 2   **D.** 3   **E.** 4

15. Which element is in the same group as the element whose electronic configuration is $1s^22s^22p^63s^23p^64s^1$?

    A. Ar   **B.** Se   **C.** Mg   **D.** P   **E.** Li

16. When an atom is most stable, how many electrons does it contain in its valence shell?

    A. 4   **B.** 6   **C.** 8   **D.** 10   **E.** 12

17. Which characteristic(s) is/are responsible for the changes seen in the first ionization energy when moving down a column?

    I. Increased shielding of electrons
    II. Larger atomic radii
    III. Increasing nuclear attraction for electrons

    A. I only   **C.** III only
    B. II only   **D.** I and II only
                                E. I, II and III

18. What is the maximum number of electrons that can occupy the $4f$ subshell?

    A. 6   **B.** 8   **C.** 14   **D.** 16   **E.** 18

19. The property that describes the energy released by a gas phase atom from adding an electron is:

    A. ionization energy   **C.** electron affinity
    B. electronegativity   **D.** hyperconjugation
                                E. none of the above

20. The attraction of the nucleus on the outermost electron in an atom tends to:

    A. decrease from right to left and bottom to top on the periodic table
    B. decrease from left to right and bottom to top on the periodic table
    C. decrease from left to right and top to bottom of the periodic table
    D. increase from right to left and top to bottom on the periodic table
    E. decrease from right to left and top to bottom on the periodic table

===

Practice Set 2: Questions 21–40

===

21. How many protons and neutrons are in ^{35}Cl, respectively?

A. 35; 18   **C.** 17; 18
B. 18; 17   **D.** 17; 17
                                 E. 35; 17

22. What must be the same if two atoms represent the same element?

A. number of neutrons   **C.** number of electron shells
B. atomic mass   **D.** atomic number
                                 E. number of valence electrons

23. Referring to the periodic table, which of the following is NOT a solid metal under normal conditions?

A. Ce   **B.** Os   **C.** Ba   **D.** Cr   **E.** Hg

24. Which of the following describe electron affinity?

    I. Ability of an atom to attract electrons when it bonds with another atom
    II. Energy needed to remove an electron from a neutral atom of the element in the gas phase
    III. Energy liberated when an electron is added to a gaseous neutral atom converting it to an anion

A. I only   **C.** III only
B. II only   **D.** I and II only
                                 E. I, II and III

25. Metalloids are elements:

A. larger than nonmetals
B. found in asteroids
C. smaller than metals
D. that have some properties like metals and some like nonmetals
E. that have properties different from either the metals or the nonmetals

26. Which of the following represent(s) halogens?

    I. Br   II. H   III. I

A. I only   **C.** III only
B. II only   **D.** I and III
                                 E. I and II

---

27. According to John Dalton, atoms of an element:

    A. are divisible   **C.** are identical

    B. have the same shape   **D.** have different masses

                                                        E. none of the above

28. Which of the following represents a pair of isotopes?

    A. $^{32}_{16}S$, $^{32}_{16}S^{2-}$ **C.** $^{14}_{6}C$, $^{14}_{7}N$

    B. O_2, O_3 **D.** $^{1}_{1}H$, $^{2}_{1}H$ **E.** None of the above

29. Early investigators proposed that the ray of the cathode tube was a negatively charged particle because the ray was:

    A. not seen from the positively charged anode

    B. diverted by a magnetic field

    C. observed in the presence or absence of a gas

    D. able to change colors depending on which gas was within the tube

    E. attracted to positively charged electric plates

30. What is the value of quantum numbers n and l in the highest occupied orbital for the element carbon with an atomic number of 6?

    A. $n = 1, l = 1$ **C.** $n = 1, l = 2$

    B. $n = 2, l = 1$ **D.** $n = 2, l = 2$ **E.** $n = 3, l = 3$

31. Which type of subshell is filled by the distinguishing electron in an alkaline earth metal?

    A. s **B.** p **C.** f **D.** d **E.** both s and p

32. If an element has an electron configuration ending in $3p^4$, which statements about the element's electron configuration is NOT correct?

    A. There are six electrons in the 3rd shell

    B. Five different subshells contain electrons

    C. There are eight electrons in the 2nd shell

    D. The 3rd shell needs two more electrons to be filled

    E. All are correct statements

33. Halogens form anions by:

    A. gaining two electrons   **C.** gaining one neutron

    B. gaining one electron   **D.** losing one electron

                                                      E. losing two electrons

34. What is the term for a broad, uninterrupted band of radiant energy?

    A. Ultraviolet spectrum   **C.** Continuous spectrum

    B. Visible spectrum   **D.** Radiant energy spectrum

                                     E. None of the above

35. Which of the following is the correct order of increasing atomic radius?

    A. Te < Sb < In < Sr < Rb   **C.** Te < Sb < In < Rb < Sr

    B. In < Sb < Te < Sr < Rb   **D.** Rb < Sr < In < Sb < Te

                                     E. In < Sb < Te < Rb < Sr

36. Which of the following electron configurations represents an excited state of an atom?

    A. $1s^2 2s^2 2p^6 3s^2 3p^3$ **C.** $1s^2 2s^2 2p^6 3d^1$

    B. $1s^2 2s^2 2p^6 3s^2 3p^6 4s^1$ **D.** $1s^2 2s^2 2p^6 3s^2 3p^6 4s^2 3d^1$

                                     E. $1s^2 2s^2 2p^6 3s^2 3p^6 4s^2 3d^{10} 4p^1$

37. Which of the following elements has the greatest ionization energy?

    A. rubidium   **C.** potassium

    B. neon   **D.** calcium

                                     E. magnesium

38. Which of the following elements is a nonmetal?

    A. Sodium   **C.** Aluminum

    B. Chlorine   **D.** Magnesium

                                     E. Palladium

39. An atom that contains 47 protons, 47 electrons, and 60 neutrons is an isotope of:

    A. Nd   **C.** Ag

    B. Bh   **D.** Al

                                   E. cannot be determined

40. The shell level of an electron is defined by which quantum number?

    A. electron spin quantum number

    B. magnetic quantum number

    C. azimuthal quantum number

    D. principal quantum number

    E. principal quantum number and electron spin quantum number

==

Practice Set 3: Questions 41–60

==

41. How many neutrons are in a Beryllium atom with an atomic number of 4 and an atomic mass of 9?

   A. 4   **B.** 5   **C.** 9   **D.** 13   **E.** 18

42. Which characteristics describe the mass, charge, and location of an electron, respectively?

   A. approximate mass 5×10^{-4} amu; charge –1; outside nucleus
   B. approximate mass 5×10^{-4} amu; charge 0; inside nucleus
   C. approximate mass 1 amu; charge –1; inside nucleus
   D. approximate mass 1 amu; charge +1; outside nucleus
   E. approximate mass 1 amu; charge 0; outside nucleus

43. What is the name of the compound $CaCl_2$?

   A. dichloromethane   **C.** carbon chloride
   B. dichlorocalcium   **D.** calcium chloride
                                       E. dicalcium chloride

44. What is the name for elements in the same column of the periodic table with similar chemical properties?

   A. congeners   **C.** diastereomers
   B. stereoisomers   **D.** epimers
                                       E. anomers

45. Which of the following sets of elements consist of members of the same group on the periodic table?

   A. ^{14}Si, ^{15}P and ^{16}S **C.** ^{9}F, ^{10}Ne and ^{11}Na
   B. ^{20}Ca, ^{26}Fe and ^{34}Se **D.** ^{31}Ga, ^{49}In and ^{81}Tl
                                                         E. ^{11}Na, ^{20}Ca and ^{39}Y

46. Which element would most likely be a metal with a low melting point?

   A. K   **B.** B   **C.** N   **D.** C   **E.** Cl

47. Mg is an example of a(n):

   A. transition metal   **C.** alkali metal
   B. noble gas   **D.** halogen
                                       E. alkaline earth metal

48. Metalloids:

      I. have some metallic and some nonmetallic properties

      II. may have low electrical conductivities

      III. contain elements in group IIIB

A. I and III only

B. II only

C. II and III only

D. I and II only

E. I, II and III

49. Which element is a halogen?

A. Os   **B.** I   **C.** O   **D.** Te   **E.** Se

50. Silicon exists as three isotopes: ^{28}Si, ^{29}Si, and ^{30}Si with atomic masses of 27.98 amu, 28.98 amu, and 29.97 amu, respectively. Which isotope is the most abundant in nature?

A. ^{28}Si

B. ^{29}Si

C. ^{30}Si

D. ^{28}Si and ^{30}Si are equally abundant

E. All are equally abundant

51. Which statement supports why early investigators proposed that the ray of the cathode ray tube was due to the cathode?

    A. The ray was diverted by a magnetic field

    B. The ray was not seen from the positively charged anode

    C. The ray was attracted to the electric plates that were positively charged

    D. The ray changed color depending on the gas used within the tube

    E. The ray was observed in the presence or absence of a gas

52. Which of the following is implied by the spin quantum number?

      I. The two spinning electrons generate magnetic fields

      II. The values are $+\frac{1}{2}$ or $-\frac{1}{2}$

      III. Orbital electrons have opposite spins

A. II only

B. I and II only

C. I and III only

D. II and III only

E. I, II and III

53. How many quantum numbers are needed to describe a single electron in an atom?

A. 1   **B.** 2   **C.** 3   **D.** 4   **E.** 5

54. Which of the following has the correct order of increasing energy in a subshell?

   A. $3s, 3p, 4s, 3d, 4p, 5s, 4d$

   B. $3s, 3p, 4s, 3d, 4p, 4d, 5s$

   C. $3s, 3p, 3d, 4s, 4p, 4d, 5s$

   D. $3s, 3p, 3d, 4s, 4p, 5s, 4d$

   E. $3s, 3p, 4s, 3d, 4d, 4p, 5s$

55. Rank the elements below in the order of decreasing atomic radius.

   A. $Al > P > Cl > Na > Mg$

   B. $Cl > Al > P > Na > Mg$

   C. $Mg > Na > P > Al > Cl$

   D. $Na > Mg > Al > P > Cl$

   E. $P > Al > Cl > Mg > Na$

56. Which of the following is the electron configuration of a boron atom?

   A. $1s^22s^12p^2$

   B. $1s^22p^3$

   C. $1s^22s^22p^2$

   D. $1s^22s^22p^1$

   E. $1s^22s^12p^1$

57. Which element has the electron configuration $1s^22s^22p^63s^23p^64s^23d^{10}4p^65s^24d^{10}5p^2$?

   A. Sn   **B.** As   **C.** Pb   **D.** Sb   **E.** In

58. Which of the following elements has the greatest ionization energy?

   A. Ar   **B.** Sr   **C.** Br   **D.** In   **E.** Sn

59. Which element has the lowest electronegativity?

   A. Mg   **B.** Al   **C.** Cl   **D.** Br   **E.** I

60. Refer to the periodic table and predict which is a solid nonmetal under normal conditions.

   A. Cl   **B.** F   **C.** Se   **D.** As   **E.** Ar

==

Practice Set 4: Questions 61–80

==

61. Lines were observed in the spectrum of uranium ore (i.e., naturally occurring solid material) identical to those of helium in the spectrum of the Sun. Which of the following produced the lines in the helium spectrum?

  A. Excited protons jumping to a higher energy level
  B. Excited protons dropping to a lower energy level
  C. Excited electrons jumping to a higher energy level
  D. Excited electrons dropping to a lower energy level
  E. None of the above

62. Which set of quantum numbers is possible?

  A. $n = 1$; $l = 2$; $m_l = 3$; $m_s = -\frac{1}{2}$
  B. $n = 4$; $l = 2$; $m_l = 2$; $m_s = -\frac{1}{2}$

  C. $n = 2$; $l = 1$; $m_l = 2$; $m_s = -\frac{1}{2}$
  D. $n = 3$; $l = 3$; $m_l = 2$; $m_s = -\frac{1}{2}$
  E. $n = 2$; $l = 3$; $m_l = 2$; $m_s = -\frac{1}{2}$

63. The number of neutrons in an atom is equal to:

  A. the mass number
  B. the atomic number

  C. mass number minus the atomic number
  D. atomic number minus the mass number
  E. mass number plus the atomic number

64. Which of the following represent(s) a compound rather than an element?

      I. O_3 II. CCl_4 III. S_8

  A. I and III only
  B. II only

  C. I and II only
  D. III only
  E. I only

65. Which of the following elements is NOT correctly classified?

  A. Mo – transition element
  B. Sr – alkaline earth metal

  C. K – representative element
  D. Ar – noble gas
  E. Po – halogen

66. Which is the name of the elements that have properties of both metals and nonmetals?

  A. alkaline earth metals
  B. metalloids

  C. nonmetals
  D. metals
  E. halogens

67. Which of the following is the correct sequence of atomic radii from smallest to largest?

    A. $Al < S < Al^{3+} < S^{2-}$ **C.** $Al^{3+} < Al < S^{2-} < S$

    B. $S < S^{2-} < Al < Al^{3+}$ **D.** $Al^{3+} < S < S^{2-} < Al$ **E.** $Al < Al^{3+} < S^{2-} < S$

68. Which of the following elements contains 6 valence electrons?

    A. S   **B.** Cl   **C.** Si   **D.** P   **E.** Ca^{2+}

69. From the periodic table, which of the following elements is a semimetal?

    A. Ar   **B.** As   **C.** Al   **D.** Ac   **E.** Am

70. Which statement does NOT describe the noble gases?

    A. The more massive noble gases react with other elements

    B. They belong to group VIIIA (or 18)

    C. They contain at least one metalloid

    D. He, Ne, Ar, Kr, Xe and Rn are included in the group

    E. They were once known as the inert gases

71. Which is a good experimental method to distinguish between ordinary hydrogen and deuterium, the rare isotope of hydrogen?

        I. Measure the density of the gas at STP

        II. Measure the rate at which the gas effuses

        III. Infrared spectroscopy

    A. I only   **C.** I and II only

    B. II only   **D.** I, II and III

                        E. II and III only

72. Which statement is true regarding the relative abundances of the ^{6}lithium or ^{7}lithium isotopes?

    A. The relative proportions change as neutrons move between the nuclei

    B. The isotopes are in roughly equal proportions

    C. The relative ratio depends on the temperature of the element

    D. ^{6}Lithium is much more abundant

    E. ^{7}Lithium is much more abundant

73. Which represents the charge on 1 mole of electrons?

    A. 96,485 C   **C.** 6.02×10^{23} grams

    B. 6.02×10^{23} C   **D.** 1 C

                      E. 1 e

74. Which of the following statement(s) is/are true?

       I. The *f* subshell contains 7 orbitals

       II. The *d* subshell contains 5 orbitals

       III. The third energy shell ($n = 3$) has no *f* orbitals

A. I only   **C.** I and II only

B. II only   **D.** II and III only

                               E. I, II and III

75. Which element has the greatest ionization energy?

A. Fr   **B.** Cl   **C.** Ga   **D.** I   **E.** Cs

76. Electrons fill up subshells in order of:

       I. decreasing distance from the nucleus

       II. increasing distance from the nucleus

       III. increasing energy

A. I only   **C.** I and II only

B. II only   **D.** II and III only

                               E. I, II and III

77. Which of the following produces the "atomic fingerprint" of an element?

A. Excited protons dropping to a lower energy level

B. Excited protons jumping to a higher energy level

C. Excited electrons dropping to a lower energy level

D. Excited electrons jumping to a higher energy level

E. None of the above

78. Which of the following is the electron configuration for manganese (Mn)?

A. $1s^2 2s^2 2p^6 3s^2 3p^6$ **C.** $1s^2 2s^2 2p^6 3s^2 3p^6 4s^2 3d^6$

B. $1s^2 2s^2 2p^6 3s^2 3p^6 4s^2 3d^{10} 4p^1$ **D.** $1s^2 2s^2 2p^6 3s^2 3p^6 4s^2 3d^8$

                                   E. $1s^2 2s^2 2p^6 3s^2 3p^6 4s^2 3d^5$

79. Which element listed below has the greatest electronegativity?

A. I   **B.** Fr   **C.** H   **D.** He   **E.** F

80. Which of the following is NOT an alkali metal?

A. Fr   **B.** Cs   **C.** Ca   **D.** Na   **E.** Rb

==

Practice Set 5: Questions 81–100

==

81. What are the atomic number and the mass number of ^{79}Br, respectively?

    A. 35; 44   **C.** 35; 79

    B. 44; 35   **D.** 79; 35

                               E. 35; 114

82. ^{65}Cu and ^{65}Zn have the same:

    A. mass number   **C.** number of ions

    B. number of neutrons   **D.** number of electrons

                               E. number of protons

83. Which of the following statements best describes an element?

    A. has consistent physical properties   **C.** consists of only one type of atom

    B. consists of more than one type of atom   **D.** material that is pure

                               E. material that has consistent chemical properties

84. Which element(s) is/are alkali metal(s)?

    I. Na   II. Sr   III. Cs

    A. I and II only   **C.** I only

    B. I and III only   **D.** II only

                                E. I, II and III

85. Which of the following is/are a general characteristic of a nonmetallic element?

    I. reacts with metals   II. pliable   III. shiny luster

    A. I only   **C.** III only

    B. II only   **D.** I and II only

                                E. I, II and III

86. Which group of the periodic table has 3 nonmetals?

    A. Group IIIA   **C.** Group VA

    B. Group IVA   **D.** Group IA

                               E. Group VIA

87. The transition metals occur in which period(s) on the periodic table?

      I. 2   II. 3   III. 4

A. I only
B. II only
C. III only
D. I and III only
E. I, II and III

88. Which of Dalton's original proposals is/are still valid?

      I. Compounds contain atoms in small whole-number ratios

      II. Atoms of different elements combine to form compounds

      III. An element is composed of tiny particles called atoms

A. I only
B. II only
C. III only
D. I and III only
E. I, II and III

89. Given that parent and daughter nuclei are isotopes of the same element, the ratio of α to β decay produced by the parent must be:

A. 1 to 1
B. 1 to 2
C. 2 to 1
D. 2 to 3
E. 3 to 2

90. ^{63}Cu isotope makes up 69% of the naturally occurring Cu. If only one other isotope is present for natural copper, what is it?

A. ^{59}Cu
B. ^{65}Cu
C. ^{61}Cu
D. ^{62}Cu
E. ^{60}Cu

91. Which does NOT contain cathode ray particles?

A. H_2O
B. K
C. H
D. He
E. All contain cathode ray particles

92. Which principle or rule states that only two electrons can occupy an orbital?

A. Pauli exclusion principle
B. Hund's rule
C. Heisenberg uncertainty principle
D. Newton's principle
E. None of the above

93. What is the maximum number of electrons to fill the atom's second electron shell?

    A. 18   **B.** 2   **C.** 4   **D.** 12   **E.** 8

94. How many electrons can occupy the $4d$ subshell?

    A. 2   **B.** 6   **C.** 8   **D.** 10   **E.** 12

95. The f subshell contains:

    A. 1 orbital   **C.** 5 orbitals

    B. 3 orbitals   **D.** 7 orbitals

                                       E. 9 orbitals

96. When an excited electron returns to the ground state, it releases:

    A. photons   **C.** beta particles

    B. protons   **D.** alpha particles

                                         E. gamma rays

97. Which statement is true of the energy levels for an electron in a hydrogen atom?

    A. The energy levels are identical to the levels in the He^+ ion

    B. The energy of each level can be computed from a known formula

    C. The distance between energy levels for $n = 1$ and $n = 2$ is the same as the distance between the $n = 3$ and $n = 4$ energy levels

    D. Since there is only one electron, the electron must be located in the lowest energy level

    E. The distance between the $n = 3$ and $n = 4$ energy levels is the same as the distance between the $n = 4$ and $n = 5$ energy levels

98. An ion is represented by which of the following electron configurations?

        I. $1s^2 2s^2 2p^6$ II. $1s^2 2s^2 2p^6 3s^2$ III. $1s^2 2s^2 2p^2 3s^2 3p^6$

    A. I only   **C.** I and II only

    B. II only   **D.** II and III only

                                         E. I, II and III

99. Consider an atom with the electron configuration $1s^2 2s^2 2p^6 3s^2 3p^6$. Which of the following is an accurate statement concerning this atom?

    A. This atom would probably be chemically inert

    B. This atom has a non-zero angular momentum

    C. This atom is in an excited state

    D. The atomic number of this atom is $Z = 11$

    E. This atom is most likely to give rise to an ion with charge +2

100. A halogen is expected to have [ ] ionization energy and [ ] electron affinity?

    A. low; small   **C.** high; large

    B. low; large   **D.** high; small

                                    E. high; neutral

101. Why does a chlorine atom form an anion more easily than a cation?

    A. Chlorine has a high electronegativity value

    B. Chlorine has a large positive electron affinity

    C. Chlorine donates one electron to complete its outer shell

    D. Chlorine gains one electron to complete its outer shell

    E. Chlorine has a low electronegativity value

102. What is the approximate mass number of an element if one mole weighs 12 grams?

    A. 6.02×10^{23} **C.** 1

    B. 12   **D.** $12 \times 6.02 \times 10^{23}$

                                    E. 24

103. How many neutrons are in the most common isotope of hydrogen?

    A. 0   **B.** 1   **C.** 2   **D.** 3   **E.** 4

104. Which of the following molecules does NOT exist?

    A. OF_5 **C.** ICl

    B. $KLiCO_3$ **D.** UF_6

                                    E. All of the above exist

Chemical Bonding

==

Practice Set 1: Questions 1–20

==

1. What is the number of valence electrons in tin (Sn)?

    A. 14   **B.** 8   **C.** 2   **D.** 4   **E.** 5

2. Unhybridized *p* orbitals participate in π bonds as double and triple bonds. How many distinct and degenerate *p* orbitals exist in the second electron shell, where n = 2?

    A. 3   **B.** 2   **C.** 1   **D.** 0   **E.** 4

3. What is the number of valence electrons in a sulfite ion, SO_3^{2-}?

    A. 22   **B.** 24   **C.** 26   **D.** 34   **E.** none of the above

4. Which type of attractive forces occurs in molecules regardless of the atoms they possess?

    A. Dipole–ion interactions   **C.** Dipole–dipole attractions

    B. Ion–ion interactions   **D.** Hydrogen bonding

                                      E. London dispersion forces

5. Given the structure of glucose, which statement explains the hydrogen bonding between glucose and water?

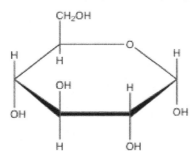

    A. Due to the cyclic structure of glucose, there is no H–bonding with water

    B. Each glucose molecule could H–bond with up to 17 water molecules

    C. H–bonds form, with water being the H–bond donor

    D. H–bonds form, with glucose being the H–bond donor

    E. Each glucose molecule could H–bond with up to 5 water molecules

6. How many valence electrons are in the Lewis dot structure of C_2H_6?

    A. 2   **B.** 14   **C.** 6   **D.** 8   **E.** 12

7. The term for a bond where the electrons are shared unequally is:

    A. ionic

    B. nonpolar covalent

    C. coordinate covalent

    D. nonpolar ionic

    E. polar covalent

8. What is the name for the weak forces of attraction between nonpolar molecules due to temporary dipoles between adjacent nonpolar molecules?

    A. van der Waals forces

    B. hydrophobic forces

    C. hydrogen bonding forces

    D. nonpolar covalent forces

    E. hydrophilic forces

9. Nitrogen has five valence electrons; which of the following types of bonding is/are possible?

        I. one single and one double bond

        II. three single bonds

        III. one triple bond

    A. I only

    B. II only

    C. I and III only

    D. I, II and III

    E. none of the above

10. What is the number of valence electrons in antimony (Sb)?

    A. 1   **B.** 2   **C.** 3   **D.** 4   **E.** 5

11. How many resonance structures, if any, can be drawn for a nitrite ion?

    A. 1   **B.** 2   **C.** 3   **D.** 4   **E.** 5

12. What is the formula of the ammonium ion?

    A. NH_4^- **B.** N_4H^+ **C.** NH_4^{2-} **D.** NH_4^{2+} **E.** NH_4^+

13. What is the type of bond that forms between oppositely charged ions?

    A. dipole

    B. covalent

    C. London

    D. induced dipole

    E. ionic

14. Which of the following series of elements are arranged in the order of increasing electronegativity?

A. Fr, Mg, Si, O

B. Br, Cl, S, P

C. F, B, O, Li

D. Cl, S, Se, Te

E. Br, Mg, Si, N

15. The attraction due to London dispersion forces between molecules depends on what two factors?

A. Volatility and shape

B. Molar mass and volatility

C. Vapor pressure and size

D. Molar mass and shape

E. Molar mass and vapor pressure

16. Which of the substances below would have the largest dipole?

A. CO_2

B. SO_2

C. H_2O

D. CCl_4

E. CH_4

17. What is the name for the attraction between H_2O molecules?

A. adhesion

B. polarity

C. cohesion

D. van der Waals

E. hydrophilicity

18. Which compound contains only covalent bonds?

A. $HC_2H_3O_2$

B. $NaCl$

C. NH_4OH

D. $Ca_3(PO_4)_2$

E. LiF

19. The distance between two atomic nuclei in a chemical bond is determined by the:

A. size of the valence electrons

B. size of the nucleus

C. size of the protons

D. balance between the repulsion of the nuclei and the attraction of the nuclei for the bonding electrons

E. size of the neutrons

20. Carbonic acid has the chemical formula of H_2CO_3. The carbonate ion has the molecular formula of CO_3^{2-}. From the Lewis structure for the CO_3^{2-} ion, what is the number of reasonable resonance structures for the anion?

A. original structure only

B. 2

C. 3

D. 4

E. 5

==

Practice Set 2: Questions 21–40

==

21. Which of the following molecules would contain a dipole?

    A. F–F   **C.** Cl–Cl
    B. H–H   **D.** H–F
                                        E. *trans*-dichloroethene

22. Which is the correct formula for the ionic compound formed between Ca and I?

    A. Ca_3I_2 **B.** Ca_2I_3 **C.** CaI_2 **D.** Ca_2I **E.** Ca_3I_5

23. Which bonding is NOT possible for a carbon atom that has four valence electrons?

    A. 1 single and 1 triple bond   **C.** 4 single bonds
    B. 1 double and 1 triple bond   **D.** 2 single and 1 double bond
                                               E. 2 double bonds

24. During strenuous exercise, why does perspiration on a person's skin form droplets?

    A. Ability of H_2O to dissipate heat   **C.** Adhesive properties of H_2O
    B. High specific heat of H_2O **D.** Cohesive properties of H_2O
                                                       E. High NaCl content of perspiration

25. If an ionic bond is stronger than a dipole–dipole interaction, why does water dissolve an ionic compound?

    A. Ions do not overcome their interatomic attraction and therefore are not soluble
    B. Ion–dipole interaction causes the ions to heat up and vibrate free of the crystal
    C. Ionic bond is weakened by the ion–dipole interactions, and ionic repulsion ejects the ions from the crystal
    D. Ion-dipole interactions of several water molecules aggregate with the ionic bond and dissociate it into the solution
    E. None of the above

26. Which one of these molecules can act as a hydrogen bond acceptor but not a donor?

    A. CH_3NH_2 **C.** H_2O
    B. CH_3CO_2H **D.** C_2H_5OH
                               E. CH_3–O–CH_3

27. When NaCl dissolves in water, what is the force of attraction between Na^+ and H_2O?

A. ion–dipole

B. hydrogen bonding

C. ion–ion

D. dipole–dipole

E. van der Waals

28. Based on the Lewis structure, how many polar and nonpolar bonds are present in H_2CO?

A. 3 polar bonds and 0 nonpolar bonds

B. 2 polar bonds and 1 nonpolar bond

C. 1 polar bond and 2 nonpolar bonds

D. 0 polar bonds and 3 nonpolar bonds

E. 2 polar bonds and 2 nonpolar bonds

29. How many more electrons can fit within the valence shell of a hydrogen atom?

A. 1　　　**B.** 2　　　**C.** 7　　　**D.** 0　　　**E.** 3

30. Which element likely forms a cation with a +2 charge?

A. Na　　　**B.** S　　　**C.** Si　　　**D.** Mg　　　**E.** Br

31. Which of the statements is an accurate description of the structure of the ionic compound NaCl?

A. Alternating rows of Na^+ and Cl^- ions are present

B. Each ion present is surrounded by six ions of opposite charge

C. Alternating layers of Na and Cl atoms are present

D. Alternating layers of Na^+ and Cl^- ions are present

E. Repeating layers of Na^+ and Cl^- ions are present

32. What is the name for the force holding two atoms together in a chemical bond?

A. gravitational force

B. strong nuclear force

C. weak hydrophobic force

D. weak nuclear force

E. electrostatic force

33. Which of the following diatomic molecules contains the bond of greatest polarity?

A. CH_4　　　**B.** BrI　　　**C.** Cl–F　　　**D.** P_4　　　**E.** Te–F

34. Why does H_2O have an unusually high boiling point compared to H_2S?

A. Hydrogen bonding

B. Van der Waals forces

C. H_2O molecules pack more closely than H_2S

D. Covalent bonds are stronger in H_2O

E. This is a false statement because H_2O has a similar boiling point to H_2S

35. What is the shape of a molecule in which the central atom has 2 bonding electron pairs and 2 nonbonding electron pairs?

    A. trigonal planar

    B. trigonal pyramidal

    C. linear

    D. bent

    E. tetrahedral

36. Which of the following represents the breaking of a noncovalent interaction?

    A. Ionization of water

    B. Decomposition of hydrogen peroxide

    C. Hydrolysis of an ester

    D. Dissolving of salt crystals

    E. None of the above

37. Which pair of elements is most likely to form an ionic compound when reacted?

    A. C and Cl

    B. K and I

    C. Ga and Si

    D. Fe and Mg

    E. H and O

38. Which of the following statements concerning coordinate covalent bonds is correct?

    A. Once formed, they are indistinguishable from any other covalent bond

    B. They are single bonds

    C. One of the atoms involved must be a metal and the other a nonmetal

    D. Both atoms involved in the bond contribute an equal number of electrons to the bond

    E. The bond is formed between two Lewis bases

39. The greatest dipole moment within a bond is when:

    A. both bonding elements have a low electronegativity

    B. one bonding element has a high electronegativity, and the other has a low electronegativity

    C. both bonding elements have a high electronegativity

    D. one bonding element has a high electronegativity, and the other has a moderate electronegativity

    E. both bonding elements have moderate electronegativity

40. Which of the following must occur for an atom to obtain the noble gas configuration?

    A. lose, gain or share an electron

    B. lose or gain an electron

    C. lose an electron

    D. share an electron

    E. share or gain an electron

===

Practice Set 3: Questions 41–60

===

41. Based on the Lewis structure for hydrogen peroxide, H_2O_2, how many polar bonds and nonpolar bonds are present?

    A. 3 polar and 0 nonpolar bonds   **C.** 1 polar and 2 nonpolar bonds

    B. 2 polar and 2 nonpolar bonds   **D.** 0 polar and 3 nonpolar bonds

                                    E. 2 polar and 1 nonpolar bond

42. To form an octet, an atom of selenium must:

    A. gain 2 electrons   **C.** gain 6 electrons

    B. lose 2 electrons   **D.** lose 6 electrons

                                      E. gain 4 electrons

43. Which of the following is an example of a chemical reaction?

    A. two solids mix to form a heterogeneous mixture

    B. two liquids mix to form a homogeneous mixture

    C. one or more new compounds are formed by rearranging atoms

    D. a new element is formed by rearranging nucleons

    E. a liquid undergoes a phase change and produces a solid

44. Which compound is NOT correctly matched with the predominant intermolecular force associated with that compound in the liquid state?

Compound	Intermolecular force
A. CH_3OH	hydrogen bonding
B. HF	hydrogen bonding
C. Cl_2O	dipole–dipole interactions
D. HBr	van der Waals interactions
E. CH_4	van der Waals interactions

45. Which element forms an ion with the greatest positive charge?

    A. Mg   **B.** Ca   **C.** Al   **D.** Na   **E.** Rb

46. What is the difference between a dipole–dipole and an ion–dipole interaction?

    A. One interaction involves dipole attraction between neutral molecules, while the other involves dipole interactions with ions

    B. One interaction involves ionic molecules interacting with other ionic molecules, while the other deals with polar molecules

    C. One interaction involves salts and water, while the other does not involve water

    D. One interaction involves hydrogen bonding, while the others do not

    E. None of the above

47. Which is a true statement about H_2O as it begins to freeze?

    A. Hydrogen bonds break

    B. Number of hydrogen bonds decreases

    C. Covalent bond strength increases

    D. Molecules move closer together

    E. Number of hydrogen bonds increases

48. In the process of forming sodium nitride (Na_3N) from its elements, what happens to the electrons of each sodium and the electrons of a nitrogen atom, respectively?

    A. one lost; three gained

    B. three lost; three gained

    C. three lost; one gained

    D. one lost; two gained

    E. two lost; three gained

49. Which species below has the least number of valence electrons in its Lewis symbol?

    A. S^{2-} **B.** Ga^+ **C.** Ar^+ **D.** Mg^{2+} **E.** F^-

50. Which of the following occur(s) naturally as nonpolar diatomic molecules?

    I. sulfur   II. chlorine   III. argon

    A. I only

    B. II only

    C. I and III only

    D. I and II only

    E. I, II and III

51. In a chemical reaction, the bonds being formed are:

    A. more energetic than the ones broken

    B. less energetic than the ones broken

    C. the same as the bonds broken

    D. different from the ones broken

    E. none of the above

52. The charge on a sulfide ion is:

    A. +1   **B.** +2   **C.** 0   **D.** –2   **E.** –3

53. Which formula for an ionic compound is NOT correct?

    A. $Al_2(CO_3)_3$ **B.** Li_2SO_4 **C.** Na_2S **D.** $MgHCO_3$ **E.** K_2O

54. What term best describes the smallest whole number repeating ratio of ions in an ionic compound?

    A. lattice

    B. unit cell

    C. formula unit

    D. covalent unit

    E. ionic unit

55. In the nitrogen monoxide molecule, the dipole moment is 0.16 D, and the bond length is 115 pm. What is the sign and magnitude of the charge on the oxygen atom? (Use the conversion factor of 1 D = 3.34×10^{-30} C·m and the charge of 1 electron = 1.602×10^{-19} C)

    A. $-0.098\ e$ **B.** $-0.71\ e$ **C.** $-1.3\ e$ **D.** $-0.029\ e$ **E.** $+1.3\ e$

56. From the electronegativity below, which single covalent bond is the most polar?

Element:	H	C	N	O
Electronegativity	2.1	2.5	3.0	3.5

    A. O–C   **B.** O–N   **C.** N–C   **D.** C–H   **E.** C–C

57. Which of the following pairs is NOT correctly matched?

Formula	Molecular Geometry		Formula	Molecular Geometry
A. CH_4	tetrahedral		**C.** PCl_3	trigonal planar
B. OF_2	bent		**D.** Cl_2CO	trigonal planar
			E. $^{+}CH_3$	trigonal planar

58. Why are adjacent water molecules attracted to each other?

    A. Ionic bonding between the hydrogens of H_2O

    B. Covalent bonding between adjacent oxygens

    C. Electrostatic attraction between the H of one H_2O and the O of another

    D. Covalent bonding between the H of one H_2O and the O of another

    E. Electrostatic attraction between the O of one H_2O and the O of another

59. What is the chemical formula for a compound that contains K^+ and CO_3^{2-} ions?

    A. $K(CO_3)_3$ **B.** $K_3(CO_3)_2$ **C.** $K_3(CO_3)_3$ **D.** KCO_3 **E.** K_2CO_3

60. Under what conditions is graphite converted to diamond?

    A. low temperature, high-pressure

    B. high temperature, low-pressure

    C. high temperature, high-pressure

    D. low temperature, low-pressure

    E. none of the above

==

Practice Set 4: Questions 61–88

==

61. Which of the following molecules is a Lewis acid?

A. NO_3^- **B.** NH_3 **C.** NH_4^+ **D.** CH_3COOH **E.** BH_3

62. Which molecule(s) is/are most likely to show a dipole-dipole interaction?

     I. $H-C{\equiv}C-H$ II. CH_4 III. CH_3SH IV. CH_3CH_2OH

A. I only   **C.** III only

B. II only   **D.** III and IV only

                               E. I and IV only

63. C=C, C=O, C=N and N=N bonds are observed in many organic compounds. However, C=S, C=P, C=Si, and other similar bonds are not often found. What is the most probable explanation for this observation?

A. The comparative sizes of $3p$ atomic orbitals make effective overlap between them less likely than between two $2p$ orbitals

B. S, P and Si do not undergo hybridization of orbitals

C. S, P and Si do not form π bonds due to the lack of occupied p orbitals in their ground state electron configurations

D. Carbon does not combine with elements found below the second row of the periodic table

E. None of the above

64. Write the formula for the ionic compound formed from magnesium and sulfur:

A. Mg_2S **B.** Mg_3S_2 **C.** MgS_2 **D.** MgS_3 **E.** MgS

65. Explain why chlorine, Cl_2, is a gas at room temperature, while bromine, Br_2, is a liquid.

A. Bromine ions are held by ionic bonds

B. Chlorine molecules are smaller and, therefore, pack tighter in their physical orientation

C. Bromine atoms are larger, which causes the formation of a stronger induced dipole induced dipole attraction

D. Chlorine atoms are larger, which causes the formation of a stronger induced dipole induced dipole attraction

E. Bromine molecules are smaller and therefore pack tighter in their physical orientation

66. What is the major intermolecular force in $(CH_3)_2NH$?

   A. hydrogen bonding   **C.** London dispersion forces

   B. dipole–dipole attractions   **D.** ion–dipole attractions

                                                  E. van der Waals forces

67. Which of the following correctly describes the molecule potassium oxide and its bond?

   A. It is a weak electrolyte with an ionic bond

   B. It is a strong electrolyte with an ionic bond

   C. It is a non-electrolyte with a covalent bond

   D. It is a strong electrolyte with a covalent bond

   E. It is a non-electrolyte with a hydrogen bond

68. An ion with an atomic number of 34 and 36 electrons has what charge?

   A. +2   **B.** −36   **C.** +34   **D.** −2   **E.** neutral

69. Based on the Lewis structure for $H_3C–NH_2$, the formal charge on N is:

   A. −1   **B.** 0   **C.** +2   **D.** +1   **E.** −2

70. The name of S^{2-} is:

   A. sulfite ion   **C.** sulfur

   B. sulfide ion   **D.** sulfate ion

                                                  E. sulfurous acid

71. An ionic bond forms between two atoms when:

   A. protons are transferred from the nucleus of the nonmetal to the nucleus of the metal

   B. each atom acquires a negative charge

   C. electron pairs are shared

   D. four electrons are shared

   E. electrons are transferred from metallic to nonmetallic atoms

72. Which of the following pairings of ions is NOT consistent with the formula?

   A. Co_2S_3 (Co^{3+} and S^{2-})   **C.** Na_3P (Na^+ and P^{3-})

   B. K_2O (K^+ and O^-)   **D.** BaF_2 (Ba^{2+} and F^-)

                                                    E. KCl (K^+ and Cl^-)

73. Which of the following solids is likely to have the smallest exothermic lattice energy?

   A. Al_2O_3 **B.** KF   **C.** NaCl   **D.** LiF   **E.** NaOH

74. In chlorine monoxide, chlorine has a charge of +0.167 *e*. If the bond length is 154 pm, what is the dipole moment of the molecule? (Use the conversion of 1 meter = 1×10^{-12} m/pm and the value of 1 electron = 1.602 $\times 10^{-19}$ C)

    A. 2.30 D   **B.** 0.167 D   **C.** 1.24 D   **D.** 3.11 D   **E.** 1.65 D

75. All of the following are examples of polar molecules, EXCEPT:

    A. H_2O **B.** CCl_4 **C.** CH_2Cl_2 **D.** HF   **E.** CO

76. Which is the most likely noncovalent interaction between an alcohol and a carboxylic acid at pH = 10?

    A. formation of an anhydride bond   **C.** dipole–charge interaction

    B. dipole–dipole interaction   **D.** charge–charge interaction

                                     E. induced dipole–dipole interaction

77. Which electron geometry is characteristic of an sp^2 hybridized atom?

    A. linear   **C.** trigonal bipyramidal

    B. tetrahedral   **D.** bent

                                     E. trigonal planar

78. Which of the following statements about noble gases is NOT correct?

    A. They have very stable electron arrangements

    B. They are the most reactive of gases

    C. They exist in nature as individual atoms rather than in the molecular form

    D. They have 8 valence electrons

    E. They have a complete octet

79. How are intermolecular forces and solubility related?

    A. Solubility is a measure of how weak the intermolecular forces in the solute are

    B. Solubility is a measure of how strong a solvent's intermolecular forces are

    C. Solubility depends on the solute's ability to overcome the intermolecular forces in the solvent

    D. Solubility depends on the solvent's ability to overcome the intermolecular forces in a solute

    E. None of the above

80. In a bond between any two of the following atoms, the bonding electrons would be most strongly attracted to:

    A. I   **B.** He   **C.** Cs   **D.** Cl   **E.** Fr

81. For small molecules of comparable molecular weight, which choice lists the intermolecular forces in increasing order?

    A. dipole–dipole forces < hydrogen bonds < London forces
    B. London forces < hydrogen bonds < dipole–dipole forces
    C. London forces < dipole–dipole forces < hydrogen bonds
    D. hydrogen bonds < dipole–dipole forces < London forces
    E. London forces < dipole–dipole forces = hydrogen bonds

82. How many electrons are in the outermost electron shell of oxygen if it has an atomic number of 8?

    A. 2   **B.** 6   **C.** 8   **D.** 16   **E.** 18

83. Based on the Lewis structure, how many resonance structures are reasonable for the nitrate ion?

    A. 1   **B.** 2   **C.** 3   **D.** 4   **E.** 5

84. Which of the compounds is most likely to be ionic?

    A. CBr_4 **B.** H_2O **C.** CH_2Cl_2 **D.** NO_2 **E.** $SrBr_2$

85. The ability of an atom in a molecule to attract electrons to itself is:

    A. ionization energy   **C.** electronegativity
    B. paramagnetism   **D.** electron affinity
                               E. hyperconjugation

86. The "octet rule" relates to the number eight because:

    A. all orbitals can hold 8 electrons
    B. electron arrangements involving 8 valence electrons are highly stable
    C. all atoms have 8 valence electrons
    D. only atoms with 8 valence electrons undergo a chemical reaction
    E. each element can accommodate a number of electrons as an integer of 8

87. Which of the following molecules is polar?

    A. SO_3 **B.** SO_2 **C.** CO_2 **D.** CH_4 **E.** CCl_4

88. Which force is intermolecular?

    A. polar covalent bond   **C.** dipole–dipole interactions
    B. ionic bond   **D.** nonpolar covalent bond
                               E. all are intermolecular forces

Notes for active learning

Phases and Phase Equilibria

===

Practice Set 1: Questions 1–20

===

1. Which statement is NOT true regarding vapor pressure?

     I. Solids do not have a vapor pressure

     II. Vapor pressure of a pure liquid does not depend on the amount of vapor present

     III. Vapor pressure of a pure liquid does not depend on the amount of liquid present

  A. I only   **C.** III only

  B. II only   **D.** I and II only

                     E. I, II and III

2. How does the volume of a fixed sample of gas change if the temperature is doubled at constant pressure?

  A. Decreases by a factor of 2   **C.** Doubles

  B. Increases by a factor of 4   **D.** Remains the same

                     E. Requires more information

3. When a solute is added to a pure solvent, the boiling point [   ] and freezing point [   ]?

  A. decreases ... decreases   **C.** increases ... decreases

  B. decreases ... increases   **D.** increases ... increases

                     E. remains the same ... remains the same

4. Consider the phase diagram for H_2O. The termination of the gas-liquid transition at which distinct gas or liquid phases do NOT exist is the:

  A. critical point   **C.** triple point

  B. endpoint   **D.** condensation point

                     E. inflection point

5. The van der Waals equation of state for a real gas is expressed as $[P + n^2a / V^2] \cdot (V - nb) = nRT$. The van der Waals constant, *a*, represents a correction for:

    A. negative deviation in the measured value of P from that of an ideal gas due to the attractive forces between the molecules of a real gas

    B. positive deviation in the measured value of P from that of an ideal gas due to the attractive forces between the molecules of a real gas

    C. negative deviation in the measured value of P from that of an ideal gas due to the finite volume of space occupied by molecules of a real gas

    D. positive deviation in the measured value of P from that of an ideal gas due to the finite volume of space occupied by molecules of a real gas

    E. positive deviation in the measured value of P from that of an ideal gas due to the finite mass of the molecules of a real gas

6. What is the value of the ideal gas constant, expressed in units (torr × mL) / mole × K?

    A. 62.4   **C.** 0.0821

    B. 62,400   **D.** 1 / 0.0821

                                        E. 8.21

7. If both the pressure and the temperature of a gas are halved, the volume is:

    A. halved   **C.** doubled

    B. the same   **D.** quadrupled

                                          E. decreased by a factor of 4

8. If container X is occupied by 2.0 moles of O_2 gas, while container Y is occupied by 10.0 grams of N_2 gas, and both containers are maintained at 5.0 °C and 760 torrs, then:

    A. container X must have a volume of 22.4 L

    B. the average kinetic energy of molecules in X equals the average kinetic energy of molecules in Y

    C. container Y must be larger than container X

    D. the average speed of the molecules in container X is greater than that of the molecules in container Y

    E. the number of atoms in container Y is greater than the number of atoms in container X

9. Among the following choices, how tall should a properly designed Torricelli mercury barometer be?

    A. 100 in   **C.** 76 mm

    B. 380 mm   **D.** 400 mm

                                        E. 800 mm

10. When volatile solvents X and Y are mixed in equal proportions, heat is released to the surroundings. If pure X has a higher boiling point than pure Y, which of the following statements is NOT true?

    A. The vapor pressure of the mixture is lower than that of pure Y

    B. The vapor pressure of the mixture is lower than that of pure X

    C. The boiling point of the mixture is lower than that of pure X

    D. The boiling point of the mixture is lower than that of pure Y

    E. Not enough information is provided

11. For the balanced reaction $2 Na + Cl_2 \rightarrow 2 NaCl$, which of the following is a gas?

    I. Na   II. Cl_2 III. NaCl

    A. I only   **C.** III only

    B. II only   **D.** I and II only

                                E. I, II and III

12. How does a real gas deviate from an ideal gas?

    I. Molecules occupy a significant amount of space

    II. Intermolecular forces may exist

    III. Pressure is created from molecular collisions with the walls of the container

    A. I only   **C.** I and II only

    B. II only   **D.** II and III only

                                E. I, II and III

13. A 2.75 L sample of He gas has a pressure of 0.950 atm. What is the pressure of the gas if the volume is reduced to 0.450 L?

    A. 5.80 atm   **C.** 0.230 atm

    B. 0.520 atm   **D.** 0.960 atm

                                E. 3.40 atm

14. Avogadro's law, in its alternate form, is very similar in form to:

    I. Boyle's law

    II. Charles' law

    III. Gay-Lussac's law

    A. I only   **C.** III only

    B. II only   **D.** II and III only

                                E. I, II and III

15. What is the relationship between the pressure and volume of a fixed amount of gas at constant temperature?

A. directly proportional

B. equal

C. inversely proportional

D. decreased by a factor of 2

E. none of the above

16. What is the term for a change of state from a liquid to a gas?

A. vaporization

B. melting

C. deposition

D. condensing

E. sublimation

17. The combined gas law can NOT be written as:

A. $V_2 = V_1 \times P_1 / P_2 \times T_2 / T_1$

B. $P_1 = P_2 \times V_2 / V_1 \times T_1 / T_2$

C. $T_2 = T_1 \times P_1 / P_2 \times V_2 / V_1$

D. $V_1 = V_2 \times P_2 / P_1 \times T_1 / T_2$

E. none of the above

18. Which singular molecule is most likely to show a dipole-dipole interaction?

A. CH_4 **B.** $H–C≡C–H$ **C.** SO_2 **D.** CO_2 **E.** CCl_4

19. What happens if the pressure of a gas above a liquid increases, such as by pressing a piston above a liquid?

A. Pressure goes down, and the gas moves out of the solvent

B. Pressure goes down, and the gas goes into the solvent

C. The gas is forced into solution, and the solubility increases

D. The solution is compressed, and the gas is forced out of the solvent

E. The amount of gas in the solution remains constant

20. Which of the following two variables are present in mathematical statements of Avogadro's law?

A. P and V

B. n and P

C. n and T

D. V and T

E. n and V

===

Practice Set 2: Questions 21–40

===

21. Which of the following acids has the lowest boiling point elevation?

Monoprotic Acids	Ka
Acid I	1.4×10^{-8}
Acid II	1.6×10^{-9}
Acid III	3.9×10^{-10}
Acid IV	2.1×10^{-8}

A. I   **C.** III

B. II   **D.** IV

                    E. Requires more information

22. An ideal gas differs from a real gas because the molecules of an ideal gas have:

A. no attraction to each other   **C.** molecular weight equal to zero

B. no kinetic energy   **D.** appreciable volumes

                    E. none of the above

23. Which of the following is the definition of standard temperature and pressure?

A. 0 K and 1 atm   **C.** 298.15 K and 1 atm

B. 298.15 K and 750 mmHg   **D.** 273 °C and 750 torr

                    E. 273.15 K and 10^5 Pa

24. At constant volume, as the temperature of a gas sample is decreased, the gas deviates from ideal behavior. Compared to the pressure predicted by the ideal gas law, actual pressure would be:

A. higher, because of the volume of the gas molecules

B. higher, because of intermolecular attractions between gas molecules

C. lower, because of the volume of the gas molecules

D. lower, because of intermolecular attractions among gas molecules

E. higher, because of intramolecular attractions between gas molecules

25. A flask contains a mixture of O_2, N_2 and CO_2. The pressure exerted by N_2 is 320 torr and by CO_2 is 240 torr. If the pressure of the gas mixture is 740 torr, what is the percent pressure of O_2?

A. 14%   **B.** 18%   **C.** 21%   **D.** 24%   **E.** 29%

26. A balloon contains 40 grams of He with a pressure of 1,000 torrs. When He is released from the balloon, the new pressure is 900 torr, and the volume is half of the original. If the temperature is the same, how many grams of He remains in the balloon?

A. 18 grams   **C.** 10 grams

B. 22 grams   **D.** 40 grams

                                   E. 28 grams

27. Which of the following is a true statement regarding evaporation?

A. Increasing the surface area of the liquid decreases the rate of evaporation

B. The temperature of the liquid changes during evaporation

C. Decreasing the surface area of the liquid increases the rate of evaporation

D. Molecules with greater kinetic energy escape from the liquid

E. Not enough information is provided to make any conclusions

28. Under which conditions does a real gas behave most nearly like an ideal gas?

A. High temperature and high pressure

B. High temperature and low pressure

C. Low temperature and low pressure

D. Low temperature and high pressure

E. If it remains in the gaseous state regardless of temperature or pressure

29. Which of the following compounds has the highest boiling point?

A. CH_3OH **C.** $CH_3OCH_2CH_2CH_2CH_3$

B. $CH_3CH_2CH_2CH_2CH_2OH$ **D.** $CH_3CH_2OCH_2CH_2CH_3$

                                            E. $CH_3CH_2CH_2C(OH)HOH$

30. According to the kinetic theory of gases, which of the following is the average kinetic energy of the gas particles directly proportional to?

A. temperature   **C.** volume

B. molar mass   **D.** pressure

                                    E. number of moles of gas

31. Under ideal conditions, which of the following gases is least likely to behave as an ideal gas?

A. CF_4 **B.** CH_3OH **C.** N_2 **D.** O_3 **E.** CO_2

32. Which statement is true regarding gases when compared to liquids?

   A. Gases have lower compressibility and higher density
   B. Gases have lower compressibility and lower density
   C. Gases have higher compressibility and higher density
   D. Gases have higher compressibility and lower density
   E. None of the above

33. When nonvolatile solute molecules are added to a solution, the vapor pressure of the solution:

   A. stays the same
   B. increases
   C. decreases
   D. is directly proportional to the second power of the amount added
   E. is directly proportional to the square root of the amount added

34. Which of the following laws states that the pressure exerted by a mixture of gases is equal to the sum of the individual gas pressures?

   A. Gay-Lussac's law
   B. Dalton's law
   C. Charles's law
   D. Boyle's law
   E. Avogadro's law

35. Which characteristics best describe a solid?

   A. Definite volume; the shape of the container; no intermolecular attractions
   B. Volume and shape of the container; no intermolecular attractions
   C. Definite shape and volume; strong intermolecular attractions
   D. Definite volume; the shape of the container; moderate intermolecular attractions
   E. Volume and shape of the container; strong intermolecular attractions

36. Which of the following statements is true, if three 2.0 L flasks are filled with H_2, O_2 and He, respectively, at STP?

   A. There are twice as many He atoms as H_2 or O_2 molecules
   B. There are four times as many H_2 or O_2 molecules as He atoms
   C. Each flask contains the same number of atoms
   D. There are twice as many H_2 or O_2 molecules as He atoms
   E. The number of H_2 or O_2 molecules is the same as the number of He atoms

37. Which of the following atoms could interact through a hydrogen bond?

   A. The hydrogen of an amine and the oxygen of an alcohol
   B. The hydrogen on an aromatic ring and the oxygen of carbon dioxide
   C. The oxygen of a ketone and the hydrogen of an aldehyde
   D. The oxygen of methanol and hydrogen on the methyl carbon of methanol
   E. None of the above

38. Which of the following laws states that the pressure and volume are inversely proportional for gas at a constant temperature?

   A. Dalton's law
   B. Gay-Lussac's law
   C. Boyle's law
   D. Charles's law
   E. Avogadro's law

39. As an automobile travels the highway, the temperature of the air inside the tires [   ] and the pressure [   ]?

   A. increases … decreases
   B. increases … increases
   C. decreases … decreases
   D. decreases … increases
   E. none of the above

40. Using the following unbalanced chemical reaction, what volume of H_2 gas at 780 mmHg and 23 °C is required to produce 12.5 L of NH_3 gas at the same temperature and pressure?

$$N_2 \,(g) + H_2 \,(g) \rightarrow NH_3 \,(g)$$

   A. 21.4 L   **B.** 15.0 L   **C.** 18.8 L   **D.** 13.0 L   **E.** 12.5 L

==

Practice Set 3: Questions 41–60

==

41. A mixture of gases containing 16 g of O_2, 14 g of N_2 and 88 g of CO_2 is collected above water at a temperature of 23 °C. The pressure is 1 atm, and the vapor pressure of water is 38 torr. What is the partial pressure exerted by CO_2?

    A. 283 torr   **B.** 367 torr   **C.** 481 torr   **D.** 549 torr   **E.** 583 torr

42. Which of the following demonstrate colligative properties?

    I. Freezing point
    II. Boiling point
    III. Vapor pressure

    A. I only   **C.** III only
    B. II only   **D.** I and II only
                                    E. I, II and III

43. What is the proportionality relationship between the pressure of a gas and its volume?

    A. directly   **C.** pressure is raised to the 2nd power
    B. inversely   **D.** pressure raised to the $\sqrt{2}$ power
                                    E. none of the above

44. Which of the following statements about gases is correct?

    A. Formation of homogeneous mixtures, regardless of the nature of non-reacting gas components
    B. Relatively long distances between molecules
    C. High compressibility
    D. No attractive forces between gas molecules
    E. All of the above

45. Which of the following describes a substance in the solid physical state?

    I. It compresses negligibly
    II. It has a fixed volume
    III. It has a fixed shape

    A. I only   **C.** I and III only
    B. II only   **D.** II and III only
                                      E. I, II and III

46. Which of the following is NOT a unit used in measuring pressure?

A. kilometers Hg

B. millimeters Hg

C. atmosphere

D. Pascal

E. torr

47. Which of the following compounds has the highest boiling point?

A. CH_4

B. $CHCl_3$

C. CH_3COOH

D. NH_3

E. CH_2Cl_2

48. Identify the decreasing ordering of attractions among particles in the three states of matter.

A. gas > liquid > solid

B. gas > solid > liquid

C. solid > liquid > gas

D. liquid > solid > gas

E. solid > gas > liquid

49. A sample of N_2 gas occupies a volume of 190 mL at STP. What volume will it occupy at 660 mmHg and 295 K?

A. 1.15 L

B. 0.760 L

C. 0.214 L

D. 1.84 L

E. 0.197 L

50. A nonvolatile liquid would have:

A. a highly explosive propensity

B. strong attractive forces between molecules

C. weak attractive forces between molecules

D. a high vapor pressure at room temperature

E. weak attractive forces within molecules

51. A closed-end manometer was constructed from a U-shaped glass tube. It was loaded with mercury so that the closed side was filled to the top, which was 820 mm above the neck, while the open end was 160 mm above the neck. The manometer was taken into a chamber used for training astronauts. What is the highest pressure that can be read with assurance on this manometer?

A. 66.0 torr

B. 660 torr

C. 220 torr

D. 760 torr

E. 5.13 torr

52. Vessels X and Y each contain 1.00 L of a gas at STP, but vessel X contains oxygen, while vessel Y contains nitrogen. Assuming the gases behave as ideal, they have the same:

I. number of molecules   II. density   III. kinetic energy

A. II only
B. III only
C. I and III only
D. I, II and III
E. I only

53. What condition must be satisfied for the noble gas Xe to exist in the liquid phase at 180 K, which is a temperature significantly greater than its normal boiling point?

A. external pressure > vapor pressure of xenon
B. external pressure < vapor pressure of xenon
C. external pressure = partial pressure of water
D. temperature is increased quickly
E. temperature is increased slowly

54. Matter is nearly incompressible in which of these states?

I. solid   II. Liquid   III. gas

A. I only
B. II only
C. III only
D. I and II only
E. I, II and III

55. Which of the following laws states that volume and temperature are directly proportional for a gas at constant pressure?

A. Gay-Lussac's law
B. Dalton's law
C. Charles' law
D. Boyle's law
E. Avogadro's law

56. Which transformation describes sublimation?

A. solid → liquid
B. solid → gas
C. liquid → solid
D. liquid → gas
E. gas → liquid

57. What is the ratio of the diffusion rate of O_2 molecules to the diffusion rate of H_2 molecules if six moles of O_2 gas and six moles of H_2 gas are placed in a large vessel, and the gases and vessels are at the same temperature?

    A. 4:1   **C.** 12:1

    B. 1:4   **D.** 1:1

                                              E. 2:1

58. How many molecules of neon gas are present in 6 liters at 10 °C and 320 mmHg? (Use the ideal gas constant R = 0.0821 L·atm K^{-1} mol^{-1})

    A. (320 mmHg / 760 atm)·(0.821)·(6 L) / (6 × 10^{23})·(283 K)

    B. (320 mmHg / 760 atm)·(6 L)·(6 × 10^{23}) / (0.0821)·(283 K)

    C. (320 mmHg)·(6 L)·(283 K)·(6 × 10^{23})

    D. (320 mmHg / 760 atm)·(6 L)·(283 K)·(6 × 10^{23}) / (0.821)

    E. (320 mmHg / 283 K)·(6 L)·(6 × 10^{23}) / (0.0821)·(760 atm)

59. Which gas has the greatest density at STP:

    A. CO_2 **B.** O_2 **C.** N_2 **D.** NO   **E.** more than one

60. A hydrogen bond is a special type of:

    A. dipole–dipole attraction involving hydrogen bonded to another hydrogen atom

    B. attraction involving any molecules that contain hydrogens

    C. dipole-dipole attraction involving hydrogen bonded to a highly electronegative atom

    D. dipole–dipole attraction involving hydrogen bonded to any other atom

    E. London dispersion force involving hydrogen bonded to an electropositive atom

==

Practice Set 4: Questions 61–80

==

61. What is the mole fraction of H_2 in a gaseous mixture that consists of 9.50 g of H_2 and 14.0 g of Ne in a 4.50-liter container maintained at 37.5 °C?

    A. 0.13   **B.** 0.43   **C.** 0.67   **D.** 0.87   **E.** 1.2

62. Which of the following is true about Liquid A if the vapor pressure of Liquid A is greater than Liquid B?

    A. Liquid A has a higher heat of fusion   **C.** Liquid A forms stronger bonds

    B. Liquid A has a higher heat of vaporization   **D.** Liquid A boils at a higher temperature

                                         E. Liquid A boils at a lower temperature

63. When 25 g of non-ionizable compound X is dissolved in 1 kg of camphor, the freezing point of the camphor falls 2.0 K. What is the approximate molecular weight of compound X? (Use $K_{camphor} = 40$)

    A. 50 g/mol   **C.** 5,000 g/mol

    B. 500 g/mol   **D.** 5,500 g/mol

                                       E. 5,750 g/mol

64. The boiling point of a liquid is the temperature:

    A. where sublimation occurs

    B. where the vapor pressure of the liquid is less than the atmospheric pressure over the liquid

    C. equal to or greater than 100 °C

    D. where the rate of sublimation equals evaporation

    E. where the vapor pressure of the liquid equals the atmospheric pressure over the liquid

65. The van der Waals equation $[(P + n^2a / v^2) \cdot (V - nb) = nRT]$ is used to describe nonideal gases. The terms n^2a / v^2 and nb stand for, respectively:

    A. volume of gas molecules and intermolecular forces

    B. nonrandom movement and intermolecular forces between gas molecules

    C. nonelastic collisions and volume of gas molecules

    D. intermolecular forces and volume of gas molecules

    E. nonrandom movement and volume of gas molecules

66. What are the units of the gas constant R?

    A. atm·K/L·mol

    B. atm·K/mol

    C. mol·L/atm·K

    D. mol·K/L·atm

    E. L·atm/mol·K

67. A sample of a gas occupying a volume of 120.0 mL at STP was placed in a different vessel with a volume of 155.0 mL, in which the pressure was measured at 0.80 atm. What was its temperature?

    A. 9.1 °C

    B. 43.1 °C

    C. 4.1 °C

    D. 93.6 °C

    E. 108.3 °C

68. Why does a beaker of water begin to boil at 22 °C when placed in a closed chamber, and a vacuum pump is used to evacuate the air from the chamber?

    A. The vapor pressure decreases

    B. Air is released from the water

    C. The atmospheric pressure decreases

    D. The vapor pressure increases

    E. The atmospheric pressure increases

69. According to the kinetic theory, what happens to the kinetic energy of gaseous molecules when the temperature of a gas decreases?

    A. Increase as does velocity

    B. Remains constant, as does velocity

    C. Increases and velocity decreases

    D. Decreases and velocity increases

    E. Decreases as does velocity

70. A sample of SO_3 gas is decomposed into SO_2 and O_2.

$$2\ SO_3\ (g) \rightarrow 2\ SO_2\ (g) + O_2\ (g)$$

If the pressure of SO_2 and O_2 is 1,250 torrs, what is the partial pressure of O_2 in torr?

    A. 417 torr

    B. 1,040 torr

    C. 1,250 torr

    D. 884 torr

    E. 12.50 torr

71. For the balanced reaction $2\ Na + Cl_2 \rightarrow 2\ NaCl$, which of the following is solid?

    I. Na   II. Cl   III. NaCl

    A. I only

    B. II only

    C. III only

    D. I and III only

    E. I, II and III

72. According to Charles's law, what happens to gas as temperature increases?

    A. volume decreases

    B. volume increases

    C. pressure decreases

    D. pressure increases

    E. mole fraction increases

73. How does the pressure of a sample of gas change if the moles of gas remain constant while the volume is halved and the temperature is quadrupled?

    A. decrease by a factor of 8

    B. quadruple

    C. decrease by a factor of 4

    D. decrease by a factor of 2

    E. increase by a factor of 8

74. For a fixed quantity of gas, gas laws describe the relationships between pressure and which two variables?

    A. chemical identity; mass

    B. volume; chemical identity

    C. temperature; volume

    D. temperature; size

    E. volume; size

75. What is the term for a direct change of state from a solid to a gas?

    A. sublimation

    B. vaporization

    C. condensation

    D. deposition

    E. melting

76. The average speed at which a methane molecule effuses at 28.5 °C is 631 m/s. The average speed at which a krypton molecule effuses at the same temperature is:

    A. 123 m/s

    B. 276 m/s

    C. 312 m/s

    D. 421 m/s

    E. 633 m/s

77. A chemical reaction A (*s*) → B (*s*) + C (*g*) occurs when substance A is vigorously heated. The molecular mass of the gaseous product was determined from the following experimental data:

Mass of A before reaction: 5.2 g

Mass of A after reaction: 0 g

Mass of residue B after cooling and weighing when no more gas evolved: 3.8 g

When all of the gas C evolved, it was collected and stored in a 668.5 mL glass vessel at 32.0 °C, and the gas exerted a pressure of 745.5 torrs.

Use the ideal gas constant R equals 0.0821 L·atm K^{-1} mol^{-1}

From this data, determine the apparent molecular mass of *C*, assuming it behaves as an ideal gas:

A. 6.46 g/mol   **C.** 53.9 g/mol

B. 46.3 g/mol   **D.** 72.2 g/mol

                                E. 142.7 g/mol

78. Which of the following describes a substance in the liquid physical state?

I. It has a variable shape
II. It compresses negligibly
III. It has a fixed volume

A. I only   **C.** I and III only

B. II only   **D.** II and III only

                            E. I, II and III

79. Which is the strongest form of intermolecular attraction between water molecules?

A. ion-dipole   **C.** induced dipole-induced dipole

B. covalent bonding   **D.** hydrogen bonding

                            E. polar-induced dipolar

80. When a helium balloon is placed in a freezer, the temperature in the balloon [ ] and the volume [ ]?

A. increases … increases   **C.** decreases … increases

B. increases … decreases   **D.** decreases … decreases

                               E. none of the above

==

Practice Set 5: Questions 81–100

==

81. How does the volume of a fixed sample of gas change if the pressure is doubled?

    A. Decreases by a factor of 2   **C.** Doubles

    B. Increases by a factor of 4   **D.** Remains the same

                                        E. Requires more information

82. What is the term that refers to the frequency and energy of gas molecules colliding with the walls of the container?

      I. partial pressure   II. vapor pressure   III. gas pressure

    A. I only   **C.** III only

    B. II only   **D.** I and II only

                                      E. I, II and III

83. The freezing point changes by 10 K when an unknown amount of toluene is added to 100 g of benzene. Find the number of moles of toluene added from the given data. (Use the $K_{benzene} = 5.0$ and $K_{toluene} = 8.4$)

    A. 0.14   **B.** 0.20   **C.** 0.23   **D.** 0.27   **E.** 0.31

84. A container is labeled "Ne, 5.0 moles" but has no pressure gauge. By measuring the temperature and determining the volume of the container, a chemist uses the ideal gas law to estimate the pressure inside the container. If the container was mislabeled and contained 5.0 moles of He, not Ne, how would this affect the scientist's estimate?

    A. The estimate is correct, but only because both gases are monatomic

    B. The estimate is too high

    C. The estimate is slightly too low

    D. The estimate is significantly lower because of the large difference in molecular mass

    E. The estimate is correct because the identity of the gas is irrelevant

85. According to the kinetic theory, which is NOT true of ideal gases?

    A. For a sample of gas molecules, average kinetic energy is directly proportional to temperature

    B. There are no attractive or repulsive forces between gas molecules

    C. Collisions among gas molecules are perfectly elastic

    D. There is no transfer of kinetic energy during collisions between gas molecules

    E. None of the above

---

86. At what temperature is degrees Celsius equivalent to degrees Fahrenheit?

A. 0   **B.** −10   **C.** −25   **D.** −40   **E.** −50

87. Which statement about the boiling point of water is NOT correct?

A. At sea level and a pressure of 760 mmHg, the boiling point is 100 °C
B. In a pressure cooker, shorter cooking times are achieved due to the change in boiling point
C. The boiling point is greater than 100 °C in a pressure cooker
D. The boiling point is less than 100 °C for locations at low elevations
E. The boiling point is greater than 100 °C for locations at low elevations

88. Which of the following compounds has the lowest boiling point?

A. CH_4 **C.** CH_3CH_2OH

B. $CHCl_3$ **D.** NH_3

                                                         E. CH_2Cl_2

89. As the pressure is increased on solid CO_2, the melting point is:

A. decreased   **C.** increased

B. unchanged   **D.** inversely proportional to the square root of the change

                                                         E. cannot be determined without further information

90. A vessel contains 32.0 g of CH_4 gas and 12.75 g of NH_3 gas at a combined pressure of 2.4 atm. What is the partial pressure of the NH_3 gas?

A. 0.30 atm   **C.** 0.44 atm

B. 1.22 atm   **D.** 1.88 atm

                                                         E. 0.66 atm

91. Which of the following statements bests describes a liquid?

A. Definite shape, but an indefinite volume
B. Indefinite shape, but a definite volume
C. Indefinite shape and volume
D. Definite shape and volume
E. Definite shape, but indefinite mass

92. Assuming constant pressure, if a volume of nitrogen gas at 420 K decreases from 100 mL to 50 mL, what is the final temperature in Kelvin?

A. 630 K   **C.** 910 K

B. 420 K   **D.** 150 K

                                                         E. 210 K

93. Which of the following laws states that pressure and Kelvin temperature are directly proportional for a gas at constant volume?

A. Gay-Lussac's law

B. Dalton's law

C. Charles' law

D. Boyle's law

E. Avogadro's law

94. The Gay–Lussac's law of increased pressure due to increased temperature is explained using kinetic molecular theory stating that the pressure must increase because the molecules:

A. increase in size

B. move slower

C. decrease in size

D. strike the container walls less often

E. strike the container walls more often

95. Which of the following terms does NOT involve the solid state?

A. solidification

B. sublimation

C. evaporation

D. melting

E. deposition

96. Which of the following compounds exhibits primarily dipole-dipole intermolecular forces?

A. CO_2 B. F_2 C. $CH_3–O–CH_3$ D. CH_3CH_3 E. CCl_4

97. Which of the following increases the pressure of a gas?

A. Decreasing the volume

B. Increasing the number of molecules

C. Increasing temperature

D. None of the above

E. All of the above

98. According to Avogadro's law, the volume of a gas [   ] as the [   ] increases while [   ] is held constant.

A. increases… temperature… pressure and number of moles

B. decreases… pressure… temperature and number of moles

C. increases… pressure… temperature and number of moles

D. decreases… number of moles… pressure and temperature

E. increases… number of moles… pressure and temperature

99. A gas initially filled a 3.0 L container. Heat was then added to the gas, which raised its temperature from 100 K to 150 K while increasing its pressure from 3.0 atm to 4.5 atm. What is the new volume of the gas?

A. 1.4 L   B. 3.0 L   C. 2.0 L   D. 4.5 L   E. 6.0 L

100. Consider a 10.0 liters sample of helium and a 10.0 liters sample of neon, both at 23 °C, 2.0 atm. Which statement regarding these samples is NOT true?

- **A.** The density of the neon sample is greater than the density of the helium sample
- **B.** Each sample contains the same number of moles of gas
- **C.** Each sample weighs the same amount
- **D.** Each sample contains the same number of atoms of gas
- **E.** All statements are true

Stoichiometry

==

Practice Set 1: Questions 1–20

==

1. What is the volume of three moles of O_2 at STP?

   A. 11.20 L   **B.** 22.71 L   **C.** 68.13 L   **D.** 32.00   **E.** 5.510 L

2. In the following reaction, which of the following describes H_2SO_4?

   $$H_2SO_4 + HI \rightarrow I_2 + SO_2 + H_2O$$

   A. reducing agent and is reduced   **C.** oxidizing agent and is reduced

   B. reducing agent and is oxidized   **D.** oxidizing agent and is oxidized

                                              E. neither an oxidizing nor reducing agent

3. Select the balanced chemical equation for the reaction: $C_6H_{14} + O_2 \rightarrow CO_2 + H_2O$

   A. $3\ C_6H_{14} + O_2 \rightarrow 18\ CO_2 + 22\ H_2O$ **C.** $2\ C_6H_{14} + 19\ O_2 \rightarrow 12\ CO_2 + 14\ H_2O$

   B. $2\ C_6H_{14} + 12\ O_2 \rightarrow 12\ CO_2 + 14\ H_2O$ **D.** $2\ C_6H_{14} + 9\ O_2 \rightarrow 12\ CO_2 + 7\ H_2O$

                                                  E. $C_6H_{14} + O_2 \rightarrow CO_2 + H_2O$

4. An oxidation number is the [   ] that an atom [   ] when the electrons in each bond are assigned to the [   ] electronegative of the two atoms involved in the bond.

   A. number of protons … definitely has … more

   B. charge … definitely has … less

   C. number of electrons … definitely has … more

   D. number of electrons … appears to have … less

   E. charge … appears to have … more

5. What is the empirical formula of acetic acid (CH_3COOH)?

   A. CH_3COOH **B.** $C_2H_4O_2$ **C.** CH_2O **D.** CO_2H_2 **E.** CHO

6. In which of the following compounds does Cl have an oxidation number of +7?

   A. $NaClO_2$ **C.** $Ca(ClO_3)_2$

   B. $Al(ClO_4)_3$ **D.** $LiClO_3$

                           E. none of the above

7. What is the formula mass of a molecule of CO_2?

   A. 44 amu   **C.** 56.5 amu

   B. 52 amu   **D.** 112 amu

                                 E. None of the above

8. What is the product of heating cadmium metal and powdered sulfur?

   A. CdS_2 **C.** CdS

   B. Cd_2S_3 **D.** Cd_2S

                                 E. Cd_3S_2

9. After balancing the following redox reaction, what is the coefficient of NaCl?

$$Cl_2\ (g) + NaI\ (aq) \rightarrow I_2\ (s) + NaCl\ (aq)$$

   A. 1   **B.** 2   **C.** 3   **D.** 5   **E.** none of the above

10. Which substance listed below is the strongest reducing agent, given the following *spontaneous* redox reaction?

$$FeCl_3\ (aq) + NaI\ (aq) \rightarrow I_2\ (s) + FeCl_2\ (aq) + NaCl\ (aq)$$

   A. $FeCl_2$ **B.** I_2 **C.** NaI   **D.** $FeCl_3$ **E.** NaCl

11. 14.5 moles of N_2 gas is mixed with 34 moles of H_2 gas in the following reaction:

$$N_2\ (g) + 3\ H_2\ (g) \rightarrow 2\ NH_3\ (g)$$

How many moles of N_2 gas remain if the reaction produces 18 moles of NH_3 gas when performed at 600 K?

   A. 0.6 moles   **C.** 5.5 moles

   B. 1.4 moles   **D.** 7.4 moles

                                 E. 9.6 moles

12. Which equation is NOT correctly classified by the type of chemical reaction?

   A. $AgNO_3 + NaCl \rightarrow AgCl + NaNO_3$ (double-replacement/non-redox)

   B. $Cl_2 + F_2 \rightarrow 2\ ClF$ (synthesis/redox)

   C. $H_2O + SO_2 \rightarrow H_2SO_3$ (synthesis/non-redox)

   D. $CaCO_3 \rightarrow CaO + CO_2$ (decomposition/redox)

   E. All are correctly classified

13. How many grams of H_2O can be formed from a reaction between 10 grams of oxygen and 1 gram of hydrogen?

    A. 11 grams of H_2O are formed since mass must be conserved

    B. 10 grams of H_2O are formed since the mass of water produced cannot be greater than the amount of oxygen reacting

    C. 9 grams of H_2O are formed because oxygen and hydrogen react in an 8:1 mass ratio

    D. No H_2O is formed because there is insufficient hydrogen to react with the oxygen

    E. Not enough information is provided

14. How many grams of Ba^{2+} ions are in an aqueous solution of $BaCl_2$ that contains 6.8×10^{22} Cl ions?

    A. 3.2×10^{48} g   **B.** 12 g   **C.** 14.5 g   **D.** 7.8 g   **E.** 9.8 g

15. What is the oxidation number of Br in $NaBrO_3$?

    A. −1   **B.** +1   **C.** +3   **D.** +5   **E.** none of the above

16. When aluminum metal reacts with ferric oxide (Fe_2O_3), a displacement reaction yields two products, with one being metallic iron. What is the sum of the coefficients of the products of the balanced reaction?

    A. 4   **B.** 6   **C.** 2   **D.** 5   **E.** 3

17. Which of the following reactions is NOT correctly classified?

    A. $AgNO_3$ (*aq*) + KOH (*aq*) → KNO_3 (*aq*) + AgOH (*s*) : non-redox / double replacement

    B. $2 H_2O_2$ (*s*) → $2 H_2O$ (*l*) + O_2 (*g*) : non-redox / decomposition

    C. $Pb(NO_3)_2$ (*aq*) + 2 Na (*s*) → Pb (*s*) + 2 $NaNO_3$ (*aq*) : redox / single-replacement

    D. HNO_3 (*aq*) + LiOH (*aq*) → $LiNO_3$ (*aq*) + H_2O (*l*) : non-redox / double-replacement

    E. All are correctly classified

18. Calculate the number of O_2 molecules if a 15.0 L cylinder was filled with O_2 gas at STP. (Use the conversion factor of 1 mole of $O_2 = 6.02 \times 10^{23}$ O_2 molecules)

    A. 443 molecules   **C.** 4.03×10^{23} molecules

    B. 6.59×10^{24} molecules   **D.** 2.77×10^{22} molecules

                                    E. 4,430 molecules

19. Which of the following is a guideline for balancing redox equations by the oxidation number method?

    A. Verify that the number of atoms and the ionic charge are the same for reactants and products

    B. In front of the substance reduced, place a coefficient that corresponds to the number of electrons lost by the substance oxidized

    C. In front of the substance oxidized, place a coefficient that corresponds to the number of electrons gained by the substance reduced

    D. Determine the electrons lost by the substance oxidized and gained by the substance reduced

    E. All of the above

20. Which of the following represents the oxidation of Co^{2+}?

    A. $Co \rightarrow Co^{2+} + 2\,e^-$ **C.** $Co^{2+} + 2\,e^- \rightarrow Co$

    B. $Co^{3+} + e^- \rightarrow Co^{2+}$ **D.** $Co^{2+} \rightarrow Co^{3+} + e^-$

                                          E. $Co^{3+} + 2\,e^- \rightarrow Co^+$

==

Practice Set 2: Questions 21–40

==

21. How many moles of phosphorous trichloride are required to produce 365 grams of HCl when the reaction yields 75%?

$$PCl_3 (g) + 3 NH_3 (g) \rightarrow P(NH_2)_3 + 3 HCl (g)$$

    A. 1 mol   **B.** 2.5 mol   **C.** 3.5 mol   **D.** 4.5 mol   **E.** 5 mol

22. What is the oxidation number of liquid bromine in the elemental state?

    A. 0   **B.** –1   **C.** –2   **D.** –3   **E.** None of the above

23. Propane burners are used by campers for cooking. What volume of H_2O is produced by the complete combustion, as shown in the unbalanced equation, of 2.6 L of propane (C_3H_8) gas when measured at the same temperature and pressure?

$$C_3H_8 (g) + O_2 (g) \rightarrow CO_2 (g) + H_2O (g)$$

    A. 0.65 L   **B.** 10.4 L   **C.** 5.2 L   **D.** 2.6 L   **E.** 26.0 L

24. Which substance is reduced in the following redox reaction?

$$HgCl_2 (aq) + Sn^{2+} (aq) \rightarrow Sn^{4+} (aq) + Hg_2Cl_2 (s) + Cl^- (aq)$$

    A. Sn^{4+} **C.** $HgCl_2$

    B. Hg_2Cl_2 **D.** Sn^{2+}

                                             E. None of the above

25. What is the coefficient (n) of P for the balanced equation: $nP (s) + nO_2 (g) \rightarrow nP_2O_5 (s)$?

    A. 1   **B.** 2   **C.** 4   **D.** 5   **E.** none of the above

26. In basic solution, which of the following are guidelines for balancing a redox equation by the half-reaction method?

    I.   Add the two half-reactions and cancel identical species on each side of the equation

    II.   Multiply each half-reaction by a whole number so that the number of electrons lost by the substance oxidized is equal to the electrons gained by the substance reduced

    III.   Write a balanced half-reaction for the substance oxidized and the substance reduced

    A. II only   **C.** I and III only

    B. III only   **D.** II and III only

                                             E. I, II and III

27. What is the molecular formula of galactose if the empirical formula is CH_2O, and the approximate molar mass is 180 g/mol?

A. $C_6H_{12}O_6$

B. CH_2O_6

C. CH_2O

D. CHO

E. $C_{12}H_{22}O_{11}$

28. How many formula units of lithium iodide (LiI) have a mass equal to 6.45 g? (Use the molecular mass of LiI = 133.85 g)

A. 3.45×10^{23} formula units

B. 6.43×10^{23} formula units

C. 1.65×10^{24} formula units

D. 7.74×10^{25} formula units

E. 2.90×10^{22} formula units

29. What is/are the product(s) of the reaction of N_2 and O_2 gases in a combustion engine?

    I. NO   II. NO_2 III. N_2O

A. I only

B. II only

C. III only

D. I and II only

E. I, II and III

30. What is the coefficient (n) of O_2 gas for the balanced equation?

    $n\text{P }(s) + n\text{O}_2 (g) \rightarrow n\text{P}_2\text{O}_3 (s)$

A. 1   B. 2   C. 3   D. 5   E. None of the above

31. Which substance is the weakest reducing agent given the spontaneous redox reaction?

    $\text{Mg }(s) + \text{Sn}^{2+} (aq) \rightarrow \text{Mg}^{2+} (aq) + \text{Sn }(s)$

A. Sn   B. Mg^{2+} C. Sn^{2+} D. Mg   E. None of the above

32. After balancing the following redox reaction in acidic solution, what is the coefficient of H^+?

    $\text{Mg }(s) + \text{NO}_3^- (aq) \rightarrow \text{Mg}^{2+} (aq) + \text{NO}_2 (aq)$

A. 1   B. 2   C. 4   D. 6   E. None of the above

33. Which substance contains the greatest number of moles in a 10 g sample?

A. SiO_2 B. SO_2 C. CBr_4 D. CO_2 E. CH_4

34. Which chemistry law is illustrated when ethyl alcohol always contains 52% carbon, 13% hydrogen, and 35% oxygen by mass?

   A. law of constant composition   **C.** law of multiple proportions

   B. law of constant percentages   **D.** law of conservation of mass

                                                     E. none of the above

35. Which could NOT be true for the following reaction: $N_2 (g) + 3 H_2 (g) \rightarrow 2 NH_3 (g)$?

   A. 25 grams of N_2 gas reacts with 75 grams of H_2 gas to form 50 grams of NH_3 gas

   B. 28 grams of N_2 gas reacts with 6 grams of H_2 gas to form 34 grams of NH_3 gas

   C. 15 moles of N_2 gas reacts with 45 moles of H_2 gas to form 30 moles of NH_3 gas

   D. 5 molecules of N_2 gas reacts with 15 molecules of H_2 gas to form 10 molecules of NH_3 gas

   E. None of the above

36. From the following reaction, if 0.2 moles of Al is allowed to react with 0.4 moles of Fe_2O_3, how many grams of aluminum oxide is produced?

        $2 Al + Fe_2O_3 \rightarrow 2 Fe + Al_2O_3$

   A. 2.8 g   **B.** 5.1 g   **C.** 10.2 g   **D.** 14.2 g   **E.** 18.6 g

37. In all of the following compounds, the oxidation number of hydrogen is +1, EXCEPT:

   A. NH_3 **B.** $HClO_2$ **C.** H_2SO_4 **D.** NaH   **E.** none of the above

38. Which equation is NOT correctly classified by the type of chemical reaction?

   A. $PbO + C \rightarrow Pb + CO$: single-replacement/non-redox

   B. $2 Na + 2HCl \rightarrow 2 NaCl + H_2$: single-replacement/redox

   C. $NaHCO_3 + HCl \rightarrow NaCl + H_2O + CO_2$: double-replacement/non-redox

   D. $2 Na + H_2 \rightarrow 2 NaH$: synthesis/redox

   E. All are correctly classified

39. What are the oxidation states of sulfur in H_2SO_4 and H_2SO_3, respectively?

   A. +4 and +4   **C.** +2 and +4

   B. +6 and +4   **D.** +4 and +2

                                                E. +4 and +6

40. A latex balloon has a volume of 500 mL when filled with gas at a pressure of 780 torr and a temperature of 320 K. How many moles of gas does the balloon contain? (Use the ideal gas constant $R = 0.08206$ L·atm K^{-1} mol^{-1})

   A. 0.0195   **B.** 0.822   **C.** 3.156   **D.** 18.87   **E.** 1.282

==

Practice Set 3: Questions 41–60

==

41. What is the term for a substance that causes the reduction of another substance in a redox reaction?

A. oxidizing agent   C. anode

B. reducing agent   D. cathode

                                          E. none of the above

42. Which of the following is a method for balancing a redox equation in acidic solution by the half-reaction method?

A. Multiply each half-reaction by a whole number so that the number of electrons lost by the substance oxidized is equal to the electrons gained by the substance reduced

B. Add the two half-reactions and cancel identical species from each side of the equation

C. Write a half-reaction for the substance oxidized and the substance reduced

D. Balance the atoms in each half-reaction; balance oxygen with water and hydrogen with H^+

E. All of the above

43. What is the formula mass of a molecule of $C_6H_{12}O_6$?

A. 148 amu   B. 27 amu   C. 91 amu   D. 180 amu   E. None of the above

44. Is it possible to have a macroscopic sample of oxygen that has a mass of 12 atomic mass units?

A. No, because oxygen is a gas at room temperature

B. Yes, because it would have the same density as nitrogen

C. No, because this is less than a macroscopic quantity

D. Yes, but it would need to be made of isotopes of oxygen atoms

E. No, because this is less than the mass of a single oxygen atom

45. What is the coefficient for O_2 when balanced with the lowest whole number coefficients?

$$C_2H_6 + O_2 \rightarrow CO_2 + H_2O$$

A. 3   B. 4   C. 6   D. 7   E. 9

46. Which element is reduced in the following redox reaction?

$$BaSO_4 + 4\ C \rightarrow BaS + 4\ CO$$

A. O in CO   C. S in BaS

B. Ba in BaS   D. C in CO

                                          E. S in $BaSO_4$

47. Ethanol (C_2H_5OH) is blended with gasoline as a fuel additive. If the combustion of ethanol produces carbon dioxide and water, what is the coefficient of oxygen in the balanced equation?

$$\underset{}{\_\_}\, C_2H_5OH\,(g) + \underset{}{\_\_}\, O_2\,(g) \xrightarrow{\text{Spark}} \underset{}{\_\_}\, CO_2\,(g) + \underset{}{\_\_}\, H_2O\,(g)$$

A. 1   **B.** 2   **C.** 3   **D.** 6   **E.** None of the above

48. How many moles of C are in a 4.50 g sample if the atomic mass of C is 12.011 amu?

A. 5.40 moles   **C.** 1.00 moles

B. 2.67 moles   **D.** 0.54 moles   **E.** 0.375 moles

49. Upon combustion analysis, a 6.84 g sample of a hydrocarbon yielded 8.98 grams of carbon dioxide. The percent, by mass, of carbon in the hydrocarbon is:

A. 18.6%   **B.** 23.7%   **C.** 35.8%   **D.** 11.4%   **E.** 52.8%

50. Select the balanced chemical equation:

A. $2\,C_2H_5OH + 2\,Na_2Cr_2O_7 + 8\,H_2SO_4 \rightarrow 2\,HC_2H_3O_2 + 2\,Cr_2(SO_4)_3 + 4\,Na_2SO_4 + 11\,H_2O$

B. $2\,C_2H_5OH + Na_2Cr_2O_7 + 8\,H_2SO_4 \rightarrow 3\,HC_2H_3O_2 + 2\,Cr_2(SO_4)_3 + 2\,Na_2SO_4 + 11\,H_2O$

C. $C_2H_5OH + 2\,Na_2Cr_2O_7 + 8\,H_2SO_4 \rightarrow HC_2H_3O_2 + 2\,Cr_2(SO_4)_3 + 2\,Na_2SO_4 + 11\,H_2O$

D. $C_2H_5OH + Na_2Cr_2O_7 + 2\,H_2SO_4 \rightarrow HC_2H_3O_2 + Cr_2(SO_4)_3 + 2\,Na_2SO_4 + 11\,H_2O$

E. $3\,C_2H_5OH + 2\,Na_2Cr_2O_7 + 8\,H_2SO_4 \rightarrow 3\,HC_2H_3O_2 + 2\,Cr_2(SO_4)_3 + 2\,Na_2SO_4 + 11\,H_2O$

51. Under acidic conditions, what is the sum of the coefficients in the balanced reaction below?

$$Fe^{2+} + Cr_2O_7^{2-} \rightarrow Fe^{3+} + Cr^{3+}$$

A. 8   **B.** 14   **C.** 17   **D.** 36   **E.** none of the above

52. Assuming STP, if 49 g of H_2SO_4 are produced in the following reaction, what volume of O_2 must be used in the reaction?

$$RuS\,(s) + O_2 + H_2O \rightarrow Ru_2O_3\,(s) + H_2SO_4$$

A. 20.6 liters   **C.** 28.3 liters

B. 31.2 liters   **D.** 29.1 liters   **E.** 25.2 liters

53. From the following reaction, if 0.20 mole of Al is allowed to react with 0.40 mole of Fe_2O_3, how many moles of iron are produced?

$$2\,Al + Fe_2O_3 \rightarrow 2\,Fe + Al_2O_3$$

A. 0.05 mole   **C.** 0.20 mole

B. 0.075 mole   **D.** 0.10 mole   **E.** 0.25 mole

54. What is the oxidation number of sulfur in the $S_2O_8^{2-}$ ion?

   A. −1   **B.** +7   **C.** +2   **D.** +6   **E.** +1

55. What are the oxidation numbers for the elements in Na_2CrO_4?

   A. +2 for Na, +5 for Cr and −6 for O   **C.** +1 for Na, +4 for Cr and −6 for O

   B. +2 for Na, +3 for Cr and −2 for O   **D.** +1 for Na, +6 for Cr and −2 for O

                                         E. +1 for Na, +5 for Cr and −2 for O

56. What is the oxidation state of sulfur in sulfuric acid?

   A. +8   **B.** +6   **C.** −2   **D.** −6   **E.** +4

57. Which substance is oxidized in the following redox reaction?

$$HgCl_2\,(aq) + Sn^{2+}\,(aq) \rightarrow Sn^{4+}\,(aq) + Hg_2Cl_2\,(s) + Cl^-\,(aq)$$

   A. Hg_2Cl_2 **C.** Sn^{2+}

   B. Sn^{4+} **D.** $HgCl_2$ **E.** None of the above

58. Which of the following statements that is NOT true with respect to the balanced equation?

$$Na_2SO_4\,(aq) + BaCl_2\,(aq) \rightarrow 2\,NaCl\,(aq) + BaSO_4\,(s)$$

   A. Barium sulfate and sodium chloride are products

   B. Barium chloride is dissolved in water

   C. Barium sulfate is a solid

   D. $2\,NaCl\,(aq)$ could be written as $Na_2Cl_2\,(aq)$

   E. Sodium sulfate has a coefficient of one

59. Which of the following represents 1 mol of phosphine gas (PH_3)?

   I. 22.71 L phosphine gas at STP

   II. 34.00 g phosphine gas

   III. 6.02×10^{23} phosphine molecules

   A. I only   **C.** III only

   B. II only   **D.** I and II only

                                   E. II and III only

60. Which of the following represents the oxidation of Co^{2+}?

   A. $Co \rightarrow Co^{2+} + 2\,e^-$ **C.** $Co^{2+} + 2\,e^- \rightarrow Co$

   B. $Co^{3+} + e^- \rightarrow Co^{2+}$ **D.** $Co^{2+} \rightarrow Co^{3+} + e^-$

                                     E. $Co^{3+} + 2\,e^- \rightarrow Co^+$

Practice Set 4: Questions 61–80

61. What is the term for a substance that causes oxidation in a redox reaction?

A. oxidized
B. reducing agent

C. anode
D. cathode
E. oxidizing agent

62. Carbon tetrachloride (CCl_4) is a potent hepatotoxin (toxic to the liver) commonly used in the past in fire extinguishers and as a refrigerant. What is the percent by mass of Cl in carbon tetrachloride?

A. 25%   **B.** 66%   **C.** 78%   **D.** 92%   **E.** 33%

63. For a redox reaction to be balanced, which of the following is true?

I. Ionic charge of reactants must equal ionic charge of products

II. Atoms of each reactant must equal atoms of the product

III. Electron gain must equal electron loss

A. I only
B. II only

C. I and II only
D. I and III only
E. I, II and III

64. The mass percent of a compound is approximately 71.8% Cl, 24.2% C, and 4.0% H. If the molecular weight of the compound is 99 g/mol, what is the molecular formula of the compound?

A. $Cl_2C_3H_6$
B. $Cl_2C_2H_4$

C. $ClCH_3$
D. ClC_2H_2
E. Cl_3CH_3

65. In the early 1980s, benzene was used as a solvent for waxes and oils but is now listed as a carcinogen by the EPA. What is the molecular formula of benzene if the empirical formula is C_1H_1 and the approximate molar mass is 78 g/mol?

A. CH_{12} **B.** $C_{12}H_{12}$ **C.** CH   **D.** CH_6 **E.** C_6H_6

66. How many O atoms are in the formula unit $GaO(NO_3)_2$?

A. 3   **B.** 4   **C.** 5   **D.** 7   **E.** 8

67. Which reaction represents the balanced reaction for the combustion of ethanol?

A. $4 C_2H_5OH + 13 O_2 \rightarrow 8 CO_2 + 10 H_2$ **C.** $C_2H_5OH + 2 O_2 \rightarrow 2CO_2 + 2 H_2O$

B. $C_2H_5OH + 3 O_2 \rightarrow 2 CO_2 + 3 H_2O$ **D.** $C_2H_5OH + O_2 \rightarrow CO_2 + H_2O$

   E. $C_2H_5OH + \frac{1}{2} O_2 \rightarrow 2 CO_2 + 3 H_2O$

68. What is the coefficient for O_2 when the following equation is balanced with the lowest whole number coefficients?

$$\_C_3H_7OH + \_O_2 \rightarrow \_CO_2 + \_H_2O$$

A. 3   **B.** 6   `C. 9   **D.** 13/2   **E.** 12

69. After balancing the following redox reaction, what is the coefficient of CO_2?

$$\_Co_2O_3 (s) + \_CO (g) \rightarrow \_Co (s) + \_CO_2 (g)$$

A. 1   **B.** 2   **C.** 3   **D.** 4   **E.** 5

70. What is the term for the volume occupied by 1 mol of any gas at STP?

A. STP volume   **C.** standard volume

B. molar volume   **D.** Avogadro's volume

   E. none of the above

71. Which substance listed below is the weakest oxidizing agent given the following spontaneous redox reaction?

$$Mg (s) + Sn^{2+} (aq) \rightarrow Mg^{2+} (aq) + Sn (s)$$

A. Mg^{2+} **B.** Sn   **C.** Mg   **D.** Sn^{2+} **E.** none of the above

72. In the following reaction performed at 500 K, 18.0 moles of N_2 gas is mixed with 24.0 moles of H_2 gas. What is the percent yield of NH_3 if the reaction produces 13.5 moles of NH_3?

$$N_2 (g) + 3 H_2 (g) \rightarrow 2 NH_3 (g)$$

A. 16%   **B.** 66%   **C.** 72%   **D.** 84%   **E.** 100%

73. How many grams of H_2O can be produced from the reaction of 25.0 grams of H_2 and 225 grams of O_2?

A. 266 grams   **C.** 184 grams

B. 223 grams   **D.** 27 grams

   E. 2.5 grams

74. What is the oxidation number of Cl in $LiClO_2$?

   A. −1   **B.** +1   **C.** +3   **D.** +5   **E.** None of the above

75. In acidic conditions, what is the sum of the coefficients in the products of the balanced reaction?

$$MnO_4^- + C_3H_7OH \rightarrow Mn^{2+} + C_2H_5COOH$$

   A. 12   **B.** 16   **C.** 18   **D.** 20   **E.** 24

76. Which reaction is NOT correctly classified?

   A. PbO (*s*) + C (*s*) → Pb (*s*) + CO (*g*) : (double-replacement)
   B. CaO (*s*) + H_2O (*l*) → $Ca(OH)_2$ (*aq*) : (synthesis)
   C. $Pb(NO_3)_2$ (*aq*) + 2LiCl (*aq*) → 2 $LiNO_3$ (*aq*) + $PbCl_2$ (*s*) : (double-replacement)
   D. Mg (*s*) + 2 HCl (*aq*) → $MgCl_2$ (*aq*) + H_2 (*g*) : (single-replacement)
   E. All are classified correctly

77. The reactants for this chemical reaction are:

$$C_6H_{12}O_6 + 6\ H_2O + 6\ O_2 \rightarrow 6\ CO_2 + 12\ H_2O$$

   A. $C_6H_{12}O_6$, H_2O, O_2 and CO_2 **C.** $C_6H_{12}O_6$
   B. $C_6H_{12}O_6$ and H_2O **D.** $C_6H_{12}O_6$ and CO_2
                                   E. $C_6H_{12}O_6$, H_2O and O_2

78. What is the charge of the electrons in 4 grams of He? (Use Faraday constant F = 96,500 C/mol)

   A. 48,250 C   **C.** 193,000 C
   B. 96,500 C   **D.** 386,000 C
                                   E. Cannot be determined

79. What is the oxidation number of iron in the compound $FeBr_3$?

   A. −2   **B.** +1   **C.** +2   **D.** +3   **E.** −1

80. What is the term for the amount of substance that contains 6.02×10^{23} particles?

   A. molar mass   **C.** Avogadro's number
   B. mole   **D.** formula mass
                                   E. none of the above

===

Practice Set 5: Questions 81–100

===

81. Which of the following species undergoes oxidation in 2 CuBr → 2 Cu + Br$_2$?

    A. Cu$^+$ **B.** Br$^-$ **C.** Cu   **D.** CuBr   **E.** Br$_2$

82. How many electrons are lost or gained by each formula unit of CuBr$_2$ in this reaction?

            Zn + CuBr$_2$ → ZnBr$_2$ + Cu

    A. loses 1 electron   **C.** gains 2 electrons
    B. gains 6 electrons   **D.** loses 2 electrons
                                              E. gains 4 electrons

83. How many molecules of CO$_2$ are in 168.0 grams of CO$_2$?

    A. 3.96×10^{23} **C.** 4.24×10^{22}
    B. 2.30×10^{24} **D.** 6.82×10^{24}
                                              E. 4.60×10^{23}

84. What is the empirical formula of a compound that, by mass, contains 64% silver, 8% nitrogen and 28% oxygen?

    A. Ag$_3$NO   **C.** AgNO$_2$
    B. Ag$_3$NO$_3$ **D.** Ag$_3$NO$_2$
                                               E. AgNO$_3$

85. What is the mass of one mole of a gas that has a density of 1.34 g/L at STP?

    A. 48.0 g   **C.** 18.3 g
    B. 56.4 g   **D.** 30.1 g
                                               E. 4.39 g

86. Propane (C$_3$H$_8$) is flammable and used as a substitute for natural gas. What is the coefficient of oxygen in the balanced equation for the combustion of propane?

                     Spark

           __C$_3$H$_8$ (g) + __O$_2$ (g) → __CO$_2$ (g) + __H$_2$O (g)

    A. 1   **B.** 7   **C.** 5   **D.** 10   **E.** None of the above

87. This reaction yields how many atoms of oxygen?

$$C_6H_{12}O_6 + 6\ H_2O + 6\ O_2 \rightarrow 6\ CO_2 + 12\ H_2O$$

A. 3   **B.** 12   **C.** 14   **D.** 24   **E.** 36

88. After balancing the following redox reaction, what is the coefficient of O_2?

$$\_Al_2O_3\ (s) + \_Cl_2\ (g) \rightarrow \_AlCl_3\ (aq) + \_O_2\ (g)$$

A. 1   **B.** 2   **C.** 3   **D.** 5   **E.** none of the above

89. Which reaction is the correctly balanced half-reaction (in acid solution) for the process below?

$$Cr_2O_7^{2-}\ (aq) \rightarrow Cr^{3+}\ (aq)$$

A. $8\ H^+ + Cr_2O_7 \rightarrow 2\ Cr^{3+} + 4\ H_2O + 3\ e^-$

B. $12\ H^+ + Cr_2O_7^{2-} + 3\ e^- \rightarrow 2\ Cr^{3+} + 6\ H_2O$

C. $14\ H^+ + Cr_2O_7^{2-} + 6\ e^- \rightarrow 2\ Cr^{3+} + 7\ H_2O$

D. $8\ H^+ + Cr_2O_7 + 3\ e^- \rightarrow 2\ Cr^{3+} + 4\ H_2O$

E. None of the above

90. Determine the oxidation number of C in $NaHCO_3$:

A. +5   **B.** +4   **C.** +12   **D.** +6   **E.** +2

91. What principle states that equal volumes of gases at the same temperature and pressure contain equal numbers of molecules?

A. Dalton's theory   **C.** Boyle's theory

B. Charles' theory   **D.** Avogadro's theory

E. None of the above

92. What are the products for this double-replacement reaction?

$$BaCl_2\ (aq) + K_2SO_4\ (aq) \rightarrow$$

A. $BaSO_3$ and $KClO_4$ **C.** BaS and $KClO_4$

B. $BaSO_4$ and $2\ KCl$ **D.** $BaSO_3$ and KCl

E. $BaSO_4$ and $KClO_4$

93. The formula for mustard gas used in chemical warfare is $C_4H_8SCl_2$. What is the percent by mass of chlorine in mustard gas? (Use the molecular mass of mustard gas = 159.09 g/mol)

A. 18.4%   **B.** 44.6%   **C.** 14.6%   **D.** 31.2%   **E.** 28.8%

94. Which substance listed is the weakest oxidizing agent given the following *spontaneous* redox reaction?

$$FeCl_3 \ (aq) + NaI \ (aq) \rightarrow I_2 \ (s) + FeCl_2 \ (aq) + NaCl \ (aq)$$

A. I_2 **B.** $FeCl_2$ **C.** $FeCl_3$ **D.** NaI   **E.** NaCl

95. If the reaction below is run at STP with excess H_2O, and 22.4 liters of O_2 reacts with 67 g of RuS, how many grams of H_2SO_4 are produced?

$$RuS \ (s) + O_2 + H_2O \rightarrow Ru_2O_3 \ (s) + H_2SO_4$$

A. 28 g   **B.** 32 g   **C.** 44 g   **D.** 54 g   **E.** 58 g

96. What is the oxidation number of sulfur in calcium sulfate, $CaSO_4$?

A. +6   **B.** +4   **C.** +2   **D.** 0   **E.** −2

97. How many moles of O_2 gas are required for combustion with one mole of $C_6H_{12}O_6$ in the unbalanced reaction?

$$C_6H_{12}O_6 \ (s) + O_2 \ (g) \rightarrow CO_2 \ (g) + H_2O \ (g)$$

A. 1   **B.** 2.5   **C.** 6   **D.** 10   **E.** 12

98. Which reaction below is a *synthesis* reaction?

A. $3 \ CuSO_4 + 2 \ Al \rightarrow Al_2(SO_4)_3 + 3 \ Cu$ **C.** $2 \ NaHCO_3 \rightarrow Na_2CO_3 + CO_2 + H_2O$

B. $SO_3 + H_2O \rightarrow H_2SO_4$ **D.** $C_3H_8 + 5 \ O_2 \rightarrow 3 \ CO_2 + 4 \ H_2O$

E. None of the above

99. Which statement is true regarding the coefficients in a chemical equation?

    I. They appear before the chemical formulas
    II. They appear as subscripts
    III. Coefficients of reactants sum to those of the products

A. I only   **C.** I and II only

B. II only   **D.** I and III only

                               E. I, II and III

100. What is the volume occupied by 17.0 g of NO gas at STP?

A. 23.4 L   **B.** 58.4 L   **C.** 0.58 L   **D.** 12.7 L   **E.** 46.8 L

==

Practice Set 6: Questions 101–120

==

101. Which of the following is/are likely to act as an oxidizing agent?

   I. Cl^- II. Cl_2 III. Na^+

   A. I only   **C.** III only
   B. II only   **D.** I and II only
                                          E. I, II and III

102. When a substance loses electrons, it is [   ], while the substance itself acts as [   ] agent.

   A. reduced… a reducing   **C.** reduced… an oxidizing
   B. oxidized… a reducing   **D.** oxidized… an oxidizing
                                          E. dissolved… a neutralizing

103. What is the term for the chemical formula of a compound that expresses the actual number of atoms of each element in a molecule?

   A. molecular formula   **C.** elemental formula
   B. empirical formula   **D.** atomic formula
                                          E. none of the above

104. Which substance is the weakest reducing agent given the spontaneous redox reaction?

   $FeCl_3$ (*aq*) + NaI (*aq*) → I_2 (*s*) + $FeCl_2$ (*aq*) + NaCl (*aq*)

   A. NaCl   **B.** I_2 **C.** NaI   **D.** $FeCl_3$ **E.** $FeCl_2$

105. What is the mass of 3.5 moles of glucose which has a molecular formula of $C_6H_{12}O_6$?

   A. 180 g   **C.** 630 g
   B. 90 g   **D.** 51.4 g
                                          E. 520 g

106. Butane (C_4H_{10}) is flammable and used in butane lighters. What is the coefficient of oxygen in the balanced equation for the combustion of butane?

                     Spark
   __C_4H_{10} (*g*) + __O_2 (*g*) → __CO_2 (*g*) + __H_2O (*g*)

   A. 9   **B.** 13   **C.** 18   **D.** 26   **E.** none of the above

---

107. Which of the following is an accurate statement for Avogadro's number?

    I. Equal to the number of atoms in 1 mole of an element

    II. Equal to the number of molecules in a compound

    III. Equal to approximately 6.022×10^{23}

A. I only

B. II only

C. III only

D. I and III only

E. I, II and III

108. What is the sum of the coefficients in the balanced reaction (no fractional coefficients)?

    __RuS (s) + __O_2 + __H_2O → __Ru_2O_3 (s) + __H_2SO_4

A. 23   **B.** 18   **C.** 13   **D.** 25   **E.** 29

109. Which is sufficient for determining the molecular formula of a compound?

    I. The % by mass of a compound

    II. The molecular weight of a compound

    III. The amount of a compound in a sample

A. I only

B. II only

C. I and II only

D. I and III only

E. I, II and III

110. Which equation(s) is/are balanced?

    I. Mg (s) + 2 HCl (aq) → $MgCl_2$ (aq) + H_2 (g)

    II. 3 Al (s) + 3 Br_2 (l) → Al_2Br_3 (s)

    III. 2 HgO (s) → 2 Hg (l) + O_2 (g)

A. I only

B. II only

C. III only

D. I and II only

E. I and III only

111. How many moles are in 60.2 g of $MgSO_4$?

A. 0.25 mole

B. 0.5 mole

C. 0.45 mole

D. 0.65 mole

E. 0.75 mole

112. How many atoms of Mg are in a solid 72.9 g sample of magnesium? (Use the molecular mass of Mg = 24.3 g/mol)

  A. 1.81×10^{24} atoms   **C.** 3.01×10^{23} atoms

  B. 4,800 atoms   **D.** 2.42×10^{25} atoms

                                         E. 6.02×10^{23} atoms

113. The formula for the illegal drug cocaine is $C_{17}H_{21}NO_4$ (303.39 g/mol). What is the percent by mass of oxygen in cocaine?

  A. 21.1%   **B.** 6.5%   **C.** 457%   **D.** 4.4%   **E.** 62.8%

114. If 7 moles of RuS are used in the following reaction, what is the maximum number of moles of Ru_2O_3 that can be produced?

$$RuS\ (s) + O_2 + H_2O \rightarrow Ru_2O_3\ (s) + H_2SO_4$$

  A. 3.5 moles   **C.** 5.8 moles

  B. 1.8 moles   **D.** 2.2 moles

                                E. 1.6 moles

115. In which sequence are sulfur-containing species arranged in the order of *decreasing* oxidation numbers for S?

  A. SO_4^{2-}, S^{2-}, $S_2O_3^{2-}$ **C.** SO_4^{2-}, $S_2O_3^{2-}$, S^{2-}

  B. $S_2O_3^{2-}$, SO_3^{2-}, S^{2-} **D.** SO_3^{2-}, SO_4^{2-}, S^{2-}

                                  E. S^{2-}, SO_4^{2-}, $S_2O_3^{2-}$

116. In the following compounds, the oxidation number of oxygen is −2, EXCEPT:

  A. $NaClO_2$ **C.** $Ba(OH)_2$

  B. Li_2O_2 **D.** Na_2SO_4

                                  E. none of the above

117. Which of the following is a *double-replacement* reaction?

  A. $2\ HI \rightarrow H_2 + I_2$ **C.** $HBr + KOH \rightarrow H_2O + KBr$

  B. $SO_2 + H_2O \rightarrow H_2SO_3$ **D.** $CuO + H_2 \rightarrow Cu + H_2O$

                                    E. none of the above

118. Which of the following substances has the lowest density?

  A. A mass of 750 g and a volume of 70 dL   **C.** A mass of 1.5 kg and a volume of 1.2 L

  B. A mass of 5 μg and a volume of 25 μL   **D.** A mass of 25 g and a volume of 20 mL

                                              E. A mass of 15 mg and a volume of 50 μL

119. How many moles of O_2 gas is required for combustion with 2 moles of hexane in the unbalanced reaction:
$C_6H_{14}\ (g) + O_2\ (g) \rightarrow CO_2\ (g) + H_2O\ (g)$

    A. 11   **B.** 14   **C.** 12   **D.** 20   **E.** 19

120. How is Avogadro's number related to the numbers on the periodic table?

    A. The atomic mass listed is the mass of Avogadro's number of atoms

    B. The periodic table provides the mass of one atom, while Avogadro's number provides the number of moles

    C. The masses are divisible by Avogadro's number, which provides the weight of one mole

    D. The periodic table only provides atomic numbers, not atomic mass

    E. The mass listed is in the units of Avogadro's number

==

Practice Set 7: Questions 121–140

==

121. Which of the following is/are likely to act as an oxidizing agent?

      I. Cl_2 II. Cl^- III. Na^+

  A. I only   **C.** III only

  B. II only   **D.** I and II only

                               E. I, II and III

122. In a redox reaction, the substance that is reduced:

  A. gains electrons

  B. loses electrons

  C. contains an element that increases in oxidation number

  D. is the reducing agent

  E. either gains or loses electrons

123. What is the term for a chemical formula that expresses the simplest whole-number ratio of atoms of each element in a molecule?

  A. empirical formula   **C.** atomic formula

  B. molecular formula   **D.** elemental formula

                                 E. none of the above

124. How many H atoms are in the molecule $C_6H_3(C_3H_7)_2(C_2H_5)$?

  A. 27   **B.** 15   **C.** 29   **D.** 20   **E.** 22

125. Which scientific principle is the basis for balancing chemical equations?

  A. Law of Conservation of Mass and Energy   **C.** Law of Conservation of Energy

  B. Law of Definite Proportions   **D.** Law of Conservation of Mass

                                 E. Avogadro's Law

126. What coefficient is needed for the O_2 molecule to balance the following equation?

               Spark

    __C_4H_{10} *(g)* + __O_2 *(g)* → __CO_2 *(g)* + __H_2O *(g)*

  A. 5   **B.** 1   **C.** 8   **D.** 15   **E.** 13

127. Which substance is reduced in the following redox reaction?

$$F_2 \, (g) + 2 \, Br^- \, (aq) \rightarrow 2 \, F^- \, (aq) + Br_2 \, (l)$$

A. F^- **B.** Br_2 **C.** F_2 **D.** Br^- **E.** None of the above

128. What is the total of the coefficients when balanced with the lowest whole number coefficients?

$$N_2H_4 + H_2O_2 \rightarrow N_2 + H_2O$$

A. 4   **B.** 8   **C.** 10   **D.** 12   **E.** 14

129. How many molecules of CH_4 gas have a mass equal to 6.40 g?

A. 3.33×10^{23} molecules   **C.** 1.15×10^{24} molecules

B. 1.85×10^{24} molecules   **D.** 3.75×10^{24} molecules

      E. 2.40×10^{23} molecules

130. If 0.2 moles of Al are allowed to react with 0.4 moles of Fe_2O_3, what is the limiting reactant in the following reaction?

$$2 \, Al + Fe_2O_3 \rightarrow 2 \, Fe + Al_2O_3$$

A. Fe_2O_3 **C.** Fe_2O_3 and Al_2O_3

B. Al_2O_3 **D.** Fe

      E. Al

131. Balance the equation: $\_\_H_2 \, (g) + \_\_N_2 \, (g) \rightarrow \_\_NH_3 \, (g)$

A. 3, 2, 2   **C.** 2, 2, 3

B. 3, 1, 2   **D.** 2, 2, 5

      E. 3, 2, 1

132. How many oxygen atoms are in 5.40 g of fructose ($C_6H_{12}O_6$)?

A. 2.34×10^{22} **C.** 3.34×10^{24}

B. 1.08×10^{23} **D.** 3.16×10^{23}

      E. 4.85×10^{22}

133. Which is the properly balanced equation for the reaction occurring when solid iron (III) oxide is reduced with carbon to produce carbon dioxide and molten iron?

A. $2 \, Fe_2O_3 + 3 \, C \, (s) \rightarrow 4 \, Fe \, (l) + 3 \, CO_2 \, (g)$ **C.** $2 \, FeO_3 + 3 \, C \, (s) \rightarrow 2 \, Fe \, (l) + 3 \, CO_2 \, (g)$

B. $4 \, Fe_2O_3 + 6 \, C \, (s) \rightarrow 8 \, Fe \, (l) + 6 \, CO_2 \, (g)$ **D.** $2 \, Fe_3O + C \, (s) \rightarrow 6 \, Fe \, (l) + CO_2 \, (g)$

      E. $2 \, FeO + C \, (s) \rightarrow 2 \, Fe \, (l) + CO_2 \, (g)$

134. How many moles of Sn^{4+} are produced from one mole of Sn^{2+} and excess O_2 in the following reaction?

$$Sn^{2+} + O_2 \rightarrow Sn^{4+}$$

A. 2 moles

B. 2.5 moles

C. 1 ½ moles

D. ½ mole

E. 1 mole

135. The oxidation number +7 is for the element:

A. Mn in $KMnO_4$

B. Br in $NaBrO_3$

C. C in MgC_2O_4

D. S in H_2SO_4

E. K in $KMnO_4$

136. Which of the following reactions is NOT correctly classified?

A. $AgNO_3$ (*aq*) + KOH (*aq*) → KNO_3 (*aq*) + AgOH (*s*) : non-redox / double replacement

B. 2 H_2O_2 (*s*) → 2 H_2O (*l*) + O_2 (*g*) : non-redox / decomposition

C. $Pb(NO_3)_2$ (*aq*) + 2 Na (*s*) → Pb (*s*) + 2 $NaNO_3$ (*aq*) : redox / single-replacement

D. HNO_3 (*aq*) + LiOH (*aq*) → $LiNO_3$ (*aq*) + H_2O (*l*) : non-redox / double-replacement

E. All are correctly classified

137. What is the molecular formula of a compound with an empirical formula of CHCl with a molar mass of 194 g/mol?

A. $C_4H_4Cl_4$

B. $C_2H_4Cl_3$

C. $C_3H_5Cl_3$

D. CHCl

E. $C_4H_6Cl_4$

138. How many moles of Al are in an 8.52 g sample if the atomic mass of Al is 26.98 amu?

A. 0.316 moles

B. 0.233 moles

C. 1.46 moles

D. 4.34 moles

E. 5.87 moles

139. What is the oxidation number of Cr in the compound HCr_2O_4Cl?

A. +3   **B.** +2   **C.** +6   **D.** +5   **E.** +4

140. What general term refers to the mass of 1 mol of any substance?

A. gram-formula mass

B. molar mass

C. gram-atomic mass

D. gram-molecular mass

E. none of the above

Notes for active learning

Thermochemistry

===

Practice Set 1: Questions 1–20

===

1. Which of the following statement regarding the symbol ΔG is NOT true?

   A. Specifies the enthalpy of the reaction
   B. Refers to the free energy of the reaction
   C. Predicts the spontaneity of a reaction
   D. Describes the effect of enthalpy and entropy on a reaction
   E. None of the above

2. What happens to the kinetic energy of a gas molecule when the gas is heated?

   A. Depends on the gas
   B. Kinetic energy increases
   C. Kinetic energy decreases
   D. Kinetic energy remains constant
   E. None of the above

3. How much heat energy (in Joules) is required to heat 21.0 g of copper from 21.0 °C to 68.5 °C? (Use the specific heat c of Cu = 0.382 J/g·°C)

   A. 462 J   B. 188 J   C. 522 J   D. 662 J   E. 381 J

4. For n moles of gas, which term expresses the kinetic energy?

   A. nPA, where n = number of moles of gas, P = total pressure, and A = surface area of the container walls
   B. $\frac{1}{2}n$PA, where n = number of moles of gas, P = total pressure, and A = surface area of the container walls
   C. $\frac{1}{2}MV^2$, where M = molar mass of the gas and V = volume of the container
   D. MV^2, where M = molar mass of the gas and V = volume of the container
   E. 3/2 nRT, where n = number of moles of gas, R = ideal gas constant, and T = absolute temperature

5. Which of the following is NOT an endothermic process?

   A. Condensation of water vapor
   B. Boiling liquid
   C. Water evaporating
   D. Ice melting
   E. All are endothermic

---

6. What is true of an endothermic reaction if it causes a decrease in the entropy (S) of the system?

    A. Only occurs at low temperatures when ΔS is insignificant

    B. Occurs if coupled to an endergonic reaction

    C. Never occurs because it decreases ΔS of the system

    D. Never occurs because ΔG is positive

    E. None of the above

7. Which of the following terms describe(s) energy contained in an object or transferred to an object?

    I. chemical   II. electrical   III. heat

    A. I only   **C.** I and II only

    B. II only   **D.** I and III only

                                        E. I, II and III

8. The greatest entropy is observed for which 10 g sample of CO_2?

    A. CO_2 (*g*)   **C.** CO_2 (*s*)

    B. CO_2 (*aq*)   **D.** CO_2 (*l*)

                                        E. All are equivalent

9. What is the term for a reaction that proceeds by absorbing heat energy?

    A. Isothermal reaction   **C.** Endothermic reaction

    B. Exothermic reaction   **D.** Spontaneous

                                          E. None of the above

10. If a chemical reaction has $\Delta H = X$, $\Delta S = Y$, $\Delta G = X - RY$ and occurs at R K, the reaction is:

    A. spontaneous   **C.** nonspontaneous

    B. at equilibrium   **D.** irreversible

                                            E. cannot be determined

11. The thermodynamic systems that have high stability tend to demonstrate:

    A. maximum ΔH and maximum ΔS **C.** minimum ΔH and maximum ΔS

    B. maximum ΔH and minimum ΔS **D.** minimum ΔH and minimum ΔS

                                                 E. none of the above

12. Whether a reaction is endothermic or exothermic is determined by:

  A. an energy balance between bond breaking and bond forming, resulting in a net loss or gain of energy
  B. the presence of a catalyst
  C. the activation energy
  D. the physical state of the reaction system
  E. none of the above

13. What role does entropy play in chemical reactions?

  A. The entropy change determines whether the reaction occurs spontaneously
  B. The entropy change determines whether the chemical reaction is favorable
  C. The entropy determines how much product is produced
  D. The entropy change determines whether the reaction is exothermic or endothermic
  E. The entropy determines how much reactant remains

14. Calculate the value of $\Delta H°$ of reaction using the provided bond energies.

   $H_2C{=}CH_2\,(g) + H_2\,(g) \rightarrow H_3C{-}CH_3\,(g)$

   C–C: 348 kJ   C≡C: 960 kJ

   C=C: 612 kJ   C–H: 412 kJ   H–H: 436 kJ

   A. –348 kJ   **B.** +134 kJ   **C.** –546 kJ   **D.** –124 kJ   **E.** –238 kJ

15. The bond dissociation energy is:

   I. useful in estimating the enthalpy change in a reaction
   II. the energy required to break a bond between two gaseous atoms
   III. the energy released when a bond between two gaseous atoms is broken

  A. I only
  B. II only
  C. I and II only
  D. I and III only
  E. I, II and III

16. Based on the following reaction, which statement is true?

   $N_2 + O_2 \rightarrow 2\,NO$ (Use the value for enthalpy, $\Delta H = 43.3$ kcal)

  A. 43.3 kcal are consumed when 2.0 mole of O_2 reacts
  B. 43.3 kcal are consumed when 2.0 moles of NO are produced
  C. 43.3 kcal are produced when 1.0 g of N_2 reacts
  D. 43.3 kcal are consumed when 2.0 g of O_2 reacts
  E. 43.3 kcal are produced when 2.0 g of NO are produced

17. Which of the following properties of gas is/are a state function

      I. temperature   II. heat   III. work

A. I only   **C.** II and III only

B. I and II only   **D.** I, II and III

                                     E. II only

18. The following statements concerning temperature change as a substance is heated are correct, EXCEPT:

A. As a liquid is heated, its temperature rises until its boiling point is reached

B. During the time a liquid is changing to the gaseous state, the temperature gradually increases until all the liquid is changed

C. As a solid is heated, its temperature rises until its melting point is reached

D. During the time for a solid to melt to a liquid, the temperature remains constant

E. The temperature remains the same during the phase change

19. Calculate the value of ΔH° of reaction for:

    O=C=O (g) + 3 H$_2$ (g) → CH$_3$–O–H (g) + H–O–H (g)

Use the following bond energies, ΔH°:

        C–C: 348 kJ   C=C: 612 kJ   C≡C: 960 kJ   C–H: 412 kJ

        C–O: 360 kJ   C=O: 743 kJ   H–H: 436 kJ   H–O: 463 kJ

A. –348 kJ   **C.** –191 kJ

B. +612 kJ   **D.** –769 kJ

                                     E. +5,779 kJ

20. Which statement(s) is/are true for ΔS?

      I. ΔS of the universe is conserved

      II. ΔS of a system is conserved

      III. ΔS of the universe increases with each reaction

A. I only   **C.** III only

B. II only   **D.** I and II only

                                     E. I and III only

Practice Set 2: Questions 21–40

21. Which of the following reaction energies is the most endothermic?

    A. 360 kJ/mole
    B. –360 kJ/mole
    C. 88 kJ/mole
    D. –88 kJ/mole
    E. 0 kJ/mole

22. A fuel cell contains hydrogen and oxygen gas that react explosively, and the energy converts water to steam which drives a turbine to turn a generator that produces electricity. The fuel cell and the steam represent which forms of energy, respectively?

    A. Electrical and heat energy
    B. Electrical and chemical energy
    C. Chemical and heat energy
    D. Chemical and mechanical energy
    E. Nuclear and mechanical energy

23. Which of the following is true for the ΔG of formation for $N_2 (g)$ at 25 °C?

    A. 0 kJ/mol
    B. positive
    C. negative
    D. 1 kJ/mol
    E. more information is needed

24. Which of the statements best describes the following reaction?

    $$HC_2H_3O_2 (aq) + NaOH (aq) \rightarrow NaC_2H_3O_2 (aq) + H_2O (l)$$

    A. Acetic acid and NaOH solutions produce sodium acetate and H_2O
    B. Aqueous solutions of acetic acid and NaOH produce aqueous sodium acetate and H_2O
    C. Acetic acid and NaOH solutions produce sodium acetate solution and H_2O
    D. Acetic acid and NaOH produce sodium acetate and H_2O
    E. An acid plus a base produce H_2O and a salt

25. If a chemical reaction is spontaneous, which value must be negative?

    A. C_p B. ΔS C. ΔG D. ΔH E. K_{eq}

26. What is the purpose of the hollow walls in a closed hollow-walled container that effectively maintains the temperature inside?

    A. To trap air trying to escape from the container, which minimizes convection
    B. To act as an effective insulator, which minimizes convection
    C. To act as an effective insulator, which minimizes conduction
    D. To provide an additional source of heat for the container
    E. Reactions occur within the walls that maintain the temperature within the container

27. If the heat of reaction is exothermic, which of the following is always true?

    A. The energy of the reactants is greater than the products
    B. The energy of the reactants is less than the products
    C. The reaction rate is fast
    D. The reaction rate is slow
    E. The energy of the reactants is equal to that of the products

28. How much heat must be absorbed to evaporate 16 g of NH_3 to its condensation point at −33 °C? (Use the heat of condensation for NH_3 = 1,380 J/g)

    A. 86.5 J   **C.** 118 J
    B. 2,846 J   **D.** 22,080 J
                                            E. 1,380 J

29. Which of the reactions is the most exothermic, assuming that the following energy profiles have the same scale? (Use the notation of R = reactants and P = products)

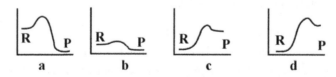

    A. a   **B.** b   **C.** c   **D.** d   **E.** Cannot be determined

30. What is the heat of formation of NH_3 (g) of the following reaction:

    $2\,NH_3\,(g) \rightarrow N_2\,(g) + 3\,H_2\,(g)$

(Use $\Delta H° = 92.4$ kJ/mol)

    A. −92.4 kJ/mol   **C.** 46.2 kJ/mol
    B. −184.4 kJ/mol   **D.** 92.4 kJ/mol
                                              E. −46.2 kJ/mol

31. If a chemical reaction has a positive ΔH and a negative ΔS, the reaction tends to be:

    A. at equilibrium   **C.** spontaneous

    B. nonspontaneous   **D.** irreversible

                                        E. unable to be determined

32. What happens to the entropy of a system as the components of the system are introduced to a larger number of possible arrangements, such as when liquid water transforms into water vapor?

    A. Entropy of a system is solely dependent upon the amount of material undergoing reaction

    B. Entropy of a system is independent of introducing the components of the system to a larger number of possible arrangements

    C. Entropy increases because there are more ways for the energy to disperse

    D. Entropy decreases because there are fewer ways in which the energy can disperse

    E. Entropy increases because there are fewer ways for the energy to disperse

33. Which of the following quantities is needed to calculate the amount of heat energy released as water turns to ice at 0 °C?

    A. The heat of condensation for water and the mass

    B. The heat of vaporization for water and the mass

    C. The heat of fusion for water and the mass

    D. The heat of solidification for water and the mass

    E. The heat of fusion for water only

34. A nuclear power plant uses ^{235}U to convert water to steam that drives a turbine which turns a generator to produce electricity. What are the initial and final forms of energy, respectively?

    A. Heat energy and electrical energy   **C.** Chemical energy and mechanical energy

    B. Nuclear energy and electrical energy   **D.** Chemical energy and heat energy

                                          E. Nuclear energy and mechanical energy

35. Which statement(s) is/are correct for the entropy?

       I. Higher for a sample of gas than for the same sample of liquid

       II. A measure of the disorder in a system

       III. Available energy for conversion into mechanical work

    A. I only   **C.** III only

    B. II only   **D.** I and II only

                                      E. I, II and III

36. Given the following data, what is the heat of formation for ethanol?

$C_2H_5OH + 3\ O_2 \rightarrow 2\ CO_2 + 3\ H_2O : \Delta H = 327.0$ kcal/mole

$H_2O \rightarrow H_2 + \frac{1}{2}\ O_2 : \Delta H = +68.3$ kcal/mole

$C + O_2 \rightarrow CO_2 : \Delta H = -94.1$ kcal/mole

A. −720.1 kcal

B. −327.0 kcal

C. +62.6 kcal

D. +720.1 kcal

E. +327.0 kcal

37. Based on the reaction shown, which statement is true?

$S + O_2 \rightarrow SO_2 + 69.8$ kcal

A. 69.8 kcal are consumed when 32.1 g of sulfur reacts

B. 69.8 kcal are produced when 32.1 g of sulfur reacts

C. 69.8 kcal are consumed when 1 g of sulfur reacts

D. 69.8 kcal are produced when 1 g of sulfur reacts

E. 69.8 kcal are produced when 1 g of sulfur dioxide is produced

38. In the reaction, $2\ H_2\ (g) + O_2\ (g) \rightarrow 2\ H_2O\ (g)$, entropy is:

A. increasing

B. the same

C. inversely proportional

D. decreasing

E. unable to be determined

39. For an isolated system, which of the following can NOT be exchanged between the system and its surroundings?

    I. Temperature   II. Matter   III. Energy

A. I only

B. II only

C. III only

D. II and III only

E. I, II and III

40. Which is a state function?

    I. ΔG II. ΔH III. ΔS

A. I only

B. II only

C. III only

D. I and II only

E. I, II and III

===

Practice Set 3: Questions 41–60

===

41. To simplify comparisons, the energy value of fuels is expressed in units of:

A. kcal/g

B. kcal/L

C. J/kcal

D. kcal/mol

E. kcal

42. Which process is slowed down when an office worker places a lid on a hot cup of coffee?

    I. Radiation   II. Conduction   III. Convection

A. I only

B. II only

C. III only

D. I and III only

E. I, II and III

43. Which statement below is always true for a spontaneous chemical reaction?

A. $\Delta S_{sys} - \Delta S_{surr} = 0$

B. $\Delta S_{sys} + \Delta S_{surr} > 0$

C. $\Delta S_{sys} + \Delta S_{surr} < 0$

D. $\Delta S_{sys} + \Delta S_{surr} = 0$

E. $\Delta S_{sys} - \Delta S_{surr} < 0$

44. A fuel cell contains hydrogen and oxygen gas that react explosively, and the energy converts water to steam, which drives a turbine to turn a generator that produces electricity. What energy changes are employed in the process?

    I. Mechanical $\rightarrow$ electrical energy

    II. Heat $\rightarrow$ mechanical energy

    III. Chemical $\rightarrow$ heat energy

A. I only

B. II only

C. I and II only

D. I and III only

E. I, II and III

45. Determine the value of $\Delta E°_{rxn}$ for this reaction, whereby the standard enthalpy of reaction ($\Delta H°_{rxn}$) is -311.5 kJ mol^{-1}:

    $C_2H_2\ (g) + 2\ H_2\ (g) \rightarrow C_2H_6\ (g)$

A. -306.5 kJ mol^{-1}

B. -318.0 kJ mol^{-1}

C. $+346.0$ kJ mol^{-1}

D. $+306.5$ kJ mol^{-1}

E. $+466$ kJ mol^{-1}

46. For a closed system, what can be exchanged between the system and its surroundings?

  I. Heat   II. Matter   III. Energy

A. I only   **C.** III only
B. II only   **D.** I and III only
                                                  E. I, II and III

47. Which reaction is accompanied by an *increase* in entropy?

  A. $Na_2CO_3 (s) + CO_2 (g) + H_2O (g) \rightarrow 2 NaHCO_3 (s)$
  B. $BaO (s) + CO_2 (g) \rightarrow BaCO_3 (s)$
  C. $CH_4 (g) + H_2O (g) \rightarrow CO (g) + 3 H_2 (g)$
  D. $ZnS (s) + 3/2 O_2 (g) \rightarrow ZnO (s) + SO_2 (g)$
  E. $N_2 (g) + 3 H_2 (g) \rightarrow 2 NH_3 (g)$

48. Under standard conditions, which reaction has the largest difference between the energy of reaction and enthalpy?

  A. C (*graphite*) $\rightarrow$ C (*diamond*)   **C.** $2 C$ (*graphite*) $+ O_2 (g) \rightarrow 2 CO (g)$
  B. C (*graphite*) $+ O_2 (g) \rightarrow$ C (*diamond*)   **D.** C (*graphite*) $+ O_2 (g) \rightarrow CO_2 (g)$
                                                    E. $CO (g) + NO_2 (g) \rightarrow CO_2 (g) + NO (g)$

49. Which type of reaction tends to be the most stable?

  I. Isothermic   II. Exergonic   III. Endergonic

A. I only   **C.** III only
B. II only   **D.** I and II only
                                                  E. I and III only

50. Which is true for the thermodynamic functions G, H and S in $\Delta G = \Delta H - T\Delta S$?

  A. G refers to the universe, H to the surroundings and S to the system
  B. G, H, and S refer to the system
  C. G and H refers to the surroundings and S to the system
  D. G and H refer to the system and S to the surroundings
  E. G and S refers to the system and H to the surroundings

51. Which of the following statements is true for the following reaction? (Use the change in enthalpy, $\Delta H° = -113.4$ kJ/mol and the change in entropy, $\Delta S° = -145.7$ J/K mol)

$$2\ NO\ (g) + O_2\ (g) \rightarrow 2\ NO_2\ (g)$$

A. The reaction is at equilibrium at 25 °C under standard conditions
B. The reaction is spontaneous at only high temperatures
C. The reaction is spontaneous only at low temperatures
D. The reaction is spontaneous at all temperatures
E. $\Delta G°$ becomes more favorable as temperature increases

52. Which of the following expressions defines enthalpy? (Use the conventions: q = heat, U = internal energy, P = pressure and V = volume)

A. $q - \Delta U$ **B.** $U + q$ **C.** q **D.** ΔU **E.** $U + PV$

53. Which statement is true regarding entropy?

    I. It is a state function
    II. It is an extensive property
    III. It has an absolute zero value

A. I only
B. III only
C. I and II only
D. I and III only
E. I, II and III

54. Where does the energy released during an exothermic reaction originate from?

A. The kinetic energy of the surrounding
B. The kinetic energy of the reacting molecules
C. The potential energy of the reacting molecules
D. The thermal energy of the reactants
E. The potential energy of the surrounding

55. The species in the reaction $KClO_3\ (s) \rightarrow KCl\ (s) + 3/2\ O_2\ (g)$ have the values for standard enthalpies of formation at 25 °C. At constant physical states, assume that the values of $\Delta H°$ and $\Delta S°$ are constant throughout a broad temperature range. Which of the following conditions may apply to the reaction? (Use $KClO_3\ (s)$ with $\Delta H_f° = -391.2$ kJ mol^{-1} and KCl (s) with $\Delta H_f° = -436.8$ kJ mol^{-1})

A. Nonspontaneous at low temperatures, but spontaneous at high temperatures
B. Spontaneous at low temperatures but nonspontaneous at high temperatures
C. Nonspontaneous at all temperatures over a broad temperature range
D. Spontaneous at all temperatures over a broad temperature range
E. No conclusion can be drawn about spontaneity based on the information

56. What is the standard enthalpy change for the reaction?

$$P_4 \, (s) + 6 \, Cl_2 \, (g) \rightarrow 4 \, PCl_3 \, (l) \qquad \Delta H° = -1,289 \, kJ$$

$$3 \, P_4 \, (s) + 18 \, Cl_2 \, (g) \rightarrow 12 \, PCl_3 \, (l)$$

A. 426 kJ

B. −1,345 kJ

C. −366 kJ

D. 1,289 kJ

E. −3,867 kJ

57. Which of the following reactions is endothermic?

A. $PCl_3 + Cl_2 \rightarrow PCl_5 + heat$

B. $2 \, NO_2 \rightarrow N_2 + 2 \, O_2 + heat$

C. $CH_4 + NH_3 + heat \rightarrow HCN + 3 \, H_2$

D. $NH_3 + HBr \rightarrow NH_4Br$

E. $PCl_3 + Cl_2 \rightarrow PCl_5 + heat$

58. When the system undergoes a spontaneous reaction, is it possible for the entropy of a system to decrease?

A. No, because this violates the second law of thermodynamics

B. No, because this violates the first law of thermodynamics

C. Yes, but only if the reaction is endothermic

D. Yes, but only if the entropy gain of the environment is greater than the entropy loss in the system

E. Yes, but only if the entropy gain of the environment is smaller than the entropy loss in the system

59. A 500 ml beaker of distilled water is placed under a bell jar, which is then covered by a layer of opaque insulation. After several days, some of the water evaporated. The contents of the bell jar are what kind of system?

A. endothermic

B. exergonic

C. closed

D. open

E. isolated

60. Which law explains the observation that the amount of heat transfer accompanying a change in one direction is equal in magnitude but opposite in sign to the amount of heat transfer in the opposite direction?

A. Law of Conservation of Mass

B. Law of Definite Proportions

C. Avogadro's Law

D. Boyle's Law

E. Law of Conservation of Energy

==

Practice Set 4: Questions 61–80

==

61. What is the term for a reaction that proceeds by releasing heat energy?

   A. Endothermic reaction   **C.** Exothermic reaction

   B. Isothermal reaction   **D.** Nonspontaneous

                                        E. None of the above

62. If a stationary gas has a kinetic energy of 500 J at 25 °C, what is its kinetic energy at 50 °C?

   A. 125 J   **B.** 450 J   **C.** 540 J   **D.** 1,120 J   **E.** 270 J

63. Which is NOT true for entropy in a closed system according to the equation $\Delta S = Q / T$?

   A. Entropy is a measure of energy dispersal of the system

   B. The equation is only valid for a reversible process

   C. Changes due to heat transfer are greater at low temperatures

   D. Disorder of the system decreases as heat is transferred out of the system

   E. Increases as temperature decreases

64. In which of the following pairs of physical changes are both processes exothermic?

   A. Melting and condensation   **C.** Sublimation and evaporation

   B. Freezing and condensation   **D.** Freezing and sublimation

                                          E. None of the above

65. A fuel cell contains hydrogen and oxygen gas that react explosively, and the energy converts water to steam which drives a turbine to turn a generator that produces electricity. What are the initial and final forms of energy, respectively?

   A. Chemical and electrical energy   **C.** Chemical and mechanical energy

   B. Nuclear and electrical energy   **D.** Chemical and heat energy

                                          E. Nuclear and mechanical energy

66. Which must be true concerning a solution that reached an equilibrium where chemicals are mixed in a redox reaction?

   A. $\Delta G° = \Delta G$ **C.** $\Delta G° < 1$

   B. $E = 0$ **D.** $K = 1$

                                      E. $K < 1$

67. A solid sample at room temperature spontaneously sublimes forming a gas. This change in state is accompanied by which of the changes in the sample?

A. Entropy decreases and energy increases   **C.** Entropy and energy decrease

B. Entropy increases and energy decreases   **D.** Entropy and energy increase

                                               E. Entropy and energy are equal

68. At constant temperature and pressure, a negative ΔG indicates that the:

A. reaction is nonspontaneous   **C.** reaction is spontaneous

B. reaction is fast   **D.** reaction is endothermic

                                               E. $\Delta S > 0$

69. The process of H_2O (*g*) $\rightarrow$ H_2O (*l*) is nonspontaneous under pressure of 760 torr and temperatures of 378 K because:

A. $\Delta H = T\Delta S$ **C.** $\Delta H > 0$

B. $\Delta G < 0$ **D.** $\Delta H < T\Delta S$

                                               E. $\Delta H > T\Delta S$

70. Consider the contribution of entropy to the spontaneity of the reaction:

      2 Al_2O_3 (*s*) $\rightarrow$ 4 Al (*s*) + 3 O_2 (*g*), ΔG = +138 kcal.

As written, the reaction is [   ] and the entropy of the system [   ].

A. non-spontaneous… decreases   **C.** spontaneous… decreases

B. non-spontaneous… increases   **D.** spontaneous… increases

                                               E. non-spontaneous… does not change

71. The ΔG of a reaction is the maximum energy that the reaction releases to do:

A. P–V work only   **C.** any type of work

B. work and release heat   **D.** non P–V work only

                                               E. work and generate heat

72. Which of the following represent forms of internal energy?

      I. bond energy

      II. thermal energy

      III. gravitational energy

A. I only   **C.** I and II only

B. II only   **D.** I and III only

                                               E. I, II and III

73. If it takes energy to break bonds and energy is gained in the formation of bonds, how can some reactions be exothermic while others are endothermic?

    A. Some products have more energy than others and require energy to be formed

    B. Some reactants have more energetic bonds than others and release energy

    C. It is the number of bonds that is determinative. Since bonds have the same amount of energy, the net gain or net loss of energy depends on the number of bonds

    D. It is the amount of energy that is determinative. Some bonds are stronger than others, so there is a net gain or net loss of energy when formed

    E. None of the above

74. Which constant is represented by A in the following calculation to determine how much heat is required to convert 60 g of ice at $-25\ °C$ to steam at $320\ °C$?

$$\text{Total heat} = [(A) \cdot (60\ g) \cdot (25\ °C)] + [(\text{heat of fusion}) \cdot (60\ g)] +$$

$$+ [(4.18\ J/g \cdot °C) \cdot (60\ g) \cdot (100\ °C)] + [(B) \cdot (60\ g)] + [(C) \cdot (60\ g) \cdot (220\ °C)]$$

    A. Specific heat of water

    B. The heat of vaporization of water

    C. The heat of condensation

    D. The heat capacity of steam

    E. Specific heat of ice

75. The heat of formation of water vapor is:

    A. positive, but greater than the heat of formation for H_2O (*l*)

    B. positive and smaller than the heat of formation for H_2O (*l*)

    C. negative, but greater than the heat of formation for H_2O (*l*)

    D. negative and smaller than the heat of formation for H_2O (*l*)

    E. positive and equal to the heat of formation for H_2O (*l*)

76. Entropy can be defined as the amount of:

    A. equilibrium in a system

    B. chemical bonds that are changed during a reaction

    C. energy required to initiate a reaction

    D. energy required to rearrange chemical bonds

    E. disorder in a system

77. Which is true of an atomic fission bomb according to the conservation of mass and energy law?

    A. The mass of the bomb and the fission products are identical

    B. A small amount of mass is converted into energy

    C. The energy of the bomb and the fission products are identical

    D. The mass of the fission bomb is greater than the mass of the products

    E. None of the above

78. Once an object enters a black hole, astronomers consider it to have left the universe which means the universe is:

A. entropic
B. isolated

C. closed
D. open
E. exergonic

79. Which ranking from lowest to highest entropy per gram of NaCl is correct?

A. NaCl (*s*) < NaCl (*l*) < NaCl (*aq*) < NaCl (*g*)
B. NaCl (*s*) < NaCl (*l*) < NaCl (*g*) < NaCl (*aq*)
C. NaCl (*g*) < NaCl (*aq*) < NaCl (*l*) < NaCl (*s*)
D. NaCl (*s*), NaCl (*aq*), NaCl (*l*), NaCl (*g*)
E. NaCl (*l*) < NaCl (*aq*) < NaCl (*g*) < NaCl (*s*)

80. From the given bond energies, how many kJ of energy are released or absorbed from the reaction of one mole of N_2 with three moles of H_2 to form two moles of NH_3?

$N{\equiv}N + H{-}H + H{-}H + H{-}H \rightarrow NH_3 + NH_3$

H–N: 389 kJ/mol   H–H: 436 kJ/mol   N≡N: 946 kJ/mol

A. −80 kJ/mol released
B. +89.5 kJ/mol absorbed

C. −946 kJ/mol released
D. +895 kJ/mol absorbed
E. +946 kJ/mol absorbed

Kinetics and Equilibrium

===

Practice Set 1: Questions 1–20

===

1. What is the general equilibrium constant (K_{eq}) expression for the following reversible reaction?

$$2\ A + 3\ B \leftrightarrow C$$

A. $K_{eq} = [C] / [A]^2 \cdot [B]^3$

B. $K_{eq} = [C] / [A] \cdot [B]$

C. $K_{eq} = [A] \cdot [B] / [C]$

D. $K_{eq} = [A]^2 \cdot [B]^3 / [C]$

E. none of the above

2. What are gases A and B likely to be, if in a mixture of these two gases gas A has twice the average velocity of gas B?

A. Ar and Kr

B. N and Fe

C. Ne and Ar

D. Mg and K

E. B and Ne

3. For a hypothetical reaction, $A + B \rightarrow C$, predict which reaction occurs at the slowest rate from the following reaction conditions.

Reaction	Activation energy	Temperature
1	103 kJ/mol	15 °C
2	46 kJ/mol	22 °C
3	103 kJ/mol	24 °C
4	46 kJ/mol	30 °C

A. 1   **B.** 2   **C.** 3   **D.** 4   **E.** requires more information

4. From the data below, what is the order of the reaction with respect to reactant A?

Determining Rate Law from Experimental Data

$$A + B \rightarrow Products$$

Exp.	Initial [A]	Initial [B]	Initial Rate M/s
1	0.015	0.022	0.125
2	0.030	0.044	0.500
3	0.060	0.044	0.500
4	0.060	0.066	1.125
5	0.085	0.088	?

A. Zero   **B.** First   **C.** Second   **D.** Third   **E.** Fourth

5. For the combustion of ethanol (C_2H_6O) to form carbon dioxide and water, what is the rate at which carbon dioxide is produced, if the ethanol is consumed at a rate of 4.0 M s^{-1}?

A. 1.5 M s^{-1}

B. 12.0 M s^{-1}

C. 8.0 M s^{-1}

D. 9.0 M s^{-1}

E. 10.0 M s^{-1}

6. Which is the correct equilibrium constant (K_{eq}) expression for the following reaction?

$$2 \text{ Ag } (s) + Cl_2 (g) \leftrightarrow 2 \text{ AgCl } (s)$$

A. $K_{eq} = [AgCl] / [Ag]^2 \times [Cl_2]$

B. $K_{eq} = [2AgCl] / [2Ag] \times [Cl_2]$

C. $K_{eq} = [AgCl]^2 / [Ag]^2 \times [Cl_2]$

D. $K_{eq} = 1 / [Cl_2]$

E. $K_{eq} = [2AgCl]^2 / [2Ag]^2 \times [Cl_2]$

7. The position of the equilibrium for a system where $K_{eq} = 6.3 \times 10^{-14}$ can be described as being favored for [   ] and the concentration of products is relatively [   ].

A. the left; large

B. the left; small

C. the right; large

D. the right; small

E. neither direction; large

8. What can be deduced about the activation energy of a reaction that takes billions of years to go to completion and a reaction that takes only a fraction of a millisecond?

A. The slow reaction has high activation energy, while the fast reaction has a low activation energy

B. The slow reaction must have low activation energy, while the fast reaction must have a high activation energy

C. The activation energy of both reactions is very low

D. The activation energy of both reactions is very high

E. The activation energy of both reactions is equal

9. Which influences the rate of a first-order chemical reaction?

I. catalyst   II. temperature   III. concentration

A. I only

B. I and II only

C. I and III only

D. II and III only

E. I, II and III

Questions **10** through **14** are based on the following:

Energy profiles for four reactions (with the same scale).

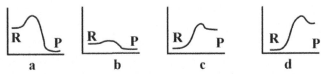

R = reactants   P = products

10. Which reaction requires the most energy?

   A. a   **B.** b   **C.** c   **D.** d   **E.** None of the above

11. Which reaction has the highest activation energy?

   A. a   **B.** b   **C.** c   **D.** d   **E.** c and d

12. Which reaction has the lowest activation energy?

   A. a   **B.** b   **C.** c   **D.** d   **E.** a and b

13. Which reaction proceeds the slowest?

   A. a   **B.** b   **C.** c   **D.** d   **E.** c and d

14. If the graphs are for the same reaction, which most likely has a catalyst?

   A. a   **B.** b   **C.** c   **D.** d   **E.** All have a catalyst

15. What is an explanation for observing that the reaction stops before all reactants are converted to products in the following reaction?

$$NH_3\,(aq) + HC_2H_3O_2\,(aq) \rightarrow NH_4^+\,(aq) + C_2H_3O_2^-\,(aq)$$

  A. The catalyst is depleted

  B. The reverse rate increases, while the forward rate decreases until they are equal

  C. As [products] increases, the acetic acid begins to dissociate, stopping the reaction

  D. As [reactants] decreases, NH_3 and $HC_2H_3O_2$ molecules stop colliding

  E. As [products] increases, NH_3 and $HC_2H_3O_2$ molecules stop colliding

16. What is the term for the principle that the rate of reaction is regulated by the frequency, energy, and orientation of molecules striking each other?

A. orientation theory

B. frequency theory

C. energy theory

D. collision theory

E. rate theory

17. What is the effect on the energy of the activated complex and on the rate of the reaction when a catalyst is added to a chemical reaction?

A. The energy of the activated complex increases and the reaction rate decreases

B. The energy of the activated complex decreases and the reaction rate increases

C. The energy of the activated complex and the reaction rate increase

D. The energy of the activated complex and the reaction rate decrease

E. The energy of the activated complex remains the same, while the reaction rate decreases

18. Which change shifts the equilibrium to the right for the reversible reaction in an aqueous solution?

$$HNO_2\,(aq) \leftrightarrow H^+\,(aq) + NO_2^-\,(aq)$$

I. Add solid NaOH

II. Decrease $[NO_2^-]$

III. Decrease $[H^+]$

IV. Increase $[HNO_2]$

A. II and III only

B. II and IV only

C. III and IV only

D. II, III and IV only

E. I, II, III and IV

19. Which conditions would favor driving the reaction to completion?

$$2\,N_2\,(g) + 6\,H_2O\,(g) + heat \leftrightarrow 4\,NH_3\,(g) + 3\,O_2\,(g)$$

A. Increasing the reaction temperature

B. Continual addition of NH_3 gas to the reaction mixture

C. Decreasing the pressure on the reaction vessel

D. Continual removal of N_2 gas

E. Decreases reaction temperature

20. The system, $H_2\,(g) + X_2\,(g) \leftrightarrow 2\,HX\,(g)$ has a value of 24.4 for K_c. A catalyst was introduced into a reaction within a 4.0-liters reactor containing 0.20 moles of H_2, 0.20 moles of X_2 and 0.800 moles of HX. The reaction proceeds in which direction?

A. to the right, $Q > K_c$

B. to the left, $Q > K_c$

C. to the right, $Q < K_c$

D. to the left, $Q < K_c$

E. requires more information

==

Practice Set 2: Questions 21–40

==

21. Which is the K_c equilibrium expression for the following reaction?

    $$4 \, CuO \,(s) + CH_4 \,(g) \leftrightarrow CO_2 \,(g) + 4 \, Cu \,(s) + 2 \, H_2O \,(g)$$

    A. $[Cu]^4 / [CuO]^4$

    B. $[CH_4]^2 / [CO_2]\cdot[H_2O]$

    C. $[CH_4] / [CO_2]\cdot[H_2O]^2$

    D. $[CuO]^4 / [Cu]^4$

    E. $[CO_2][H_2O]^2 / [CH_4]$

22. Which of the following concentrations of CH_2Cl_2 should be used in the rate law for Step 2, if CH_2Cl_2 is a product of the fast (first) step and a reactant of the slow (second) step?

    A. $[CH_2Cl_2]$ at equilibrium

    B. $[CH_2Cl_2]$ in Step 2 cannot be predicted because Step 1 is the fast step

    C. Zero moles per liter

    D. $[CH_2Cl_2]$ after Step 1 is completed

    E. None of the above

23. What is the term for a substance that allows a reaction to proceed faster by lowering the energy of activation?

    A. rate barrier

    B. energy barrier

    C. collision energy

    D. activation energy

    E. catalyst

24. Which statement is true for the grams of products present after a chemical reaction reaches equilibrium?

    A. Must equal the grams of the initial reactants

    B. May be less than, equal to, or greater than the grams of reactants present, depending upon the chemical reaction

    C. Must be greater than the grams of the initial reactants

    D. Must be less than the grams of the initial reactants

    E. None of the above

25. What is the rate law when rates were measured at different concentrations for the dissociation of hydrogen gas: $H_2(g) \rightarrow 2H(g)$?

$[H_2]$	Rate M/s s^{-1}
1.0	1.3×10^5
1.5	2.6×10^5
2.0	5.2×10^5

A. rate $= k^2[H] / [H_2]$

B. rate $= k[H]^2 / [H_2]$

C. rate $= k[H_2]^2$

D. rate $= k[H_2] / [H]^2$

E. requires more information

120. Which change to this reaction system causes the equilibrium to shift to the right?

$N_2(g) + 3H_2(g) \leftrightarrow 2NH_3(g) + heat$

A. Heating the system

B. Removal of $H_2(g)$

C. Addition of $NH_3(g)$

D. Lowering the temperature

E. Addition of a catalyst

27. Which statement is NOT correct for $aA + bB \rightarrow dD + eE$ whereby rate $= k[A]^q \times [B]^r$?

A. The overall order of the reaction is $q + r$

B. The exponents q and r are equal to the coefficients a and b, respectively

C. The exponents q and r must be determined experimentally

D. The exponents q and r are often integers

E. The symbol k represents the rate constant

28. Which is the correct equilibrium constant (K_{eq}) expression for the following reaction?

$CO(g) + 2H_2(g) \leftrightarrow CH_3OH(l)$

A. $K_{eq} = 1 / [CO] \cdot [H_2]^2$

B. $K_{eq} = [CO] \cdot [H_2]^2$

C. $K_{eq} = [CH_3OH] / [CO] \cdot [H_2]^2$

D. $K_{eq} = [CH_3OH] / [CO] \cdot [H_2]$

E. None of the above

Questions **29-32** are based on the following graph and net reaction:

The reaction proceeds in two consecutive steps.

$$XY + Z \leftrightarrow XYZ \leftrightarrow X + YZ$$

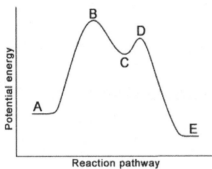

29. Where is the activated complex for this reaction?

   A. A

   B. A and C

   C. C

   D. E

   E. B and D

30. The activation energy of the slow step for the forward reaction is given by:

   A. A → B

   B. A → C

   C. B → C

   D. C → E

   E. A → E

31. The activation energy of the slow step for the reverse reaction is given by:

   A. C → A

   B. E → C

   C. C → B

   D. E → D

   E. E → A

32. The change in energy (ΔE) of the overall reaction is given by the difference between:

   A. A and B

   B. A and E

   C. A and C

   D. B and D

   E. B and C

33. Which of the following increases the collision energy of gaseous molecules?

      I. Increasing the temperature

      II. Adding a catalyst

      III. Increasing the concentration

A. I only

B. II only

C. III only

D. I and II only

E. I and III only

34. Which of the following conditions favors the formation of NO (*g*) in a closed container?

      N_2 (*g*) + O_2 (*g*) ↔ 2 NO (*g*)

      (Use ΔH = +181 kJ/mol)

A. Increasing the temperature

B. Decreasing the temperature

C. Increasing the pressure

D. Decreasing the pressure

E. Decreasing the temperature and decreasing the pressure

35. A chemical system is considered to have reached dynamic equilibrium when the:

A. activation energy of the forward reaction equals the activation energy of the reverse reaction

B. rate of production of each of the products equals the rate of their consumption by the reverse reaction

C. frequency of collisions between the reactant molecules equals the frequency of collisions between the product molecules

D. sum of the concentrations of each of the reactant species equals the sum of the concentrations of each of the product species

E. none of the above

36. Chemical equilibrium is reached in a system when:

A. complete conversion of reactants to products has occurred

B. product molecules begin reacting with each other

C. reactant concentrations steadily decrease

D. reactant concentrations steadily increase

E. product and reactant concentrations remain constant

37. At a given temperature, $K = 46.0$ for the reaction:

$$4 \text{ HCl } (g) + \text{O}_2 (g) \leftrightarrow 2 \text{ H}_2\text{O} (g) + 2 \text{ Cl}_2 (g)$$

At equilibrium, $[\text{HCl}] = 0.150$, $[\text{O}_2] = 0.395$ and $[\text{H}_2\text{O}] = 0.625$. What is the concentration of Cl_2 at equilibrium?

A. 0.153 M   **B.** 0.444 M   **C.** 1.14 M   **D.** 0.00547 M   **E.** 2.64 M

38. If at equilibrium, reactant concentrations are slightly smaller than product concentrations, the equilibrium constant would be:

A. slightly greater than 1   **C.** much lower than 1
B. slightly lower than 1   **D.** much greater than 1
                                              E. equal to zero

39. Hydrogen gas reacts with iron (III) oxide to form iron metal (which produces steel), as shown in the reaction below. Which statement is NOT correct concerning the equilibrium system?

$$\text{Fe}_2\text{O}_3 (s) + 3 \text{ H}_2 (g) + heat \leftrightarrow 2 \text{ Fe } (s) + 3 \text{ H}_2\text{O} (g)$$

A. Continually removing water from the reaction chamber increases the yield of iron
B. Decreasing the volume of hydrogen gas reduces the yield of iron
C. Lowering the reaction temperature increases the concentration of hydrogen gas
D. Increasing the pressure on the reaction chamber increases the formation of products
E. Decreasing the reaction temperature reduces the formation of iron

40. Which of the changes shifts the equilibrium to the right for the following reversible reaction?

$$\text{CO } (g) + \text{H}_2\text{O} (g) \leftrightarrow \text{CO}_2 (g) + \text{H}_2 (g) + heat$$

A. increasing volume   **C.** increasing $[\text{CO}_2]$
B. increasing temperature   **D.** adding a catalyst
                                              E. increasing $[\text{CO}]$

==

Practice Set 3: Questions 41–60

==

41. Which is the correct equilibrium constant (K_{eq}) expression for the following reaction?

$$4 \, NH_3 \, (g) + 5 \, O_2 \, (g) \leftrightarrow 4 \, NO \, (g) + 6 \, H_2O \, (g)$$

A. $K_{eq} = [NO]^4 \times [H_2O]^6 / [NH_3]^4 \times [O_2]^5$

B. $K_{eq} = [NH_3]^4 \times [O_2]^5 / [NO]^4 \times [H_2O]^6$

C. $K_{eq} = [NO] \times [H_2O] / [NH_3] \times [O_2]$

D. $K_{eq} = [NH_3] \times [O_2] / [NO] \times [H_2O]$

E. $K_{eq} = [NH_3]^2 \times [O_2]^5 / [NO]^2 \times [H_2O]^3$

42. Carbonic acid equilibrium in blood:

$$CO_2 \, (g) + H_2O \, (l) \leftrightarrow H_2CO_3 \, (aq) \leftrightarrow H^+ \, (aq) + HCO_3^- \, (aq)$$

If a person hyperventilates, the rapid breathing expels carbon dioxide gas. Which of the following decreases when a person hyperventilates?

I. $[HCO_3^-]$ II. $[H^+]$ III. $[H_2CO_3]$

A. I only

B. II only

C. III only

D. I and II only

E. I, II and III

43. What is the overall order of the reaction if the units of the rate constant for a particular reaction are min^{-1}?

A. Zero   **B.** First   **C.** Second   **D.** Third   **E.** Fourth

44. Heat is often added to chemical reactions performed in the laboratory to:

A. compensate for the natural tendency of energy to disperse

B. increase the rate at which reactants collide

C. allow a greater number of reactants to overcome the barrier of the activation energy

D. increase the energy of the reactant molecules

E. all of the above

45. Predict which reaction occurs at a faster rate for a hypothetical reaction $X + Y \rightarrow W + Z$.

Reaction	Activation energy	Temperature
1	low	low
2	low	high
3	high	high
4	high	low

A. 1   **B.** 2   **C.** 3   **D.** 4   **E.** 1 and 4

46. For the reaction, $2\ XO + O_2 \rightarrow 2\ XO_2$, data obtained from measurement of the initial rate of reaction at varying concentrations are:

Experiment	[*XO*]	[O₂]	Rate (mmol $L^{-1}\ s^{-1}$)
1	0.010	0.010	2.5
2	0.010	0.020	5.0
3	0.030	0.020	45.0

What is the expression for the rate law?

A. rate $= k[XO]\cdot[O_2]$

B. rate $= k[XO]^2\cdot[O_2]^2$

C. rate $= k[XO]^2\cdot[O_2]$

D. rate $= k[XO]\cdot[O_2]^2$

E. rate $= k[XO]^2 / [O_2]^2$

47. What is the equilibrium (K_{eq}) expression for the following reaction?

$$CaO\ (s) + CO_2\ (g) \leftrightarrow CaCO_3\ (s)$$

A. $K_{eq} = [CaCO_3] / [CaO]$

B. $K_{eq} = 1 / [CO_2]$

C. $K_{eq} = [CaCO_3] / [CaO]\cdot[CO_2]$

D. $K_{eq} = [CO_2]$

E. $K_{eq} = [CaO]\cdot[CO_2] / [CaCO_3]$

48. If a reaction does not occur extensively and gives a low concentration of products at equilibrium, which of the following is true?

A. The rate of the forward reaction is greater than the reverse reaction

B. The rate of the reverse reaction is greater than the forward reaction

C. The equilibrium constant is greater than one; that is, K_{eq} is larger than 1

D. The equilibrium constant is less than one; that is, K_{eq} is smaller than 1

E. The equilibrium constant equals 1

49. Which of the following changes most likely decreases the rate of a reaction?

A. Increasing the reaction temperature

B. Increasing the concentration of a reactant

C. Increasing the activation energy for the reaction

D. Decreasing the activation energy for the reaction

E. Increasing the reaction pressure

50. Which factors would increase the rate of a reversible chemical reaction?

    I. Increasing the temperature of the reaction

    II. Removing products as they form

    III. Adding a catalyst to the reaction vessel

A. I only

B. II only

C. I and II only

D. I and III only

E. I, II and III

51. What is the ionization equilibrium constant (K_i) expression for the following weak acid?

$$H_2S\ (aq) \leftrightarrow H^+\ (aq) + HS^-\ (aq)$$

A. $K_i = [H^+]^2 \cdot [S^{2-}] / [H_2S]$

B. $K_i = [H_2S] / [H^+] \cdot [HS^-]$

C. $K_i = [H^+] \cdot [HS^-] / [H_2S]$

D. $K_i = [H^+]^2 \cdot [HS^-] / [H_2S]$

E. $K_i = [H_2S] / [H^+]^2 \cdot [HS^-]$

52. Increasing the temperature of a chemical reaction:

A. increases the reaction rate by lowering the activation energy

B. increases the reaction rate by increasing reactant collisions per unit time

C. increases the activation energy, thus increasing the reaction rate

D. raises the activation energy, thus decreasing the reaction rate

E. causes fewer reactant collisions to take place

53. All of the following factors determine reaction rates, EXCEPT:

A. orientation of collisions between molecules

B. spontaneity of the reaction

C. force of collisions between molecules

D. number of collisions between molecules

E. the activation energy of the reaction

54. Coal burning plants release sulfur dioxide into the atmosphere, while nitrogen monoxide is released via industrial processes and combustion engines. Sulfur dioxide can be produced in the atmosphere by the following equilibrium reaction:

$$SO_3\ (g) + NO\ (g) + heat \leftrightarrow SO_2\ (g) + NO_2\ (g)$$

Which of the following does NOT shift the equilibrium to the right?

A. [NO$_2$] decrease

B. [NO] increase

C. Decrease the reaction chamber volume

D. Temperature increase

E. All of the above shift the equilibrium to the right

55. Which of the following statements can be assumed to be true about how reactions occur?

A. Reactant particles must collide with each other

B. Energy must be released as the reaction proceeds

C. Catalysts must be present in the reaction

D. Energy must be absorbed as the reaction proceeds

E. The energy of activation must have a negative value

56. At equilibrium, increasing the temperature of an exothermic reaction likely:

A. increases the heat of reaction

B. decreases the heat of reaction

C. increases the forward reaction

D. decreases the forward reaction

E. increases the heat of reaction and the forward reaction

57. Which shifts the equilibrium to the left for the reversible reaction in an aqueous solution?

$$HC_2H_3O_2\ (aq) \leftrightarrow H^+\ (aq) + C_2H_3O_2^-\ (aq)$$

I. increase pH

II. increase $[HC_2H_3O_2]$

III. add solid $KC_2H_3O_2$

A. I only

B. I and III only

C. II only

D. II and III only

E. III only

58. Which of the following statements is true concerning the equilibrium system, whereby S combines with H_2 to form hydrogen sulfide, toxic gas from the decay of organic material? (Use the equilibrium constant, $K_{eq} = 2.8 \times 10^{-21}$)

$$S\ (g) + H_2\ (g) \leftrightarrow H_2S\ (g)$$

A. Almost all the starting molecules are converted to product

B. Decreasing $[H_2S]$ shifts the equilibrium to the left

C. Decreasing $[H_2]$ shifts the equilibrium to the right

D. Increasing the volume of the sealed reaction container shifts the equilibrium to the right

E. Very little hydrogen sulfide gas is present in the equilibrium

59. Which of the following conditions characterizes a system in a state of chemical equilibrium?

A. Product concentrations are greater than reactant concentrations

B. Reactant molecules no longer react with each other

C. Concentrations of reactants and products are equal

D. The rate of the forward reaction has dropped to zero

E. Reactants are being consumed at the same rate they are being produced

60. Which statement is NOT true regarding an equilibrium constant for a particular reaction?

A. It does not change as the product is removed

B. It does not change as an additional quantity of a reactant is added

C. It changes when a catalyst is added

D. It changes as the temperature increases

E. All are true statements

==

Practice Set 4: Questions 61–80

==

Questions **61** through **63** refer to the rate data
for the conversion of reactants W, X, and Y to product Z.

Trial Number	Concentration (moles/L)			Rate of Formation of Z (moles/l·s)
	W	X	Y	
1	0.01	0.05	0.04	0.04
2	0.015	0.07	0.06	0.08
3	0.01	0.15	0.04	0.36
4	0.03	0.07	0.06	0.08
5	0.01	0.05	0.16	0.08

61. From the above data, what is the overall order of the reaction?

    A. 3½   **B.** 4   **C.** 3   **D.** 2   **E.** 2½

62. From the above data, the order with respect to W suggests that the rate of formation of Z is:

    A. dependent on [W]   **C.** semi-dependent on [W]

    B. independent of [W]   **D.** unable to be determined

                                            E. inversely proportional to [W]

63. From the above data, the magnitude of *k* for trial 1 is:

    A. 20   **B.** 40   **C.** 60   **D.** 80   **E.** 90

64. Which of the following changes shifts the equilibrium to the left for the given reversible reaction?

SO_3 (*g*) + NO (*g*) + *heat* $\leftrightarrow$ SO_2 (*g*) + NO_2 (*g*)

    A. Decrease temperature   **C.** Increase [NO]

    B. Decrease volume   **D.** Decrease [SO_2]

                                            E. Add a catalyst

65. What is the ionization equilibrium constant (K_i) expression for the following weak acid?

      H_3PO_4 (*aq*) $\leftrightarrow$ H^+ (*aq*) + $H_2PO_4^-$ (*aq*)

    A. $K_i = [H_3PO_4] / [H^+] \cdot [H_2PO_4^-]$ **C.** $K_i = [H^+]^3 \cdot [H_2PO_4^-] / [H_3PO_4]$

    B. $K_i = [H^+]^3 \cdot [PO_4^{3-}] / [H_3PO_4]$ **D.** $K_i = [H^+] \cdot [H_2PO_4^-] / [H_3PO_4]$

                                            E. $K_i = [H_3PO_4] / [H^+]^3 \cdot [PO_4^{3-}]$

66. What effect does a catalyst have on equilibrium?

    A. It increases the rate of the forward reaction

    B. It shifts the reaction to the right

    C. It increases the rate at which equilibrium is reached without changing ΔG

    D. It increases the rate at which equilibrium is reached and lowers ΔG

    E. It slows the reverse reaction

67. What is the correct ionization equilibrium constant (K_i) expression for the following weak acid?

$$H_2SO_3\,(aq) \leftrightarrow H^+\,(aq) + HSO_3^-\,(aq)$$

    A. $K_i = [H_2SO_3] / [H^+] \cdot [HSO_3^-]$

    B. $K_i = [H^+]^2 \cdot [SO_3^{2-}] / [H_2SO_3]$

    C. $K_i = [H^+]^2 \cdot [HSO_3^-] / [H_2SO_3]$

    D. $K_i = [H^+] \cdot [HSO_3^-] / [H_2SO_3]$

    E. $K_i = [H_2SO_3] / [H^+]^2 \cdot [SO_3^{2-}]$

68. Which of the following is true if a reaction occurs extensively and yields a high concentration of products at equilibrium?

    A. The rate of the reverse reaction is greater than the forward reaction

    B. The rate of the forward reaction is greater than the reverse reaction

    C. The equilibrium constant is less than one; K_{eq} is much smaller than 1

    D. The equilibrium constant is greater than one; K_{eq} is much larger than 1

    E. The equilibrium constant equals 1

69. The minimum combined kinetic energy reactants must possess for collisions to result in a reaction is:

    A. orientation energy

    B. activation energy

    C. collision energy

    D. dissociation energy

    E. bond energy

70. Which factors decrease the rate of a reaction?

    I. Lowering the temperature

    II. Increasing the concentration of reactants

    III. Adding a catalyst to the reaction vessel

    A. I only

    B. II only

    C. III only

    D. I and II only

    E. I, II and III

71. For a collision between molecules to result in a reaction, the molecules must possess a favorable orientation relative to each other and:

A. be in the gaseous state

B. have a certain minimum energy

C. adhere for at least 2 nanoseconds

D. exchange electrons

E. be in the liquid state

72. Most reactions are carried out in a liquid solution or the gaseous phase, because in such situations:

A. kinetic energies of reactants are lower

B. reactant collisions occur more frequently

C. activation energies are higher

D. reactant activation energies are lower

E. reactant collisions occur less frequently

73. Find the reaction rate for A + B → C:

Trial	$[A]_{t=0}$	$[B]_{t=0}$	Initial rate (M/s)
1	0.05 M	1.0 M	1.0×10^{-3}
2	0.05 M	4.0 M	16.0×10^{-3}
3	0.15 M	1.0 M	3.0×10^{-3}

A. rate = $k[A]^2 \cdot [B]^2$

B. rate = $k[A] \cdot [B]^2$

C. rate = $k[A]^2 \cdot [B]$

D. rate = $k[A] \cdot [B]$

E. rate = $k[A]^2 \cdot [B]^3$

74. Why does a glowing splint of wood burn only slowly in air, but rapidly in a burst of flames when placed in pure oxygen?

A. A glowing wood splint is extinguished within pure oxygen because oxygen inhibits the smoke

B. Pure oxygen is able to absorb carbon dioxide at a faster rate

C. Oxygen is a flammable gas

D. There is an increased number of collisions between the wood and oxygen molecules

E. There is a decreased number of collisions between the wood and oxygen

75. Which of the changes shift the equilibrium to the right for the following system at equilibrium?

$N_2 (g) + 3 H_2 (g) \leftrightarrow 2 NH_3 (g) + 92.94$ kJ

I. Removing NH_3

II. Adding NH_3

III. Removing N_2

IV. Adding N_2

A. I and III

B. II and III

C. II and IV

D. I and IV

E. None of the above

76. Which of the changes has no effect on the equilibrium for the reversible reaction in an aqueous solution?

$$HC_2H_3O_2\ (aq) \leftrightarrow H^+\ (aq) + C_2H_3O_2^-\ (aq)$$

A. Adding solid $NaC_2H_3O_2$

B. Adding solid $NaNO_3$

C. Increasing $[HC_2H_3O_2]$

D. Increasing $[H^+]$

E. Adding solid NaOH

77. Consider the following reaction: $H_2\ (g) + I_2\ (g) \rightarrow 2\ HI\ (g)$

At 160 K, this reaction has an equilibrium constant of 35. If at 160 K, the concentration of hydrogen gas is 0.4 M, iodine gas is 0.6 M, and hydrogen iodide gas is 3 M:

A. system is at equilibrium

B. [iodine] decreases

C. [hydrogen iodide] increases

D. [hydrogen iodide] decreases

E. [hydrogen] decreases

78. If the concentration of reactants decreases, which of the following is true?

    I. The amount of products increases

    II. The heat of reaction decreases

    III. The rate of reaction decreases

A. I only

B. II only

C. III only

D. II and III only

E. I, II and III

79. What does a chemical equilibrium expression of a reaction depend on?

    I. mechanism

    II. stoichiometry

    III. rate

A. I only

B. II only

C. III only

D. I and II only

E. I, II and III

80. For the following reaction where $\Delta H < 0$, which factor decreases the magnitude of the equilibrium constant K?

$$CO\ (g) + 2\ H_2O\ (g) \leftrightarrow CH_3OH\ (g)$$

A. Decreasing the temperature of this system

B. Decreasing volume

C. Decreasing the pressure of this system

D. All of the above

E. None of the above

==

Practice Set 5: Questions 81–100

==

81. Which changes shift the equilibrium to the product side for the following reaction at equilibrium?

$$SO_2Cl_2\ (g) \leftrightarrow SO_2\ (g) + Cl_2\ (g)$$

   I. Addition of SO_2Cl_2 III. Removal of SO_2Cl_2

  II. Addition of SO_2 IV. Removal of Cl_2

A. I, II and III   **C.** I and II

B. II and III   **D.** III and IV   **E.** I and IV

82. Calculate a value for K_c for the following reaction:

$$NOCl\ (g) + \tfrac{1}{2}\ O_2\ (g) \leftrightarrow NO_2\ (g) + \tfrac{1}{2}\ Cl_2\ (g)$$

Use the data:

$$2\ NO\ (g) +\ Cl_2\ (g) \leftrightarrow 2\ NOCl\ (g) \qquad K_c = 3.20 \times 10^{-3}$$

$$2\ NO_2\ (g) \leftrightarrow 2\ NO\ (g) + O_2\ (g) \qquad K_c = 15.5$$

A. 4.49   **C.** 4.32×10^{-4}

B. 0.343   **D.** 1.33×10^{-5} **E.** 18.4

83. Assuming vessels a, b and c are drawn to relative proportions, which of the following reactions proceeds the fastest? (Assume equal temperatures)

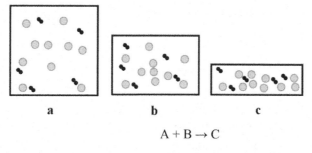

a   b   c

$$A + B \rightarrow C$$

A. a   **C.** c

B. b   **D.** All proceed at same rate

                      E. Not enough information given

84. The reaction rate is:

  A. the ratio of the masses of products and reactants

  B. the ratio of the molecular masses of the elements in a given compound

  C. the speed at which reactants are consumed or product is formed

  D. the balanced chemical formula that relates the number of product molecules to reactant molecules

  E. none of the above

85. What is the equilibrium constant (K_{eq}) expression for the reversible reaction below?

$$2\ A \leftrightarrow B + 3\ C$$

A. $K_{eq} = [B] \times [C]^3 / [A]^2$ **C.** $K_{eq} = [A]^2 / [B] \times [C]^3$

B. $K_{eq} = [B] \times [C] / [A]$ **D.** $K_{eq} = [A] / [B] \times [C]$

                                                  E. None of the above

86. With respect to A, what is the order of the reaction, if the rate law = $k[A]^3[B]^6$?

A. 2   **B.** 3   **C.** 4   **D.** 5   **E.** 6

87. Nitrogen monoxide reacts with bromine at elevated temperatures according to the equation:

$$2\ NO\ (g) + Br_2\ (g) \rightarrow 2\ NOBr\ (g)$$

What is the rate of consumption of $Br_2\ (g)$ if, in a certain reaction mixture, the rate of formation of NOBr (g) was 4.50×10^{-4} mol L^{-1} s^{-1}?

A. 3.12×10^{-4} mol L^{-1} s^{-1} **C.** 2.25×10^{-4} mol L^{-1} s^{-1}

B. 8.00×10^{-4} mol L^{-1} s^{-1} **D.** 4.50×10^{-4} mol L^{-1} s^{-1}

                                                  E. 4.00×10^{-3} mol L^{-1} s^{-1}

88. Which of the following statements about "activation energy" is correct?

A. Activation energy is the energy given off when reactants collide

B. Activation energy is high for reactions that occur rapidly

C. Activation energy is low for reactions that occur rapidly

D. Activation energy is the maximum energy a reacting molecule may possess

E. Activation energy is low for reactions that occur slowly

89. Which is the correct equilibrium constant (K_{eq}) expression for the following reaction?

$$A + 2\ B \leftrightarrow 2\ C + D$$

A. $[C]^2 \cdot [D] / [A] \cdot [B]^2$ **C.** $[A] \cdot 2[B] / 2[C] \cdot [D]$

B. $[A]^2 \cdot [B]^2 / 2[C]^2 \cdot [D]$ **D.** $2[C] \cdot [D] / [A] \cdot 2[B]$

                                              E. $[A] \cdot [B]^2 / [C]^2 \cdot [D]$

90. What is the effect on the equilibrium after adding H_2O to the equilibrium mixture if CO_2 and H_2 react until equilibrium is established?

$$CO_2\ (g) + H_2\ (g) \leftrightarrow H_2O\ (g) + CO\ (g)$$

A. $[H_2]$ decreases and $[H_2O]$ increases   **C.** $[H_2]$ decreases and $[CO_2]$ increases

B. $[CO]$ and $[CO_2]$ increase   **D.** Equilibrium shifts to the left

                                              E. Equilibrium shifts to the right

91. For a reaction that has an equilibrium constant of 4.3×10^{-17} at 25 °C, the relative position of equilibrium is described as:

A. equal amounts of reactants and products

B. significant amounts of both reactants and products

C. mostly products

D. mostly reactants

E. amount of product is slightly greater than reactants

92. What is the term for the energy necessary for reactants to achieve the transition state and form products?

A. heat of reaction

B. energy barrier

C. rate barrier

D. collision energy

E. activation energy

93. What is the equilibrium constant K_c value if, at equilibrium, the concentrations of $[NH_3] = 0.40$ M, $[H_2] = 0.12$ M and $[N_2] = 0.040$ M?

$$2\ NH_3\ (g) \leftrightarrow N_2\ (g) + 3\ H_2\ (g)$$

A. 6.3×10^{12}

B. 7.1×10^{-7}

C. 4.3×10^{-4}

D. 3.9×10^{-3}

E. 8.5×10^{-9}

94. Which of the following increases the collision frequency of molecules?

    I. Increasing the concentration

    II. Adding a catalyst

    III. Decreasing the temperature

A. I only

B. II only

C. III only

D. I and II only

E. I and III only

95. For the following reaction with $\Delta H < 0$, which factor increases the equilibrium yield for methanol?

$$CO\ (g) + 2\ H_2O\ (g) \leftrightarrow CH_3OH\ (g)$$

    I. Decreasing the volume

    II. Decreasing the pressure

    III. Lowering the temperature of the system

A. I only

B. II only

C. I and II only

D. I and III only

E. I, II and III

96. If there is too much chlorine in the water, swimmers complain that their eyes burn. Consider the equilibrium found in swimming pools. Predict which increases the chlorine concentration.

$$Cl_2 (g) + H_2O (l) \leftrightarrow HClO (aq) \leftrightarrow H^+ (aq) + ClO^- (aq)$$

A. Decreasing the pH

B. Adding hydrochloric acid, HCl (*aq*)

C. Adding hypochlorous acid, HClO (*aq*)

D. Adding sodium hypochlorite, NaClO

E. All of the above

97. What is the ratio of the diffusion rates of H_2 gas to O_2 gas?

A. 4:1   **B.** 2:1   **C.** 1:3   **D.** 1:4   **E.** 1:2

98. Which of the following increases the amount of product formed from a reaction?

    I. Using a UV light catalyst

    II. Adding an acid catalyst

    III. Adding a metal catalyst

A. I only

B. II only

C. III only

D. I and II only

E. None of the above

99. A mixture of 1.40 moles of A and 2.30 moles of B reacted. At equilibrium, 0.90 moles of A are present. How many moles of C is present at equilibrium?

$$3 A (g) + 2 B (g) \rightarrow 4 C (g)$$

A. 2.7 moles   **B.** 1.3 moles   **C.** 0.09 moles   **D.** 1.8 moles   **E.** 0.67 moles

100. What is the equilibrium constant K_c for the following reaction?

$$PCl_5 (g) + 2 NO (g) \leftrightarrow PCl_3 (g) + 2 NOCl (g)$$

$$K_1 = PCl_3 (g) + Cl_2 (g) \leftrightarrow PCl_5 (g)$$

$$K_2 = 2 NO (g) + Cl_2 (g) \leftrightarrow 2 NOCl (g)$$

A. K_1 / K_2 **B.** $(K_1K_2)^{-1}$ **C.** $K_1 \times K_2$ **D.** K_2 / K_1 **E.** $K_2 - K_1$

===

Practice Set 6: Questions 101–120

===

101. Write the mass action (K_c) expression for the following reaction?

$$4 \text{ Cr } (s) + 3 \text{ CCl}_4 (g) \leftrightarrow 4 \text{ CrCl}_3 (g) + 4 \text{ C } (s)$$

A. $K_c = [\text{C}] \cdot [\text{CrCl}_3] / [\text{Cr}] \cdot [\text{CCl}_4]$

B. $K_c = [\text{C}]^4 \cdot [\text{CrCl}_3]^4 / [\text{Cr}]^4 \cdot [\text{CCl}_4]^3$

C. $K_c = [\text{CrCl}_3]^4 / [\text{CCl}_4]^3$

D. $K_c = [\text{CrCl}_3] / [\text{CCl}_4]$

E. $K_c = [\text{CrCl}_3] + [\text{CCl}_4]$

102. Which of the changes shifts the equilibrium to the left for the reversible reaction in an aqueous solution?

$$\text{HNO}_2 (aq) \leftrightarrow \text{H}^+ (aq) + \text{NO}_2^- (aq)$$

A. Adding solid KNO_2

B. Adding solid KCl

C. Increasing $[\text{HNO}_2]$

D. Increasing pH

E. Adding solid KOH

103. Given the equation $x\text{A} + y\text{B} \rightarrow z$, the rate expression reaction in terms of the rate of change of the concentration with respect to time is:

A. $k[\text{A}]^x[\text{B}]^{yz}$

B. $k[\text{A}]^y[\text{B}]^x$

C. $k[\text{A}]^x[\text{B}]^y$

D. $k[\text{A}]^x[\text{B}]^y / [z]$

E. cannot be determined

104. What is the rate law for the reaction $3 \text{ D} + \text{E} \rightarrow \text{F} + 2 \text{ G}$, given the following experimental data?

Experiment	[D]	[E]	Rate (mol L^{-1} s^{-1})
1	0.100	0.250	0.000250
2	0.200	0.250	0.000500
3	0.100	0.500	0.00100

A. rate $= k[\text{D}]^2 \cdot [\text{E}]^2$

B. rate $= k[\text{D}]^2 \cdot [\text{E}]$

C. rate $= k[\text{D}]^3 \cdot [\text{E}]$

D. rate $= k[\text{D}] \cdot [\text{E}]^2$

E. rate $= k[\text{D}] \cdot [\text{E}]$

105. A 10-mm cube of copper metal is placed in 400 mL of 10 M nitric acid at 27 °C and the reaction below occurs:

$$Cu(s) + 4\,H^+(aq) + 2\,NO_3^-(aq) \rightarrow Cu^{2+}(aq) + 2\,NO_2(g) + 2\,H_2O\,(l)$$

If nitrogen dioxide is being produced at the rate of 2.8×10^{-4} M/min, what is the rate at which hydrogen ions are being consumed?

A. 3.2×10^{-3} M/min

B. 1.8×10^{-2} M/min

C. 1.1×10^{-4} M/min

D. 5.6×10^{-4} M/min

E. 6.7×10^{-5} M/min

106. What is the equilibrium constant (K_{eq}) expression for the following reaction?

$$2\,CO\,(g) + O_2\,(g) \leftrightarrow 2\,CO_2\,(g)$$

A. $K_{eq} = 2[CO_2] / 2[CO] \times [O_2]$

B. $K_{eq} = [CO] \times [O_2] / [CO_2]$

C. $K_{eq} = [CO_2]^2 / [CO]^2 \times [O_2]$

D. $K_{eq} = [CO_2] / [CO] + [O_2]$

E. $K_{eq} = [CO]^2 \times [O_2] / [CO_2]^2$

107. In the reaction $A + B \rightarrow AB$, which of the following does NOT increase the rate of the reaction?

I. increasing the temperature
II. decreasing the temperature
III. adding a catalyst

A. I only

B. II only

C. III only

D. I and III only

E. II and III only

108. If $K_{eq} = 6.1 \times 10^{-11}$, which statement is true?

A. Slightly more products are present

B. The number of reactants equals products

C. Mostly products are present

D. Mostly reactants are present

E. Cannot be determined

109. By convention, what is the equilibrium constant for step 1 in a reaction?

A. $k_1 k_{+2}$ B. $k_1 + k_{+2}$ C. k_{-1} D. k_1 E. k_1 / k_{-1}

110. A catalyst changes which of the following?

A. ΔS B. ΔG C. ΔH D. $E_{activation}$ E. $\Sigma(\Delta H_{products})$

111. Which changes shift(s) the equilibrium to the right for the reversible reaction in an aqueous solution?

$$HC_2H_3O_2 \ (aq) \leftrightarrow H^+ \ (aq) + C_2H_3O_2^- \ (aq)$$

I. Decreasing $[C_2H_3O_2^-]$
II. Decreasing $[H^+]$
III. Increasing $[HC_2H_3O_2]$
IV. Decreasing $[HC_2H_3O_2]$

A. I and II only
B. I and III only
C. II and III only
D. I, II and III only
E. I, II and IV only

112. For the reaction A + B → C, which proceeds the slowest? (Assume equal temperatures)

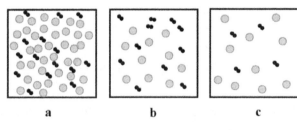

a   b   c

A. a
B. b
C. c
D. All proceed at the same rate
E. Not enough information

113. Why might increasing the concentration of a set of reactants increase the rate of reaction?

A. The rate of reaction depends only on the mass of the atoms and increases as the mass of the reactants increase
B. There is an increased probability that any two reactant molecules collide and react
C. There is an increased ratio of reactants to products
D. The concentration of reactants is unrelated to the rate of reaction
E. None of the above

114. For a chemical reaction to occur, all of the following must happen, EXCEPT:

A. reactant particles must collide with the correct orientation
B. a large enough number of collisions must occur
C. chemical bonds must break or form
D. reactant particles must collide with enough energy for change to occur
E. none of the above

115. If a catalyst is added to the reaction, which direction does the equilibrium shift toward?

$$CO + H_2O + heat \leftrightarrow CO_2 + H_2$$

A. To the left
B. To the right
C. No effect
D. Not enough information
E. Initially to the right, but settles at a ½ current equilibrium

116. For a chemical reaction at equilibrium, which of the following decreases the concentration of products?

A. Decreasing the pressure
B. Increasing the temperature and decreasing the temperature
C. Increasing the temperature
D. Decreasing the temperature
E. Decreasing the concentration of a gaseous or aqueous reactant

117. Which factor does NOT describe activated complexes?

A. May be chemically isolated
B. Decompose rapidly
C. Have specific geometry
D. Are extremely reactive
E. At the high point of the reaction profile

118. When a reaction system is at equilibrium:

A. rate in the forward and reverse directions is equal
B. rate in the forward direction is at a maximum
C. amounts of reactants and products are equal
D. rate in the reverse direction is at a minimum
E. reaction is complete and static

119. A reaction vessel contains NH_3, N_2, and H_2 at equilibrium with $[NH_3] = 0.1$ M, $[N_2] = 0.2$ M, and $[H_2] = 0.3$ M. For decomposition, what is K for $NH_3 \rightarrow N_2$ and H_2?

A. $K = (0.2) \cdot (0.3)^3 / (0.1)^2$
B. $K = (0.1) / (0.2)^2 \cdot (1.5)^3$

C. $K = (0.1)^2 / (0.2) \cdot (0.3)^3$
D. $K = (0.2) \cdot (0.3)^2 / (0.1)$
E. $K = (0.2) \cdot (0.3)^3 / (0.1)$

120. The data provides the rate of a reaction as affected by the concentration of the reactants. What is the order of the reaction rate?

Experiment	[X]	[Y]	[Z]	Rate (mol L^{-1} hr^{-1})
1	0.200 M	0.100 M	0.600 M	5.0
2	0.200 M	0.400 M	0.400 M	80.0
3	0.600 M	0.100 M	0.200 M	15.0
4	0.200 M	0.100 M	0.200 M	5.0
5	0.200 M	0.200 M	0.400 M	20.0

A. Zero order with respect to X

B. Order for X is minus one (rate proportional to 1 / [X])

C. First order with respect to X

D. Second order with respect to X

E. Order for X cannot be determined from data

Solution Chemistry

===

Practice Set 1: Questions 1–20

===

1. If the solubility of nitrogen in blood is 1.90 cc/100 cc at 1.00 atm, what is the solubility of nitrogen in a scuba diver's blood at a depth of 125 feet where the pressure is 4.5 atm?

   A. 1.90 cc/100 cc

   B. 2.36 cc/100 cc

   C. 4.5 cc/100 cc

   D. 0.236 cc/100 cc

   E. 8.55 cc/100 cc

2. All of the statements about molarity are correct, **EXCEPT**:

   A. volume = moles/molarity

   B. moles = molarity × volume

   C. molarity of a diluted solution is less than the molarity of the original solution

   D. abbreviation is M

   E. molarity equals moles of solute per mole of solvent

3. Which of the following molecules is expected to be most soluble in water?

   A. NaCl

   B. $CH_3CH_2CH_2COOH$

   C. $CH_3CH_2CH_2\ OH$

   D. $Al(OH)_3$

   E. CH_4

4. The equation for the reaction shown below can be written as an ionic equation.

   $$BaCl_2\ (aq) + K_2CrO_4\ (aq) \rightarrow BaCrO_4\ (s) + 2\ KCl\ (aq)$$

   In the ionic equation, the spectator ions are:

   A. K^+ and Cl^-

   B. Ba^{2+} and CrO_4^{2-}

   C. Ba^{2+} and K^+

   D. K^+ and CrO_4^{2-}

   E. Cl^- and CrO_4^{2-}

5. In commercially prepared soft drinks, carbon dioxide gas is injected into soda. Under what conditions are carbon dioxide gas most soluble?

   A. High temperature, high-pressure

   B. High temperature, low-pressure

   C. Low temperature, low-pressure

   D. Low temperature, high-pressure

   E. Solubility is the same for all conditions

6. A solute is a:

    A. substance that dissolves into a solvent

    B. substance containing a solid, liquid or gas

    C. solid substance that does not dissolve into water

    D. solid substance that does not dissolve at a given temperature

    E. liquid that does not dissolve into another liquid

7. How many ions are produced in solution by dissociation of one formula unit of $Co(NO_3)_2 \cdot 6H_2O$?

    A. 2   **B.** 3   **C.** 4   **D.** 6   **E.** 9

8. 15 grams of an unknown substance is dissolved in 60 grams of water. When the solution is transferred to another container, it weighs 78 grams. Which of the following is a possible explanation?

    A. The solution reacted with the second container, forming a precipitate

    B. Some of the solution remained in the first container

    C. The reaction was endothermic, which increased the average molecular speed

    D. The solution reacted with the first container, causing some byproducts to be transferred with the solution

    E. The reaction was exothermic, which increased the average molecular speed

9. Which of the following is NOT soluble in H_2O?

    A. Iron (III) hydroxide   **C.** Potassium sulfate

    B. Iron (III) nitrate   **D.** Ammonium sulfate

                                        E. Sodium chloride

10. What is the v/v% concentration of a solution made by adding 25 mL of acetone to 75 mL of water?

    A. 33% v/v   **C.** 25% v/v

    B. 0.33% v/v   **D.** 2.5% v/v

                                        E. 3.3% v/v

11. Why is octane less soluble in H_2O than in benzene?

    A. Bonds between benzene and octane are much stronger than the bonds between H_2O and octane

    B. Octane cannot dissociate in the presence of H_2O

    C. Bonds between H_2O and octane are weaker than the bonds between H_2O molecules

    D. Octane and benzene have similar molecular weights

    E. H_2O dissociates in the presence of octane

12. What is the K_{sp} for slightly soluble copper (II) phosphate in an aqueous solution?

$$Cu_3(PO_4)_2\ (s) \leftrightarrow 3\ Cu^{2+}\ (aq) + 2\ PO_4^{3-}\ (aq)$$

A. $K_{sp} = [Cu^{2+}]^3 \cdot [PO_4^{3-}]^2$

B. $K_{sp} = [Cu^{2+}] \cdot [PO_4^{3-}]^2$

C. $K_{sp} = [Cu^{2+}]^3 \cdot [PO_4^{3-}]$

D. $K_{sp} = [Cu^{2+}] \cdot [PO_4^{3-}]$

E. $K_{sp} = [Cu^{2+}]^2 \cdot [PO_4^{3-}]^3$

13. Apply the *like dissolves like* rule to predict which of the following liquids is/are miscible with water:

    I.   carbon tetrachloride, CCl_4

    II. toluene, C_7H_8

    III. ethanol, C_2H_5OH

A. I only

B. II only

C. III only

D. I and II only

E. I, II and III

14. In which of the following pairs of substances would both species in the pair be written in the molecular form in a net ionic equation?

    I. CO_2 and H_2SO_4 II. LiOH and H_2 III. HF and CO_2

A. I only

B. II only

C. III only

D. I, II and III

E. None of the above

15. Which statement best describes a supersaturated solution?

A. It contains dissolved solute in equilibrium with undissolved solid

B. It rapidly precipitates if a seed crystal is added

C. It contains as much solvent as it can accommodate

D. It contains no double bonds

E. It contains only electrolytes

16. What is the mass of a 7.50% urine sample that contains 122 g of dissolved solute?

A. 1,250 g   **B.** 935 g   **C.** 49.35 g   **D.** 155.4 g   **E.** 1,627 g

17. Calculate the molarity of a solution prepared by dissolving 15.0 g of NH_3 in 250 g of water with a final density of 0.974 g/mL.

A. 36.2 M

B. 3.23 M

C. 0.0462 M

D. 0.664 M

E. 6.80 M

18. Which of the following compounds has the highest boiling point?

A. 0.2 M $Al(NO_3)_3$ **C.** 0.2 M glucose ($C_6H_{12}O_6$)

B. 0.2 M $MgCl_2$ **D.** 0.2 M Na_2SO_4

          E. Pure H_2O

19. Which compound produces four ions per formula unit by dissociation when dissolved in water?

A. Li_3PO_4 **C.** $MgSO_4$

B. $Ca(NO_3)_2$ **D.** $(NH_4)_2SO_4$

          E. $(NH_4)_4Fe(CN)_6$

20. Which of the following would be a weak electrolyte in a solution?

A. HBr (*aq*)   **B.** KCl   **C.** KOH   **D.** $HC_2H_3O_2$ **E.** HI

===

Practice Set 2: Questions 21–40

===

21. The ions Ca^{2+}, Mg^{2+}, Fe^{2+}, Fe^{3+}, present in groundwater, can be removed by pretreating the water with:

    A. $PbSO_4$ **C.** KNO_3

    B. $Na_2CO_3 \cdot 10H_2O$ **D.** $CaCl_2$

                                           E. 0.05 M HCl

22. Choose the spectator ions: $Pb(NO_3)_2$ *(aq)* + H_2SO_4 *(aq)* → ?

    A. NO_3^- and H^+ **C.** Pb^{2+} and H^+

    B. H^+ and SO_4^{2-} **D.** Pb^{2+} and NO_3^-

                                           E. Pb^{2+} and SO_4^{2-}

23. From the *like dissolves like* rule, predict which of the following vitamins is soluble in water:

    A. α-tocopherol ($C_{29}H_{50}O_2$)   **C.** ascorbic acid ($C_6H_8O_6$)

    B. calciferol ($C_{27}H_{44}O$)   **D.** retinol ($C_{20}H_{30}O$)

                                         E. none of the above

24. How much water must be added when 125 mL of a 2.00 M solution of HCl is diluted to a final concentration of 0.400 M?

    A. 150 mL   **C.** 625 mL

    B. 850 mL   **D.** 750 mL

                                         E. 500 mL

25. Which of the following is the sulfate ion?

    A. SO_4^{2-} **B.** S^{2-} **C.** CO_3^{2-} **D.** PO_4^{3-} **E.** S^-

26. Which of the following explains why bubbles form inside a pot of water when the pot of water is heated?

    A. As temperature increases, the vapor pressure increases

    B. As temperature increases, the atmospheric pressure decreases

    C. As temperature increases, the solubility of air decreases

    D. As temperature increases, the kinetic energy decreases

    E. None of the above

27. A solution in which the rate of crystallization is equal to the rate of dissolution is:

    A. saturated

    B. supersaturated

    C. dilute

    D. unsaturated

    E. impossible to determine

28. What is the term that refers to liquids that do not dissolve in one another and separate into two layers?

    A. Soluble

    B. Miscible

    C. Insoluble

    D. Immiscible

    E. None of the above

29. Which is a correctly balanced hydration equation for the hydration of Na_2SO_4?

    A. $Na_2SO_4 (s) \xrightarrow{H_2O} Na^+ (aq) + 2SO_4^{2-} (aq)$

    B. $Na_2SO_4 (s) \xrightarrow{H_2O} 2\,Na^{2+} (aq) + S^{2-} (aq) + O_4^{2-} (aq)$

    C. $Na_2SO_4 (s) \xrightarrow{H_2O} Na_2^{2+} (aq) + SO_4^{2-} (aq)$

    D. $Na_2SO_4 (s) \xrightarrow{H_2O} 2\,Na^+ (aq) + SO_4^{2-} (aq)$

    E. $Na_2SO_4 (s) \xrightarrow{H_2O} 2\,Na^{2+} (aq) + S^{2-} (aq) + SO_4^{2-} (aq) + O_4^{2-} (aq)$

30. Soft drinks are carbonated by injection with carbon dioxide gas. Under what conditions is carbon dioxide gas least soluble?

    A. High temperature, low-pressure

    B. High temperature, high-pressure

    C. Low temperature, high-pressure

    D. Low temperature, low-pressure

    E. None of the above

31. Which type of compound is likely to dissolve in H_2O?

    I.   One with hydrogen bonds

    II. Highly polar compound

    III. Salt

    A. I only

    B. II only

    C. III only

    D. I and III only

    E. I, II and III

32. Which of the following might have the best solubility in water?

A. CH_3CH_3

B. CH_3OH

C. CCl_4

D. O_2

E. None of the above

33. What is the K_{sp} for slightly soluble gold (III) chloride in an aqueous solution for the reaction shown?

$$AuCl_3 (s) \leftrightarrow Au^{3+} (aq) + 3 \, Cl^- (aq)$$

A. $K_{sp} = [Au^{3+}]^3 \cdot [Cl^-] / [AuCl_3]$

B. $K_{sp} = [Au^{3+}] \cdot [Cl^-]^3$

C. $K_{sp} = [Au^{3+}]^3 \cdot [Cl^-]$

D. $K_{sp} = [Au^{3+}] \cdot [Cl^-]$

E. $K_{sp} = [Au^{3+}] \cdot [Cl^-]^3 / [AuCl_3]$

34. What is the net ionic equation for the reaction shown?

$$CaCO_3 + 2 \, HNO_3 \rightarrow Ca(NO_3)_2 + CO_2 + H_2O$$

A. $CO_3^{2-} + H^+ \rightarrow CO_2$

B. $CaCO_3 + 2 \, H^+ \rightarrow Ca^{2+} + CO_2 + H_2O$

C. $Ca^{2+} + 2 \, NO_3^- \rightarrow Ca(NO_3)_2$

D. $CaCO_3 + 2 \, NO_3^- \rightarrow Ca(NO_3)_2 + CO_3^{2-}$

E. None of the above

35. What is the volume of a 0.550 M $Fe(NO_3)_3$ solution needed to supply 0.950 moles of nitrate ions?

A. 265 mL

B. 0.828 mL

C. 22.2 mL

D. 576 mL

E. 384 mL

36. What is the molarity of a solution that contains 48 mEq Ca^{2+} per liter?

A. 0.024 M

B. 0.048 M

C. 1.8 M

D. 2.4 M

E. 0.96 M

37. If 36.0 g of LiOH is dissolved in water to make 975 mL of solution, what is the molarity of the LiOH solution? (Use molecular mass of LiOH = 24.0 g/mol)

A. 1.54 M

B. 2.48 M

C. 0. 844 M

D. 0.268 M

E. 0.229 M

38. What volume of 8.50% (m/v) solution contains 60.0 grams of glucose?

A. 170 mL

B. 448 mL

C. 706 mL

D. 344 mL

E. 960 mL

39. In an AgCl solution, if the K_{sp} for AgCl is A, and the concentration Cl^- in a container is B molar, what is the concentration of Ag (in moles/liter)?

I. A moles/liter

II. B moles/liter

III. A/B moles/liter

A. I only

B. II only

C. III only

D. II and III only

E. I and III only

40. Which statement below is generally true?

A. Bases are strong electrolytes and ionize completely when dissolved in water

B. Salts are strong electrolytes and dissociate completely when dissolved in water

C. Acids are strong electrolytes and ionize completely when dissolved in water

D. Bases are weak electrolytes and ionize completely when dissolved in water

E. Salts are weak electrolytes and ionize partially when dissolved in water

===

Practice Set 3: Questions 41–60

===

41. Which of the following intermolecular attractions is/are important for the formation of a solution?

     I. solute-solute

     II. solvent-solute

     III. solvent-solvent

A. I only   **C.** I and II only

B. III only   **D.** II and III only

                             E. I, II and III

42. What is the concentration of I^- ions in a 0.40 M solution of magnesium iodide?

A. 0.05 M   **C.** 0.60 M

B. 0.80 M   **D.** 0.20 M

                             E. 0.40 M

43. Which is most likely soluble in NH_3?

A. CO_2 **C.** CCl_4

B. SO_2 **D.** N_2

                             E. H_2

44. Which of the following represents the symbol for the chlorite ion?

A. ClO_2^- **B.** ClO^- **C.** ClO_4^- **D.** ClO_3^- **E.** ClO_2

45. Which of the following statements best describes what is happening in a water softening unit?

  A. Sodium is removed from the water, making the water interact less with the soap molecules

  B. Ions in the water softener are softened by chemically bonding with sodium

  C. Hard ions are trapped in the softener, which filters out the ions

  D. Hard ions in water are exchanged for ions that do not interact as strongly with soaps

  E. None of the above

46. Which of the following compounds are soluble in water?

     I. $Mn(OH)_2$ II. $Cr(NO_3)_3$ III. $Ni_3(PO_4)_2$

A. I only   **C.** III only

B. II only   **D.** I and III only

                             E. I, II and III

47. The hydration number of an ion is the number of:

   A. water molecules bonded to an ion in an aqueous solution

   B. water molecules required to dissolve one mole of ions

   C. ions bonded to one mole of water molecules

   D. ions dissolved in one liter of an aqueous solution

   E. water molecules required to dissolve the compound

48. When a solid dissolves, each molecule is removed from the crystal by interaction with the solvent. This process of surrounding each ion with solvent molecules is called:

   A. hemolysis

   B. electrolysis

   C. crenation

   D. dilution

   E. solvation

49. The term *miscible* describes which type of solution?

   A. Solid/solid

   B. Liquid/gas

   C. Liquid/solid

   D. Liquid/liquid

   E. Solid/gas

50. Apply the *like dissolves like* rule to predict which of the following vitamins is insoluble in water:

   A. niacinamide ($C_6H_6N_2O$)

   B. pyridoxine ($C_8H_{11}NO_3$)

   C. retinol ($C_{20}H_{30}O$)

   D. thiamine ($C_{12}H_{17}N_4OS$)

   E. cyanocobalamin ($C_{63}H_{88}CoN_{14}O_{14}P$)

51. Which species is NOT written as its constituent ions when the equation is expanded into the ionic equation?

$$Mg(OH)_2\ (s) + 2\ HCl\ (aq) \rightarrow MgCl_2\ (aq) + 2\ H_2O\ (l)$$

   A. $Mg(OH)_2$ only

   B. H_2O and $Mg(OH)_2$

   C. HCl

   D. $MgCl_2$

   E. HCl and $MgCl_2$

52. If x moles of $PbCl_2$ fully dissociate in 1 liter of H_2O, the K_{sp} is equivalent to:

   A. x^2 **B.** $2x^4$ **C.** $3x^2$ **D.** $2x^3$ **E.** $4x^3$

53. Which of the following are strong electrolytes?

   I. salts   II. strong bases   III. weak acids

   A. I only

   B. I and II only

   C. III only

   D. I, II and III

   E. I and III only

54. What are the spectator ions in the reaction between KOH and HNO_3?

   A. K^+ and NO_3^-

   B. H^+ and NO_3^-

   C. K^+ and H^+

   D. H^+ and ^-OH

   E. K^+ and ^-OH

55. What volume of 14 M acid must be diluted with distilled water to prepare 6.0 L of 0.20 M acid?

   A. 86 mL

   B. 62 mL

   C. 0.94 mL

   D. 6.8 mL

   E. 120 mL

56. What is the molarity of the solution obtained by diluting 160 mL of 4.50 M NaOH to 595 mL?

   A. 0.242 M

   B. 1.21 M

   C. 2.42 M

   D. 1.72 M

   E. 0.115 M

57. Which compound is most likely to be more soluble in the nonpolar solvent of benzene than in water?

   A. SO_2

   B. CO_2

   C. Silver chloride

   D. H_2S

   E. CH_2Cl_2

58. Which of the following concentrations is dependent on temperature?

   A. Mole fraction

   B. Molarity

   C. Mass percent

   D. Molality

   E. More than one of the above

59. Which of the following solutions is the most concentrated?

   A. One liter of water with 1 gram of sugar

   B. One liter of water with 2 grams of sugar

   C. One liter of water with 5 grams of sugar

   D. One liter of water with 10 grams of sugar

   E. All are the same

60. What mass of NaOH is contained in 75.0 mL of a 5.0% (w/v) NaOH solution?

   A. 6.50 g

   B. 15.0 g

   C. 7.50 g

   D. 0.65 g

   E. 3.75 g

===

Practice Set 4: Questions 61–80

===

Questions **61** through **63** are based on the following data:

	K_{sp}
$PbCl_2$	1.0×10^{-5}
AgCl	1.0×10^{-10}
$PbCO_3$	1.0×10^{-15}

61. Consider a saturated solution of $PbCl_2$. The addition of NaCl would:

    I.   decrease $[Pb^{2+}]$

    II. increase the precipitation of $PbCl_2$

    III. have no effect on the precipitation of $PbCl_2$

A. I only
B. II only
C. III only
D. I and II only
E. I and III only

62. What occurs when $AgNO_3$ is added to a saturated solution of $PbCl_2$?

    I.   AgCl precipitates

    II. $Pb(NO_3)_2$ forms a white precipitate

    III. More $PbCl_2$ forms

A. I only
B. II only
C. III only
D. I, II and III
E. I and II only

63. Comparing equal volumes of saturated solutions for $PbCl_2$ and AgCl, which solution contains a greater concentration of Cl^-?

    I. $PbCl_2$

    II. AgCl

    III. Both have the same concentration of Cl^-

A. I only
B. II only
C. III only
D. I and II only
E. Cannot be determined

64. Which of the following is the reason why hexane is significantly soluble in octane?

   A. Entropy increases for the two substances as the dominant factor in the ΔG when mixed
   B. Hexane hydrogen bonds with octane
   C. Intermolecular bonds between hexane-octane are much stronger than either hexane or octane molecular bonds
   D. ΔH for hexane-octane is greater than hexane-H_2O
   E. Hexane and octane have similar molecular weights

65. Which of the following are characteristics of an ideally dilute solution?

   I.   Solute molecules do not interact with each other
   II. Solvent molecules do not interact with each other
   III. The mole fraction of the solvent approaches 1

   A. I only
   B. II only
   C. I and III only
   D. I, II and III
   E. I and II only

66. Which principle states that the solubility of a gas in a liquid is proportional to the partial pressure of the gas above the liquid?

   A. Solubility principle
   B. Tyndall effect
   C. Colloid principle
   D. Henry's law
   E. None of the above

67. Water and methanol are two liquids that dissolve in each other. When the two are mixed they form one layer, because the liquids are:

   A. unsaturated
   B. saturated
   C. miscible
   D. immiscible
   E. supersaturated

68. What is the molarity of a glucose solution that contains 10.0 g of $C_6H_{12}O_6$ dissolved in 100.0 mL of solution? (Use the molecular mass of $C_6H_{12}O_6$ = 180.0 g/mol)

   A. 1.80 M
   B. 0.555 M
   C. 0.0555 M
   D. 0.00555 M
   E. 18.0 M

69. Which of the following solid compounds is insoluble in water?

   I. $BaSO_4$ II. Hg_2Cl_2 III. $PbCl_2$

   A. I only
   B. II only
   C. III only
   D. I and III only
   E. I, II and III

70. The net ionic equation for the reaction between zinc and hydrochloric acid solution is:

A. $Zn\ (s) + 2\ H^+\ (aq) + 2\ Cl^-\ (aq) \rightarrow Zn^{2+}\ (aq) + 2\ Cl^-\ (aq) + H_2\ (g)$

B. $ZnCl_2\ (aq) + H_2\ (g) \rightarrow Zn\ (s) + 2\ HCl\ (aq)$

C. $Zn\ (s) + 2\ H^+\ (aq) \rightarrow Zn^{2+}\ (aq) + H_2\ (g)$

D. $Zn\ (s) + 2\ HCl\ (aq) \rightarrow ZnCl_2\ (aq) + H_2\ (g)$

E. None of the above

71. Apply the *like dissolves like* rule to predict which of the following liquids is/are miscible with water:

      I.   methyl ethyl ketone, C_4H_8O

      II. glycerin, $C_3H_5(OH)_3$

      III. formic acid, $HCHO_2$

A. I only

B. II only

C. III only

D. I and II only

E. I, II and III

72. What is the K_{sp} for calcium fluoride (CaF_2) if the calcium ion concentration in a saturated solution is 0.00021 *M*?

A. $K_{sp} = 3.7 \times 10^{-11}$

B. $K_{sp} = 2.6 \times 10^{-10}$

C. $K_{sp} = 3.6 \times 10^{-9}$

D. $K_{sp} = 8.1 \times 10^{-10}$

E. $K_{sp} = 7.3 \times 10^{-11}$

73. A 4 M solution of H_3A is completely dissociated in water. How many equivalents of H^+ are found in 1/3 liter?

A. ¼   **B.** 1   **C.** 1.5   **D.** 3   **E.** 4

74. Which of the following aqueous solutions are poor conductors of electricity?

      I.   sucrose, $C_{12}H_{22}O_{11}$

      II. barium nitrate, $Ba(NO_3)_2$

      III. calcium bromide, $CaBr_2$

A. I only

B. II only

C. III only

D. I and II only

E. I, II and III

75. Which of the following solid compounds is insoluble in water?

A. $BaSO_4$

B. Na_2S

C. $(NH_4)_2CO_3$

D. K_2CrO_4

E. $Sr(OH)_2$

76. Which is true if the ion concentration product of a solution of AgCl is less than the K_{sp}?

    I. Precipitation occurs

    II. The ions are insoluble in water

    III. Precipitation does not occur

A. I only

B. II only

C. III only

D. I and II only

E. II and III only

77. Under which conditions is the expected solubility of oxygen gas in water the highest?

A. High temperature and high O_2 pressure above the solution

B. Low temperature and low O_2 pressure above the solution

C. Low temperature and high O_2 pressure above the solution

D. High temperature and low O_2 pressure above the solution

E. The O_2 solubility is independent of temperature and pressure

78. What is the molarity of an 8.60 molal solution of methanol (CH_3OH) with a density of 0.94 g/mL?

A. 0.155 M

B. 23.5 M

C. 6.34 M

D. 9.68 M

E. 2.35 M

79. Which of the following is NOT a unit factor related to a 15.0% aqueous solution of potassium iodide (KI)?

A. 100 g solution / 85.0 g water

B. 85.0 g water / 15.0 g KI

C. 85.0 g water / 100 g solution

D. 15.0 g KI / 85.0 g water

E. 15.0 g KI / 100 g water

80. What is the molar concentration of a solution containing 0.75 mol of solute in 75 cm^3 of solution?

A. 0.1 M   **B.** 1.5 M   **C.** 3 M   **D.** 10 M   **E.** 1 M

===

Practice Set 5: Questions 81–107

===

81. If 25.0 mL of seawater has a mass of 25.88 g and contains 1.35 g of solute, what is the mass/mass percent concentration of solute in the seawater sample?

    A. 1.14%
    C. 2.62%

    B. 12.84%
    D. 5.22%

        E. 9.45%

82. Which of the following statements describing solutions is NOT true?

    A. Solutions are colorless

    B. The particles in a solution are atomic or molecular

    C. Making a solution involves a physical change

    D. Solutions are homogeneous

    E. Solutions are transparent

83. Calculate the solubility product of AgCl if the solubility of AgCl in H_2O is 1.3×10^{-4} mol/L?

    A. 1.3×10^{-4}
    C. 2.6×10^{-4}

    B. 1.3×10^{-2}
    D. 3.9×10^{-5}

        E. 1.7×10^{-8}

84. Which of the following solutions is the most dilute?

    A. 0.1 liters of H_2O with 1 gram of sugar
    C. 0.5 liters of H_2O with 5 grams of sugar

    B. 0.2 liters of H_2O with 2 grams of sugar
    D. 1 liter of H_2O with 10 grams of sugar

        E. All have the same concentration

85. When salt A is dissolved into water to form a 1 molar unsaturated solution, the temperature of the solution decreases. Under these conditions, which statement is accurate when salt A is dissolved in water?

    A. $\Delta H°$ and $\Delta G°$ are positive
    C. $\Delta H°$ is negative and $\Delta G°$ is positive

    B. $\Delta H°$ is positive and $\Delta G°$ is negative
    D. $\Delta H°$ and $\Delta G°$ are negative

        E. $\Delta H°$, $\Delta S°$ and $\Delta G°$ are positive

86. The heat of a solution measures the energy absorbed during:

    I. the formation of solvent-solute bonds

    II. the breaking of solute-solute bonds

    III. the breaking of solvent-solvent bonds

A. I only

B. II only

C. I and II only

D. I and III only

E. II and III only

87. Why does the reaction proceed if, when solid potassium chloride is dissolved in H_2O, the energy of the bonds formed is less than the energy of the bonds broken?

A. The electronegativity of the H_2O increases from interaction with potassium and chloride ions

B. The reaction does not take place under standard conditions

C. The decreased disorder due to mixing decreases entropy within the system

D. Remaining potassium chloride which does not dissolve offsets the portion that dissolves

E. The increased disorder due to mixing increases entropy within the system

88. Why are salts more soluble in H_2O than in benzene?

A. Benzene is aromatic and therefore very stable

B. The dipole moment of H_2O compensates for the loss of ionic bonding when salt dissolves

C. Strong intermolecular attractions in benzene must be disrupted to dissolve salt in benzene

D. The molecular mass of H_2O is similar to the atomic mass of most ions

E. The dipole moment of H_2O compensates for the increased ionic bonding when salt dissolves

89. What is the formula of the solid formed when aqueous barium chloride is mixed with aqueous potassium chromate?

A. K_2CrO_4

B. K_2Ba

C. KCl

D. $BaCrO_4$

E. $BaCl_2$

90. Why might sodium carbonate (washing soda, Na_2CO_3) be added to hard water for cleaning?

A. The soap gets softer due to the added ions

B. The ions solubilize the soap due to ion-ion intermolecular attraction, which improves the cleaning ability

C. The hard ions in the water are more attracted to the carbonate ions' –2 charge

D. The added sodium ions dissolve the hard ions

E. None of the above

91. What volume of 0.25 M hydrochloric acid reacts completely with 0.400 g of sodium hydrogen carbonate, $NaHCO_3$? (Use the molar mass of $NaHCO_3$ = 84.0 g/mol)

$$NaHCO_3 \ (s) + HCl \ (aq) \rightarrow NaCl \ (aq) + H_2O \ (l) + CO_2 \ (g)$$

A. 142 mL   **C.** 19.0 mL

B. 34.6 mL   **D.** 54.6 mL

                                                    E. 32.6 mL

92. What is the concentration of the contaminant in ppm (m/m) if 8.8×10^{-3} g of a contaminant is present in 5,246 g of a particular solution?

A. 12.40 ppm   **C.** 4.06 ppm

B. 1.53 ppm   **D.** 1.68 ppm

                                                    E. 8.72 ppm

93. Which of the following structures represents the bicarbonate ion?

A. HCO_3^- **C.** CO_3^-

B. $H_2CO_3^{2-}$ **D.** CO_3^{2-}

                                                    E. HCO_3^{2-}

94. Of the following, which can serve as the solute in a solution?

         I. solid   II. liquid   III. gas

A. I only   **C.** III only

B. II only   **D.** I and II only

                                                    E. I, II and III

95. Hydration involves the:

A. formation of water–solute bonds

B. breaking of water–water bonds

C. breaking of water–solute bonds

D. breaking of solute–solute bonds

E. breaking of water-water bonds and formation of water-solute bonds

96. Which is the name for a substance represented by a formula written as $M_xLO_y \cdot zH_2O$?

A. Solvent   **C.** Solid hydrate

B. Solute   **D.** Colloid

                                                    E. Suspension

97. In the reaction between aqueous silver nitrate and aqueous potassium chromate, what is the identity of the soluble substance that is formed?

A. Potassium nitrate

B. Potassium chromate

C. Silver nitrate

D. Silver chromate

E. No soluble substance is formed

98. Apply the *like dissolves like* rule to predict which of the following liquids is miscible with liquid bromine (Br_2).

    I.   benzene (C_6H_6)

    II. hexane (C_6H_{14})

    III. carbon tetrachloride (CCl_4)

A. I only

B. II only

C. III only

D. I and II only

E. I, II and III

99. Hydrochloric acid contains:

A. ionic bonds and is a weak electrolyte

B. hydrogen bonds and is a weak electrolyte

C. covalent bonds and is a strong electrolyte

D. ionic bonds and is a non-electrolyte

E. covalent bonds and is a non-electrolyte

100. At 22 °C, a one-liter sample of pure water has a vapor pressure of 18.5 torr. If 7.5 g of NaCl is added to the sample, what is the result for the vapor pressure of the water?

A. It equals the vapor pressure of the NaCl added to the sample

B. It remains unchanged at 18.5 torr

C. It increases the vapor pressure

D. It decreases the vapor pressure

E. It increases to the square root of solute added

101. At room temperature, barium hydroxide ($Ba(OH)_2$) is dissolved in pure water. From the equilibrium concentrations of Ba^+ and ^-OH ions, which K_{sp} expression could be used to find the solubility product for barium hydroxide?

A. $K_{sp} = [Ba^{2+}] \cdot [^-OH]^2$

B. $K_{sp} = 2[Ba^{2+}] \cdot [^-OH]^2$

C. $K_{sp} = [Ba^{2+}] \cdot [^-OH]$

D. $K_{sp} = [Ba^{2+}] \times 2[^-OH]$

E. $K_{sp} = 2[Ba^{2+}] \cdot [^-OH]$

102. How many grams of H_3PO_4 are needed to make 200 mL of a 0.2 M H_3PO_4 solution?

A. 3.9 g   **B.** 5.8 g   **C.** 0.34 g   **D.** 2.6 g   **E.** 1.4 g

103. What is the concentration of ^-OH, if the concentration of H_3O^+ is 1×10^{-8} M? (Use the expression $[H_3O^+] \times [^-OH] = K_w = 1 \times 10^{-14}$)

A. 1×10^8

B. 1×10^6

C. 1×10^{-14}

D. 1×10^{-6}

E. 1×10^{-8}

104. What is the molarity of a solution prepared by dissolving 4.50 mol NaCl in enough water to make 1.50 L of solution?

A. 6.58 M

B. 3.0 M

C. 4.65 M

D. 7.68 M

E. 11.68 M

105. What is the net ionic equation for the reaction between HNO_3 and Na_2SO_3?

A. $H_2O + SO_2\,(g) \rightarrow H^+\,(aq) + HSO_3^-\,(aq)$

B. $2\,H^+\,(aq) + SO_3^{2-}\,(aq) \rightarrow H_2O + SO_2\,(g)$

C. $HNO_3\,(aq) + Na_2SO_3\,(aq) \rightarrow H_2O + SO_2\,(g) + NO_3^-\,(aq)$

D. $H_2SO_3\,(aq) \rightarrow H_2O + SO_2\,(g)$

E. $2\,H^+\,(aq) + Na_2SO_3\,(aq) \rightarrow 2\,Na^+\,(aq) + H_2O + SO_2\,(g)$

106. Which percentage concentration units are used by chemists?

I. % (v/v)   II. % (m/v)   III. % (m/m)

A. I only

B. II only

C. III only

D. I, II and III

E. II and III only

107. Which of the following aqueous solutions are poor conductors of electricity?

I. ethanol, CH_3CH_2OH

II. ammonium chloride, NH_4Cl

III. ethylene glycol, $HOCH_2CH_2OH$

A. I only

B. II only

C. III only

D. I and III only

E. I, II and III

===

Practice Set 1: Questions 1–20

===

1. Which is the conjugate acid–base pair in the reaction?

$$CH_3NH_2 + HCl \leftrightarrow CH_3NH_3^+ + Cl^-$$

A. HCl and Cl^-

B. $CH_3NH_3^+$ and Cl^-

C. CH_3NH_2 and Cl^-

D. CH_3NH_2 and HCl

E. HCl and H_3O^+

2. What is the pH of an aqueous solution if the $[H^+] = 0.10$ M?

A. 0.0   **B.** 1.0   **C.** 2.0   **D.** 10.0   **E.** 13.0

3. In which of the following pairs of substances are both species salts?

A. NH_4F and KCl

B. $CaCl_2$ and HCN

C. LiOH and K_2CO_3

D. NaOH and $CaCl_2$

E. HCN and K_2CO_3

4. Which reactant is a Brønsted-Lowry acid?

$$HCl\ (aq) + KHS\ (aq) \rightarrow KCl\ (aq) + H_2S\ (aq)$$

A. KCl   **B.** H_2S **C.** HCl   **D.** KHS   **E.** None of the above

5. If a light bulb in a conductivity apparatus glows brightly when testing a solution, which of the following must be true about the solution?

A. It is highly reactive

B. It is slightly reactive

C. It is highly ionized

D. It is slightly ionized

E. It is not an electrolyte

6. What is the term for a substance capable of either donating or accepting a proton in an acid–base reaction?

I. Nonprotic   II. Aprotic   III. Amphoteric

A. I only

B. II only

C. III only

D. I and III only

E. II and III only

7. Which of the following compounds is a strong acid?

        I. $HClO_4$ (*aq*)   II. H_2SO_4 (*aq*)   III. HNO_3 (*aq*)

    A. I only   **C.** II and III only

    B. II only   **D.** I and II only

                                                     E. I, II and III

8. What is the approximate pH of a solution of a strong acid where $[H_3O^+] = 8.30 \times 10^{-5}$?

    A. 9   **B.** 11   **C.** 3   **D.** 5   **E.** 4

9. Which of the following reactions represents the ionization of H_2O?

    A. $H_2O + H_2O \rightarrow 2\ H_2 + O_2$ **C.** $H_2O + H_3O^+ \rightarrow H_3O^+ + H_2O$

    B. $H_2O + H_2O \rightarrow H_3O^+ + {}^-OH$ **D.** $H_3O^+ + {}^-OH \rightarrow H_2O + H_2O$

                                                     E. None of the above

10. Which set below contains only weak electrolytes?

    A. NH_4Cl (*aq*), $HClO_2$ (*aq*), HCN (*aq*)   **C.** KOH (*aq*), H_3PO_4 (*aq*), $NaClO_4$ (*aq*)

    B. NH_3 (*aq*), HCO_3^- (*aq*), HCN (*aq*)   **D.** HNO_3 (*aq*), H_2SO_4 (*aq*), HCN (*aq*)

                                                     E. $NaOH$ (*aq*), H_2SO_4 (*aq*), $HC_2H_3O_2$ (*aq*)

11. Which of the following statements describes a neutral solution?

    A. $[H_3O^+] / [{}^-OH] = 1 \times 10^{-14}$ **C.** $[H_3O^+] < [{}^-OH]$

    B. $[H_3O^+] / [{}^-OH] = 1$ **D.** $[H_3O^+] > [{}^-OH]$

                                                      E. $[H_3O^+] \times [{}^-OH] \neq 1 \times 10^{-14}$

12. Which of the following is an example of an Arrhenius acid?

    A. H_2O (*l*)   **C.** $Ba(OH)_2$ (*aq*)

    B. $RbOH$ (*aq*)   **D.** $Al(OH)_3$ (*s*)

                                                     E. None of the above

13. In the following reaction, which reactant is a Brønsted-Lowry base?

        H_2CO_3 (*aq*) + Na_2HPO_4 (*aq*) $\rightarrow$ $NaHCO_3$ (*aq*) + NaH_2PO_4 (*aq*)

    A. $NaHCO_3$ **C.** Na_2HPO_4

    B. NaH_2PO_4 **D.** H_2CO_3

                                                     E. None of the above

14. Which of the following is the conjugate base of $^-$OH?

   A. O_2 **B.** O^{2-} **C.** H_2O **D.** O^- **E.** H_3O^+

15. What of the following describes the solution for a vinegar sample at pH of 5?

   A. Weakly basic   **C.** Weakly acidic

   B. Neutral   **D.** Strongly acidic

                                       E. Strongly basic

16. What is the pI for glutamic acid that contains two carboxylic acid groups and an amino group? (Use the carboxyl $pK_{a1} = 2.2$, carboxyl $pK_{a2} = 4.2$ and amino $pK_a = 9.7$)

   A. 3.2   **B.** 1.0   **C.** 6.4   **D.** 5.4   **E.** 5.95

17. Which of the following compounds cannot act as an acid?

   A. NH_3 **C.** HSO_4^{1-}

   B. H_2SO_4 **D.** SO_4^{2-}

                              E. CH_3CO_2H

18. A weak acid is titrated with a strong base. When the concentration of the conjugate base is equal to the concentration of the acid, the titration is at the:

   A. endpoint   **B.** equivalence point

   C. buffering region   **D.** diprotic point

                              E. indicator zone

19. If $[H_3O^+]$ in an aqueous solution is 7.5×10^{-9} M, what is the $[^-OH]$?

   A. 6.4×10^{-5} M   **C.** 7.5×10^{-23} M

   B. $3.8 \times 10^{+8}$ M   **D.** 1.3×10^{-6} M

                              E. 9.0×10^{-9} M

20. Which species has a K_a of 5.7×10^{-10} if NH_3 has a K_b of 1.8×10^{-5}?

   A. H^+ **B.** NH_2^- **C.** NH_4^+ **D.** H_2O **E.** $NaNH_2$

===

Practice Set 2: Questions 21–40

===

21. Which of the following are the conjugate bases of HSO_4^-, CH_3OH and H_3O^+, respectively:

A. SO_4^{2-}, CH_2OH^- and ^-OH

B. CH_3O^-, SO_4^{2-} and H_2O

C. SO_4^-, CH_3O^- and ^-OH

D. SO_4^-, CH_2OH^- and H_2O

E. SO_4^{2-}, CH_3O^- and H_2O

22. If 30.0 mL of 0.10 M $Ca(OH)_2$ is titrated with 0.20 M HNO_3, what volume of nitric acid is required to neutralize the base according to the following expression?

$$2\ HNO_3\ (aq) + Ca(OH)_2\ (aq) \rightarrow 2\ Ca(NO_3)_2\ (aq) + 2\ H_2O\ (l)$$

A. 30.0 mL

B. 15.0 mL

C. 10.0 mL

D. 20.0 mL

E. 45.0 mL

23. Which of the following expressions describes an acidic solution?

A. $[H_3O^+] / [^-OH] = 1 \times 10^{-14}$

B. $[H_3O^+] \times [^-OH] \neq 1 \times 10^{-14}$

C. $[H_3O^+] < [^-OH]$

D. $[H_3O^+] > [^-OH]$

E. $[H_3O^+] / [^-OH] = 1$

24. Which is incorrectly classified as an acid, a base, or an amphoteric species?

A. LiOH / base

B. H_2O / amphoteric

C. H_2S / acid

D. NH_4^+ / base

E. None of the above

25. Which of the following is the strongest weak acid?

A. CH_3COOH; $K_a = 1.8 \times 10^{-5}$

B. HF; $K_a = 6.5 \times 10^{-4}$

C. HCN; $K_a = 6.3 \times 10^{-10}$

D. HClO; $K_a = 3.0 \times 10^{-8}$

E. HNO_2; $K_a = 4.5 \times 10^{-4}$

26. What are the products from the complete neutralization of phosphoric acid with aqueous lithium hydroxide?

A. $LiHPO_4$ (aq) and H_2O (l)

B. Li_3PO_4 (aq) and H_2O (l)

C. Li_2HPO4 (aq) and H_2O (l)

D. LiH_2PO_4 (aq) and H_2O (l)

E. Li_2PO_4 (aq) and H_2O (l)

---

27. Which of the following compounds is NOT a strong base?

   A. $Ca(OH)_2$ **B.** $Fe(OH)_3$ **C.** KOH   **D.** NaOH   **E.** $^-NH_2$

28. What is the $[H^+]$ in stomach acid that registers a pH of 2.0 on a strip of pH paper?

   A. 0.2 M   **B.** 0.1 M   **C.** 0.02 M   **D.** 0.01 M   **E.** 2 M

29. Which statement is true about distinguishing between dissociation and ionization?

   A. Ionization is the separation of existing charged particles
   B. Dissociation produces new charged particles
   C. Ionization involves polar covalent compounds
   D. Dissociation involves polar covalent compounds
   E. Ionization is reversible, while dissociation is irreversible

30. Which of the following is a general property of a basic solution?

   I.   Turns litmus paper red
   II. Tastes sour
   III. Causes the skin of the fingers to feel slippery

   A. I only
   B. II only
   C. III only
   D. I and II only
   E. I, II and III

31. Which of the following compound-classification pairs is incorrectly matched?

   A. HF – weak acid
   B. $LiC_2H_3O_2$ –salt
   C. NH_3 – weak base
   D. HI – strong acid
   E. $Ca(OH)_2$ – weak base

32. What is the term for a substance that releases H^+ in H_2O?

   A. Brønsted-Lowry acid
   B. Brønsted-Lowry base
   C. Arrhenius acid
   D. Arrhenius base
   E. Lewis acid

33. Which molecule is acting as a base in the following reaction?

   $^-OH + NH_4^+ \rightarrow H_2O + NH_3$

   A. ^-OH **B.** NH_4^+ **C.** H_2O **D.** NH_3 **E.** H_3O^+

34. Citric acid is a triprotic acid with three carboxylic acid groups having pK_a values of 3.2, 4.8 and 6.4. At a pH of 5.7, what is the predominant protonation state of citric acid?

   A. Three carboxylic acid groups are deprotonated

   B. Three carboxylic acid groups are protonated

   C. One carboxylic acid group is deprotonated, while two are protonated

   D. Two carboxylic acid groups are deprotonated, while one is protonated

   E. The protonation state cannot be determined

35. When fully neutralized by treatment with barium hydroxide, a phosphoric acid yields $Ba_2P_2O_7$ as one of its products. The parent acid for the anion in this compound is:

   A. monoprotic acid   **C.** triprotic acid

   B. diprotic acid   **D.** hexaprotic acid

                                 E. tetraprotic acid

36. Does a solution become more or less acidic when a weak acid solution is added to a concentrated solution of HCl?

   A. Less acidic because the concentration of OH^- increases

   B. No change in acidity because [HCl] is too high to be changed by the weak solution

   C. Less acidic because the solution becomes more dilute with a less concentrated solution of H_3O^+ being added

   D. More acidic because more H_3O^+ is being added to the solution

   E. More acidic because the solution becomes more dilute with a less concentrated solution of H_3O^+ being added

37. Which of the following is a triprotic acid?

   A. HNO_3 **B.** H_3PO_4 **C.** H_2SO_3 **D.** $HC_2H_3O_2$ **E.** CH_2COOH

38. For which of the following pairs of substances do the two members of the pair NOT react?

   A. Na_3PO_4 and HCl   **C.** HF and LiOH

   B. KCl and NaI   **D.** $PbCl_2$ and H_2SO_4

                                 E. All react to form products

39. What happens to the pH when sodium acetate is added to a solution of acetic acid?

   A. Decreases due to the common ion effect

   B. Increases due to the common ion effect

   C. Remains constant because sodium acetate is a buffer

   D. Remains constant because sodium acetate is neither acidic nor basic

   E. Remains constant due to the common ion effect

40. Which of the following is the acidic anhydride of phosphoric acid (H_3PO_4)?

   A. P_2O **B.** P_2O_3 **C.** PO_3 **D.** PO_2 **E.** P_4O_{10}

===

Practice Set 3: Questions 41–60

===

41. Which of the following does NOT act as a Brønsted-Lowry acid?

    A. CO_3^{2-} **B.** HS^- **C.** HSO_4^- **D.** H_2O **E.** H_2SO_4

42. Which of the following is the strongest weak acid?

    A. HF; $pK_a = 3.17$ **C.** $H_2PO_4^-$; $pK_a = 7.18$
    B. HCO_3^-; $pK_a = 10.32$ **D.** NH_4^+; $pK_a = 9.20$
                                             E. $HC_2H_3O_2$; $pK_a = 4.76$

43. Why does boiler scale form on the walls of hot water pipes from groundwater?

    A. Transformation of $H_2PO_4^-$ ions to PO_4^{3-} ions, which precipitate with the "hardness ions," Ca^{2+}, Mg^{2+}, Fe^{2+}/Fe^{3+}
    B. Transformation of HSO_4^- ions to SO_4^{2-} ions, which precipitate with the "hardness ions," Ca^{2+}, Mg^{2+}, Fe^{2+}/Fe^{3+}
    C. Transformation of HSO_3^- ions to SO_3^{2-} ions, which precipitate with the "hardness ions," Ca^{2+}, Mg^{2+}, Fe^{2+}/Fe^{3+}
    D. Transformation of HCO_3^- ions to CO_3^{2-} ions, which precipitate with the "hardness ions," Ca^{2+}, Mg^{2+}, Fe^{2+}/Fe^{3+}
    E. The reaction of the CO_3^{2-} ions present in groundwater with the "hardness ions," Ca^{2+}, Mg^{2+}, Fe^{2+}/Fe^{3+}

44. Which of the following substances, when added to a solution of sulfoxylic acid (H_2SO_2), could be used to prepare a buffer solution?

    A. H_2O **C.** KCl
    B. $HC_2H_3O_2$ **D.** HCl
                                     E. $NaHSO_2$

45. Which of the following statements is NOT correct?

    A. Acidic salts are formed by partial neutralization of a diprotic acid by a diprotic base
    B. Acidic salts are formed by partial neutralization of a triprotic acid by a diprotic base
    C. Acidic salts are formed by partial neutralization of a monoprotic acid by a monoprotic base
    D. Acidic salts are formed by partial neutralization of a diprotic acid by a monoprotic base
    E. Acidic salts are formed by partial neutralization of a polyprotic acid by a monoprotic base

46. Which of the following is the acid anhydride for $HClO_4$?

    A. ClO   **B.** ClO_2 **C.** ClO_3 **D.** ClO_4 **E.** Cl_2O_7

---

47. Identify the acid/base behavior of each substance for the reaction:

$$H_3O^+ + Cl^- \rightleftharpoons H_2O + HCl$$

A. H_3O^+ acts as an acid, Cl^- acts as a base, H_2O acts as a base and HCl acts as an acid

B. H_3O^+ acts as a base, Cl^- acts as an acid, H_2O acts as a base and HCl acts as an acid

C. H_3O^+ acts as an acid, Cl^- acts as a base, H_2O acts as an acid and HCl acts as a base

D. H_3O^+ acts as a base, Cl^- acts as an acid, H_2O acts as an acid and HCl acts as a base

E. H_3O^+ acts as an acid, Cl^- acts as a base, H_2O acts as a base and HCl acts as a base

48. Given the pK_a values for phosphoric acid of 2.15, 6.87 and 12.35, what is the ratio of HPO_4^{2-} / $H_2PO_4^-$ in a typical muscle cell when the pH is 7.35?

A. 6.32×10^{-6}

B. 1.18×10^5

C. 0.46

D. 3.02

E. 3.31×10^3

49. If a light bulb in a conductivity apparatus glows dimly when testing a solution, which of the following must be true about the solution?

   I.   It is slightly reactive

   II. It is slightly ionized

   III. It is highly ionized

A. I only

B. II only

C. III only

D. I and II only

E. II and III only

50. Which of the following properties is NOT characteristic of an acid?

A. It is neutralized by a base

B. It has a slippery feel

C. It produces H^+ in water

D. It tastes sour

E. Its pH reading is less than 7

51. What is the term for a substance that releases hydroxide ions in water?

A. Brønsted-Lowry base

B. Brønsted-Lowry acid

C. Arrhenius base

D. Arrhenius acid

E. Lewis base

52. For the reaction below, which of the following is the conjugate acid of C_5H_5N?

$$C_5H_5N + H_2CO_3 \leftrightarrow C_5H_6N^+ + HCO_3^-$$

A. $C_5H_6N^+$ B. HCO_3^- C. C_5H_5N D. H_2CO_3 E. H_3O^+

53. Which of the following terms applies to Cl^- in the reaction below?

$$HCl\ (aq) \rightarrow H^+ + Cl^-$$

A. Weak conjugate base

B. Strong conjugate base

C. Weak conjugate acid

D. Strong conjugate acid

E. Strong conjugate base and weak conjugate acid

54. Which of the following compounds is a diprotic acid?

A. HCl   **B.** H_3PO_4 **C.** HNO_3 **D.** H_2SO_3 **E.** H_2O

55. Lysine contains two amine groups ($pK_a = 9.0$ and 10.0) and a carboxylic acid group ($pK_a = 2.2$). In a solution of pH 9.5, which describes the protonation and charge state of lysine?

A. Carboxylic acid is deprotonated and negative; amine (pK_a 9.0) is deprotonated and neutral, whereby the amine ($pK_a = 10.0$) is protonated and positive

B. Carboxylic acid is deprotonated and negative; amine ($pK_a = 9.0$) is protonated and positive, whereby the amine ($pK_a = 10.0$) is deprotonated and neutral

C. Carboxylic acid is deprotonated and neutral; both amines are protonated and positive

D. Carboxylic acid is deprotonated and negative; both amines are deprotonated and neutral

E. Carboxylic acid is deprotonated and neutral; amine ($pK_a = 9.0$) is deprotonated and neutral, whereby the amine ($pK_a = 10.0$) is protonated and positive

56. Which of the following is the chemical species present in acidic solutions?

A. $H_2O^+\ (aq)$

B. $H_3O^+\ (l)$

C. $H_2O\ (aq)$

D. $^-OH\ (aq)$

E. $H_3O^+\ (aq)$

57. Which compound has a value of K_a that is approximately equal to 10^{-5}?

A. $CH_3CH_2CH_2CO_2H$

B. KOH

C. NaBr

D. HNO_3

E. NH_3

58. Relative to a pH of 7, a solution with a pH of 4 has:

A. 30 times less $[H^+]$

B. 300 times less $[H^+]$

C. 1,000 times greater $[H^+]$

D. 300 times greater $[H^+]$

E. 30 times greater $[H^+]$

59. What is the pH of this buffer system if the concentration of undissociated weak acid is equal to the concentration of the conjugate base? (Use the K_a of the buffer = 4.6 ×10^{-4})

   A. 1 and 2   **C.** 5 and 6

   B. 3 and 4   **D.** 7 and 8

                            E. 9 and 10

60. Which of the following is the ionization constant expression for water?

   A. K_w = [H$_2$O] / [H$^+$]·[$^-$OH]   **C.** K_w = [H$^+$]·[$^-$OH]

   B. K_w = [H+]·[$^-$OH] / [H$_2$O]   **D.** K_w = [H$_2$O]·[H$_2$O]

                            E. None of the above

==

Practice Set 4: Questions 61–80

==

61. Which of the following statements about strong or weak acids is true?

   A. A weak acid reacts with a strong base
   B. A strong acid does not react with a strong base
   C. A weak acid readily forms ions when dissolved in water
   D. A weak acid and a strong acid at the same concentration are equally corrosive
   E. None of the above

62. What is the value of K_w at 25 °C?

   A. 1.0
   B. 1.0×10^{-7}
   C. 1.0×10^{-14}
   D. 1.0×10^7
   E. 1.0×10^{14}

63. Which of the statements below best describes the following reaction?

   $$HNO_3\ (aq) + LiOH\ (aq) \rightarrow LiNO_3\ (aq) + H_2O\ (l)$$

   A. Nitric acid and lithium hydroxide solutions produce lithium nitrate solution and H_2O
   B. Nitric acid and lithium hydroxide solutions produce lithium nitrate and H_2O
   C. Nitric acid and lithium hydroxide produce lithium nitrate and H_2O
   D. Aqueous solutions of nitric acid and lithium hydroxide produce aqueous lithium nitrate and H_2O
   E. An acid plus a base produces H_2O and a salt

64. The Brønsted-Lowry acid and base in the following reaction are, respectively:

   $$NH_4^+ + CN^- \rightarrow NH_3 + HCN$$

   A. NH_4^+ and ^-CN
   B. ^-CN and HCN
   C. NH_4^+ and HCN
   D. NH_3 and ^-CN
   E. NH_3 and NH_4^+

65. Which would NOT be used to make a buffer solution?

   A. H_2SO_4
   B. H_2CO_3
   C. NH_4OH
   D. CH_3COOH
   E. Tricine

66. Which of the following is a general property of an acidic solution?

   A. Turns litmus paper blue
   B. Neutralizes acids
   C. Tastes bitter
   D. Feels slippery
   E. None of the above

67. What is the term for a solution that is a good conductor of electricity?

A. Strong electrolyte   **C.** Non-electrolyte

B. Weak electrolyte   **D.** Aqueous electrolyte

                                              E. None of the above

68. Which of the following compounds is an acid?

A. HBr   **B.** C_2H_6 **C.** KOH   **D.** NaF   **E.** $NaNH_2$

69. A metal and a salt solution react only if the metal introduced into the solution is:

A. below the replaced metal in the activity series   **C.** below hydrogen in the activity series

B. above the replaced metal in the activity series   **D.** above hydrogen in the activity series

                                                E. equal to the replaced metal in the activity series

70. Which of the following is an example of an Arrhenius base?

           I. NaOH (*aq*)   II. $Al(OH)_3$ (*s*)   III. $Ca(OH)_2$ (*aq*)

A. I only   **C.** III only

B. II only   **D.** I and II only

                                                E. I and III only

71. Which of the following is NOT a conjugate acid/base pair?

A. S^{2-} / H_2S **C.** H_2O / ^-OH

B. HSO_4^- / SO_4^{2-} **D.** PH_4^+ / PH_3

                                              E. All are conjugate acid/base pairs

72. If a buffer is made with the pH below the pK_a of the weak acid, the [base] / [acid] ratio is:

A. equal to 0   **C.** greater than 1

B. equal to 1   **D.** less than 1

                                              E. undetermined

73. Which of the following acids listed below has the strongest conjugate base?

Monoprotic Acids	K_a
Acid I	1.3×10^{-8}
Acid II	2.9×10^{-9}
Acid III	4.2×10^{-10}
Acid IV	3.8×10^{-8}

A. I

B. II

C. III

D. IV

E. Not enough data to determine

74. Complete neutralization of phosphoric acid with barium hydroxide, when separated and dried, yields $Ba_3(PO_4)_2$ as one of the products. Therefore, which term describes phosphoric acid?

A. Monoprotic acid

B. Diprotic acid

C. Hexaprotic acid

D. Tetraprotic acid

E. Triprotic acid

75. Which is the correct net ionic equation for the hydrolysis reaction of Na_2S?

A. $Na^+ + H_2O \rightarrow NaOH + H_2$

B. $Na^+ + 2\,H_2O \rightarrow NaOH + H_2O^+$

C. $S^{2-} + H_2O \rightarrow 2\,HS^- + OH^-$

D. $S^{2-} + 2\,H_2O \rightarrow HS^- + H_3O^+$

E. $S^{2-} + H_2O \rightarrow OH^- + HS^-$

76. Calculate the pH of 0.0765 M HNO_3.

A. 1.1   **B.** 3.9   **C.** 11.7   **D.** 7.9   **E.** 5.6

77. Which of the following compounds is NOT a strong acid?

A. HBr *(aq)*

B. HNO_3

C. CH_3COOH

D. H_2SO_4

E. HCl *(aq)*

78. Which reaction produces $NiCr_2O_7$ as a product?

A. Nickel (II) hydroxide and dichromic acid

B. Nickel (II) hydroxide and chromic acid

C. Nickelic acid and chromium (II) hydroxide

D. Nickel (II) hydroxide and chromate acid

E. Nickel (II) hydroxide and trichromic acid

79. Which of the following statements describes a Brønsted-Lowry base?

A. Donates protons to other substances

B. Accepts protons from other substances

C. Produces hydrogen ions in aqueous solution

D. Produces hydroxide ions in aqueous solution

E. Accepts hydronium ions from other substances

80. When dissolved in water, the Arrhenius acid/bases KOH, H_2SO_4 and HNO_3 are, respectively:

A. base, acid and base

B. base, base and acid

C. base, acid and acid

D. acid, base and base

E. acid, acid and base

===

Practice Set 5: Questions 81–100

===

Questions **81–85** are based on the following titration graph:

The unknown acid is completely titrated with NaOH, as shown on the following titration curve:

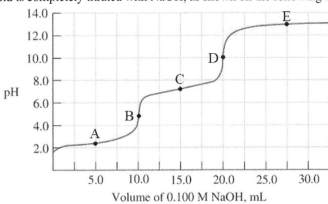

81. The unknown acid shown in the graph must be a(n):

   A. monoprotic acid   **C.** triprotic acid
   B. diprotic acid   **D.** weak acid
                                          E. aprotic acid

82. The pK_{a2} for this acid is located at point:

   A. A   **B.** B   **C.** C   **D.** D   **E.** E

83. At which point does the acid exist as 50% fully protonated and 50% singly deprotonated?

   A. A   **B.** B   **C.** C   **D.** D   **E.** E

84. At which point is the acid 100% singly deprotonated?

   A. A   **B.** B   **C.** C   **D.** D   **E.** E

85. Which points are the best buffer regions?

   A. A and B   **C.** B and D
   B. A and C   **D.** C and B
                                          E. C and D

86. Sodium hydroxide is a strong base. If a concentrated solution of NaOH spills on a latex glove, it feels like water. Why is it that if the solution were to splash directly on a person's skin, it feels very slippery?

    A. As a liquid, NaOH is slippery, but this cannot be detected through a latex glove because of the friction between the latex surfaces

    B. NaOH destroys skin cells on contact, and the remnants of skin cells feel slippery because the cells have been lysed

    C. NaOH lifts oil directly out of the skin cells, and the extruded oil causes the slippery sensation

    D. NaOH reacts with skin oils, transforming them into soap

    E. NaOH, as a liquid, causes the skin to feel slippery from low viscosity

87. Which of the following statements is/are true for a neutralization reaction?

    I.   Water is formed

    II. It is the reaction of an ^-OH with an H^+

    III. One molecule of acid neutralizes one molecule of base

    A. I only

    B. II only

    C. III. only

    D. I and II only

    E. I and III only

88. Which of the following statements is/are true for strong acids?

    I.   They form positively charged ions when dissolved in H_2O

    II. They form negatively charged ions when dissolved in H_2O

    III. They are strongly polar

    A. I only

    B. II only

    C. III only

    D. II and III only

    E. I, II and III

89. A Brønsted-Lowry base is defined as a substance that:

    A. increases $[H^+]$ when placed in water

    B. decreases $[H^+]$ when placed in water

    C. acts as a proton donor

    D. acts as a proton acceptor

    E. acts as a buffer

90. What is the salt to acid ratio needed to prepare a buffer solution with pH = 4.0 and an acid with a pK_a of 3.0?

    A. 1:1

    B. 1:1000

    C. 1:100

    D. 1:5

    E. 10:1

91. Which of the following must be true if an unknown solution is a poor conductor of electricity?

A. Solution is slightly reactive

B. Solution is highly corrosive

C. Solution is highly ionized

D. Solution is slightly ionized

E. Solution is highly reactive

92. Which of the following is a general property of a basic solution?

    I.   Turns litmus paper blue

    II. Tastes bitter

    III. Feels slippery

A. I only

B. II only

C. III only

D. I and II only

E. I, II and III

93. Since $pK_a = -\log K_a$, which of the following is a correct statement?

A. Since the pK_a for conversion of the ammonium ion to ammonia is 9.3; ammonia is a weaker base than the ammonium ion

B. For carbonic acid with pK_a values of 6.3 and 10.3, the bicarbonate ion is a stronger base than the carbonate ion

C. Acetic acid ($pK_a = 4.8$) is a weaker acid than lactic acid ($pK_a = 3.9$)

D. Lactic acid ($pK_a = 3.9$) is weaker than the forms of phosphoric acid ($pK_a = 2.1, 6.9$ and 12.4)

E. None of the above

94. Which of the following compounds is NOT a strong acid?

A. HI (*aq*)

B. $HClO_4$

C. HCl (*aq*)

D. HNO_3

E. $HC_2H_3O_2$

95. Which of the following is the acid anhydride for H_3CCOOH?

A. H_3CCOO^-

B. $H_3CCO_4CCH_3$

C. $H_3CCO_3CCH_3$

D. $H_3CCO_2CCH_3$

E. $H_3CCOCCH_3$

96. Acids and bases react to form:

A. Brønsted-Lowry acids

B. Arrhenius acids

C. Lewis acids

D. Lewis bases

E. salts

97. Which of the following does NOT act as a Brønsted-Lowry acid and base?

    A. H_2O **B.** HCO_3^- **C.** $H_2PO_4^-$ **D.** NH_4^+ **E.** HS^-

98. Which of the following is the conjugate acid of hydrogen phosphate, HPO_4^{2-}?

    A. $H_2PO_4^{2-}$ **C.** $H_2PO_4^-$

    B. H_3PO_4 **D.** $H_2PO_3^-$

                                                  E. None of the above

99. A sample of $Mg(OH)_2$ salt is dissolved in water and reaches equilibrium with its dissociated ions. The addition of the strong base NaOH increases the concentration of:

    A. H_2O^+ **C.** undissociated sodium hydroxide

    B. Mg^{2+} **D.** undissociated magnesium hydroxide

                                                  E. H_2O^+ and undissociated sodium hydroxide

100. Which of the following is the weakest acid?

    A. HCO_3^- ; $pK_a = 10.3$ **C.** NH_4^+ ; $pK_a = 9.2$

    B. $HC_2H_3O_2$; $pK_a = 4.8$ **D.** $H_2PO_4^-$; $pK_a = 7.2$

                                            E. CCl_3COOH; $pK_a = 2.9$

===

Practice Set 6: Questions 101–120

===

101. In which of the following pairs of acids are both chemical species weak acids?

    A. $HC_2H_3O_2$ and HI   **C.** HCN and H_2S

    B. H_2CO_3 and HBr   **D.** H_3PO_4 and H_2SO_4

                                        E. HCl and HBr

102. Which of the following indicators are yellow in an acidic solution and blue in a basic solution?

    I. methyl red   II. phenolphthalein   III. bromothymol blue

    A. I only   **C.** I and III only

    B. II only   **D.** III only

                                        E. I, II and III

103. Which of the following compounds is a basic anhydride?

    A. BaO   **B.** O_2 **C.** CO_2 **D.** SO_2 **E.** N_2O_5

104. What is the pK_a of an unknown acid if, in a solution at pH 7.0, 24% of the acid is in its deprotonated form?

    A. 6.0   **B.** 6.5   **C.** 7.5   **D.** 8.0   **E.** 10.0

105. When acids and bases react, which of the following, other than water, is the product?

    A. hydronium ion   **C.** hydrogen ion

    B. metal   **D.** hydroxide ion

                                        E. salt

106. Which of the following acids listed below is the strongest acid?

Monoprotic Acids	K_a
Acid I	1.0×10^{-8}
Acid II	1.8×10^{-9}
Acid III	3.7×10^{-10}
Acid IV	4.6×10^{-8}

    A. Acid I   **C.** Acid III

    B. Acid II   **D.** Acid IV

                                        E. Requires more information

107. In a basic solution, the pH is […] and the $[H_3O^+]$ is […].

A. < 7 and $> 1 \times 10^{-7}$ M

B. < 7 and $< 1 \times 10^{-7}$ M

C. $= 7$ and 1×10^{-7} M

D. < 7 and 1×10^{-7} M

E. > 7 and $< 1 \times 10^{-7}$ M

108. If a battery acid solution is a strong electrolyte, which of the following must be true of the battery acid?

A. It is highly ionized

B. It is slightly ionized

C. It is highly reactive

D. It is slightly reactive

E. It is slightly ionized and weakly reactive

109. Which species form in the second step of the dissociation of H_3PO_4?

A. PO_4^{3-} B. $H_2PO_4^{2-}$ C. $H_2PO_4^-$ D. HPO_4^{2-} E. H_3PO_3

110. Which statement concerning the Arrhenius acid–base theory is NOT correct?

A. Neutralization reactions produce H_2O, plus a salt

B. Acid–base reactions must take place in an aqueous solution

C. Arrhenius acids produce H^+ in H_2O solution

D. Arrhenius bases produce ^-OH in H_2O solution

E. All are correct

111. What is the term for a substance that donates a proton in an acid–base reaction?

A. Brønsted-Lowry acid

B. Brønsted-Lowry base

C. Arrhenius acid

D. Arrhenius base

E. Lewis acid

112. Which of the following does NOT represent a conjugate acid/base pair?

A. HCl / Cl^-

B. $HC_2H_3O_2 / {}^-OH$

C. H_3O^+ / H_2O

D. $HCN / {}^-CN$

E. All represent a conjugate acid/base pair

113. Which of the following statements is correct for a solution of 100% H_2O?

A. It contains no ions

B. It is an electrolyte

C. The $[^-OH]$ equals $[H_3O^+]$

D. The $[^-OH]$ is greater than $[H_3O^+]$

E. The $[H_3O^+]$ is greater than $[^-OH]$

114. The isoelectric point of an amino acid is defined as the pH at which the:

  A. value is equal to the pK_a **C.** amino acid exists in the basic form

  B. amino acid exists in the acidic form   **D.** amino acid exists in the zwitterion form

                                                    E. amino acid exists in the protonated form

115. In an aqueous solution, what is the term for ions that do not participate in a reaction and do not appear in the net ionic equation?

  A. Zwitterions   **C.** Nonelectrolyte ions

  B. Spectator ions   **D.** Electrolyte ions

                     E. None of the above

116. Salts that result from the reaction of strong acids with strong bases are:

  A. neutral   **B.** basic   **C.** acidic   **D.** salts   **E.** none of the above

117. Which of the following pairs of chemical species contains two polyprotic acids?

  A. HNO_3 and $H_2C_4H_4O_6$ **C.** H_3PO_4 and HCN

  B. $HC_2H_3O_2$ and $H_3C_6H_5O_7$ **D.** H_2S and H_2CO_3

                                        E. HCN and HNO_3

118. Which of the following pairs of acids and conjugate bases is NOT correctly labeled?

Acid	Conjugate Base		Acid	Conjugate Base

  A. $NH_4^+ \leftrightarrow NH_3$ **C.** $H_2SO_4 \leftrightarrow HSO_4^-$

  B. $HSO_3^- \leftrightarrow SO_3^{2-}$ **D.** $HSO_4^- \leftrightarrow SO_4^{2-}$

                               E. $HFO_2 \leftrightarrow HFO_3$

119. What happens to the respective corrosive properties of an acid and base after a neutralization reaction?

  A. The corrosive properties are doubled because the acid and base are combined in the salt

  B. The corrosive properties remain the same when the salt is mixed into water

  C. The corrosive properties are neutralized because the acid and base are transformed

  D. The corrosive properties are unaffected because salt is a corrosive agent

  E. The corrosive properties are increased because salt is a corrosive agent

120. What is the molarity of a nitric acid solution if 25.00 mL of HNO_3 is required to neutralize 0.500 g of calcium carbonate? (Use molecular mass of $CaCO_3$ = 100.09 g/mol)

$$2\ HNO_3\ (aq) + CaCO_3\ (s) \rightarrow Ca(NO_3)_2\ (aq) + H_2O\ (l) + CO_2\ (g)$$

  A. 0.200 M   **B.** 0.250 M   **C.** 0.400 M   **D.** 0.550 M   **E.** 0.700 M

===

Practice Set 7: Questions 121–140

===

121. Which of the following compounds is a weak acid?

    A. $^-$OH   **B.** $NaNH_2$ **C.** HNO_3 **D.** HF   **E.** HI

122. Which of the following indicators are yellow in a basic solution and red in an acidic solution?

    I.   phenolphthalein
    II. bromothymol blue
    III. methyl red

    A. I only   **C.** III only
    B. II only   **D.** I and II only
                                E. I, II and III

123. What are the products from the complete neutralization of carbonic acid with aqueous potassium hydroxide?

    A. $KHCO_4$ (*aq*) and H_2O (*l*)   **C.** K_2CO_3 (*aq*) and H_2O (*l*)
    B. $KC_2H_3O_2$ (*aq*) and H_2O (*l*)   **D.** $KHCO_3$ (*aq*) and H_2O (*l*)
                                          E. $K_2C_2H_3O_2$ (*aq*) and H_2O (*l*)

124. Which of the following compounds are strong acids?

    I. HNO_3 (*aq*)
    II. H_2SO_4 (*aq*)
    III. HCl (*aq*)

    A. I only   **C.** III only
    B. II only   **D.** I, II and III
                                E. I and II only

125. In an acidic solution, what is the $[H_3O^+]$?

    A. $< 1 \times 10^{-7}$ M   **C.** 1×10^{-14} M
    B. $> 1 \times 10^{-7}$ M   **D.** 1×10^{-7} M
                                E. It cannot be determined

126. Which of the following substances is NOT an electrolyte?

    A. NaOH   **B.** NH_4NO_3 **C.** HBr   **D.** KCl   **E.** CH_4

---

127. Which of the following statements about weak acids is correct?

    A. Weak acids can only be prepared as dilute solutions

    B. Weak acids have a strong affinity for acidic hydrogens

    C. Weak acids always contain carbon atoms

    D. The percentage dissociation for weak acids is usually in the range of 40-60%

    E. Weak acids have a weak affinity for acidic hydrogens

128. Which of the following is a synonym for hydrogen ion donor?

    A. Brønsted-Lowry base   **C.** Amphiprotic

    B. Proton donor   **D.** Arrhenius base

                                  E. Hydroxide donor

129. Which of the following are examples of an Arrhenius acid?

    I. $NaC_2H_3O_2$ *(aq)*   II. $HC_2H_3O_2$ *(aq)*   III. $NH_4C_2H_3O_2$ *(aq)*

    A. I only   **C.** III only

    B. II only   **D.** I and II only

                                    E. I and III only

130. Identify the correct combination of Brønsted-Lowry bases in the following equilibrium reaction:

$$H_2PO_4^- + H_2O \leftrightarrow H_3PO_4 + {}^-OH$$

    A. $H_2O + {}^-OH$ **C.** $H_2PO_4^- + H_3PO_4$

    B. $H_2O + H_3PO_4$ **D.** $H_2PO_4^- + {}^-OH$

                                  E. $H_2PO_4^- + H_2O$

131. Which of the following is the conjugate acid of water?

    A. H^+ *(aq)*   **B.** ${}^-OH$ *(aq)*   **C.** O_2^- *(aq)*   **D.** H_2O *(l)*   **E.** H_3O^+ *(aq)*

132. Calculate the $[H_3O^+]$ and the pH of a 0.045 M HNO_3 solution.

    A. 0.045 M and 1.35   **C.** 4.5×10^{-13} M and 12.55

    B. 4.51 M and −1.88   **D.** 4.5×10^{-13} M and 13.55

                                  E. 0.45 and 3.36

133. At which pH will the net charge on an amino acid be zero if the pI is 6.5?

    A. 5.5   **B.** 6.0   **C.** 7.0   **D.** 6.5   **E.** 7.5

134. What is the term for a substance that changes color according to the pH of the solution?

    A. Arrhenius acid

    B. Brønsted-Lowry acid

    C. Acid–base indicator

    D. Acid–base signal

    E. None of the above

135. Which of the following acts as a buffer system?

    A. $NH_3 + H_2O \rightleftharpoons {}^-OH + NH_4^+$

    B. $H_2PO_4 \rightleftharpoons H^+ + HPO_4^{2-}$

    C. $HC_2H_3O_2 \rightleftharpoons H^+ + C_2H_3O_2^-$

    D. $CO_2 + H_2O \rightleftharpoons H_2CO_3 \rightleftharpoons HCO_3^- + H^+$

    E. All of the above

136. What is the term for a solution in which concentration has been established accurately, usually to three or four significant digits?

    A. standard solution

    B. stock solution

    C. normal solution

    D. reference solution

    E. none of the above

137. The hydrogen sulfate ion HSO_4^- is amphoteric. In which of the following equations does it act as an acid?

    A. $HSO_4^- + {}^-OH \rightarrow H_2SO_4 + O^{2-}$

    B. $HSO_4^- + H_2O \rightarrow SO_4^{2-} + H_3O^+$

    C. $HSO_4^- + H_3O^+ \rightarrow SO_3 + 2H_2O$

    D. $HSO_4^- + H_2O \rightarrow H_2SO_4 + {}^-OH$

    E. In none of the above

138. In the following titration curve, what does the inflection point represent?

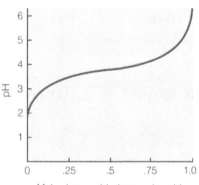

    A. The weak acid is 50% protonated and 50% deprotonated

    B. The pH where the solution functions most effectively as a buffer

    C. Equal concentration of weak acid and conjugate base

    D. pH of the solution equals pK_a of a weak acid

    E. All of the above

139. What is the sum of the coefficients for the balanced molecular equation of the following acid–base reaction?

$$LiOH\ (aq) + H_2SO_4\ (aq) \rightarrow$$

A. 6 **B.** 7 **C.** 8 **D.** 10 **E.** 12

140. A typical amino acid has a carboxylic acid and an amine with pK_a values of 2.3 and 9.6, respectively. In a solution of pH 4.5, which describes its protonation and charge state?

A. Carboxylic acid is deprotonated and negative; amine is protonated and neutral
B. Carboxylic acid is deprotonated and negative; amine is protonated and positive
C. Carboxylic acid is protonated and neutral; amine is deprotonated and negative
D. Carboxylic acid is protonated and neutral; amine is protonated and neutral
E. Carboxylic acid is deprotonated and negative; amine is deprotonated and neutral

Notes for active learning

Electrochemistry

===

Practice Set 1: Questions 1–20

===

1. Which is NOT true regarding the redox reaction occurring in a spontaneous electrochemical cell?

$$Cl_2 (g) + 2\ Br^- (aq) \rightarrow Br_2 (l) + 2\ Cl^- (aq)$$

   A. Anions flow towards the anode   **C.** Cl_2 is reduced at the cathode

   B. Electrons flow from the anode to the cathode   **D.** Br^- is oxidized at the cathode

                                                    E. Br^- is oxidized at the anode

2. What is the relationship between an element's ionization energy and its ability to function as an oxidizing agent? As ionization energy increases:

   A. the ability of an element to function as an oxidizing agent remains the same

   B. the ability of an element to function as an oxidizing agent decreases

   C. the ability of an element to function as an oxidizing agent increases

   D. the ability of an element to function as a reducing agent remains the same

   E. the ability of an element to function as a reducing agent increases

3. How many electrons are needed to balance the following half-reaction $H_2S \rightarrow S_8$ in an acidic solution?

   A. 14 electrons to the left side   **C.** 12 electrons to the left side

   B. 6 electrons to the right side   **D.** 8 electrons to the right side

                                                    E. 16 electrons to the right side

4. Which is NOT true regarding the following redox reaction occurring in an electrolytic cell?

                    Electricity

$$3\ C + 2\ Co_2O_3 \quad \rightarrow \quad 4\ Co + 3\ CO_2$$

   A. $CO_2 (g)$ is produced at the anode

   B. Co metal is produced at the anode

   C. Oxidation half-reaction: $C + 2\ O^{2-} \rightarrow CO_2 + 4\ e^-$

   D. Reduction half-reaction: $Co^{3+} + 3\ e^- \rightarrow Co$

   E. Co metal is produced at the cathode

5. The anode in a galvanic cell attracts:

A. cations

B. neutral particles

C. anions

D. both anions and neutral particles

E. both cations and neutral particles

6. The electrode with the standard reduction potential of 0 V is assigned as the standard reference electrode and uses the half-reaction:

A. $2 NH_4^+ (aq) + 2 e^- \leftrightarrow H_2 (g) + 2 NH_3 (g)$

B. $Ag^+ (aq) + e^- \leftrightarrow Ag (s)$

C. $Cu^{2+} (aq) + 2 e^- \leftrightarrow Cu (s)$

D. $Zn^{2+} (aq) + 2 e^- \leftrightarrow Zn (s)$

E. $2 H^+ (aq) + 2 e^- \leftrightarrow H_2 (g)$

7. In the reaction for a discharging nickel–cadmium (NiCd) battery, which substance is being oxidized?

$$Cd (s) + NiO_2 (s) + 2 H_2O (l) \rightarrow Cd(OH)_2 (s) + Ni(OH)_2 (s)$$

A. H_2O B. $Cd(OH)_2$ C. Cd D. NiO_2 E. $Ni(OH)_2$

8. What happens at the anode if rust forms when Fe is in contact with H_2O?

$$4 Fe + 3 O_2 \rightarrow 2 Fe_2O_3$$

A. Fe is reduced

B. Oxygen is reduced

C. Oxygen is oxidized

D. Fe is oxidized

E. None of the above

9. Which is true regarding the redox reaction occurring in the spontaneous electrochemical cell for

$$Cl_2 (g) + 2 Br^- (aq) \rightarrow Br_2 (l) + 2 Cl^- (aq)?$$

A. Electrons flow from the cathode to the anode

B. Cl_2 is oxidized at the cathode

C. Br^- is reduced at the anode

D. Cations in the salt bridge flow from the Br_2 half-cell to the Cl_2 half-cell

E. Br^- is reduced at the anode, and Cl_2 is oxidized at the cathode

10. What is the term for a chemical reaction that involves electron transfer between two reacting substances?

A. Reduction reaction

B. Electrochemical reaction

C. Oxidation reaction

D. Half-reaction

E. Redox reaction

11. Using the following metal ion/metal reaction potentials,

$$Cu^{2+} (aq)|Cu (s) \quad Ag^+ (aq)|Ag (s) \quad Co^{2+} (aq)|Co (s) \quad Zn^{2+} (aq)|Zn (s)$$

$$+0.34 \text{ V} \qquad\qquad +0.80 \text{ V} \qquad\qquad -0.28 \text{ V} \quad -0.76 \text{ V}$$

calculate the standard cell potential for the cell whose reaction is:

$$Co (s) + Cu^{2+} (aq) \rightarrow Co^{2+} (aq) + Cu (s)$$

A. +0.62 V   C. +0.48 V

B. −0.62 V   D. −0.48 V

             E. +0.68 V

12. How are photovoltaic cells different from many other forms of solar energy?

A. Light is reflected, and the coolness of the shade is used to provide a temperature differential

B. Light is passively converted into heat

C. Light is converted into heat and then into steam

D. Light is converted into heat and then into electricity

E. Light is converted directly to electricity

13. What is the term for an electrochemical cell with a single electrode where oxidation or reduction can occur?

A. Half-cell   C. Dry cell

B. Voltaic cell   D. Electrolytic cell

             E. None of the above

14. By which method could electrolysis be used to raise the hull of a sunken ship?

A. Electrolysis could only be used to raise the hull if the ship is made of iron. If so, the electrolysis of the iron metal might produce sufficient gas to lift the ship

B. The gaseous products of the electrolysis of H_2O are collected with bags attached to the hull of the ship and the inflated bags raise the ship

C. The electrolysis of the H_2O beneath the hull of the ship boils H_2O and creates upward pressure to raise the ship

D. An electric current passed through the hull of the ship produces electrolysis, and the gases trapped in compartments of the vessel would push it upwards

E. Electrolysis of the ship's hull decreases its mass, and the reduced weight causes the ship to rise

15. What are the products for the single-replacement reaction $Zn (s) + CuSO_4 (aq) \rightarrow$?

A. CuO and $ZnSO_4$ C. Cu and $ZnSO_4$

B. CuO and $ZnSO_3$ D. Cu and $ZnSO_3$

            E. No reaction

16. Which of the following is a true statement about the electrochemical reaction for an electrochemical cell with a cell potential of +0.36 V?

A. The reaction favors the formation of reactants and would be considered a galvanic cell

B. The reaction favors the formation of reactants and would be considered an electrolytic cell

C. The reaction is at equilibrium and is a galvanic cell

D. The reaction favors the formation of products and would be considered an electrolytic cell

E. The reaction favors the formation of products and would be considered a galvanic cell

17. Which of the following is a unit of electrical energy?

A. Coulomb   **B.** Joule   **C.** Watt   **D.** Ampere   **E.** Volt

18. Which observation describes the solution near the cathode when an aqueous solution of sodium chloride is electrolyzed, and hydrogen gas is evolved at the cathode?

A. Colored   **B.** Acidic   **C.** Basic   **D.** Frothy   **E.** Viscous

19. How many grams of Ag is deposited in the cathode of an electrolytic cell if a current of 3.50 A is applied to a solution of $AgNO_3$ for 12 minutes? (Use the molecular mass of Ag = 107.86 g/mol and the conversion of 1 mole $e^- = 9.65 \times 10^4$ C)

A. 0.32 g   **B.** 0.86 g   **C.** 2.82 g   **D.** 3.86 g   **E.** 4.38 g

20. In an electrochemical cell, which of the following statements is FALSE?

A. The anode is the electrode where oxidation occurs

B. A salt bridge provides electrical contact between the half-cells

C. The cathode is the electrode where reduction occurs

D. A spontaneous electrochemical cell is called a galvanic cell

E. All of the above are true

===

Practice Set 2: Questions 21–40

===

21. What is the relationship between an element's ionization energy and its ability to function as an oxidizing and reducing agent? Elements with high ionization energy are:

   A. strong oxidizing and weak reducing agents   **C.** strong oxidizing and strong reducing agents

   B. weak oxidizing and weak reducing agents   **D.** weak oxidizing and strong reducing agents

        E. weak oxidizing and neutral reducing agents

22. In a basic solution, how many electrons are needed to balance the following half-reaction?

$$C_8H_{10} \rightarrow C_8H_4O_4^{2-}$$

   A. 8 electrons on the left side   **C.** 4 electrons on the left side

   B. 12 electrons on the left side   **D.** 12 electrons on the right side

        E. 8 electrons on the right side

23. Which statement is true regarding the following redox reaction occurring in an electrolytic cell?

$$\text{Electricity}$$
$$3\,C\,(s) + 2\,Co_2O_3\,(l) \quad \rightarrow \quad 4\,Co\,(l) + 3\,CO_2\,(g)$$

   A. CO_2 gas is produced at the anode   **C.** Oxidation half-reaction: $Co^{3+} + 3\,e^- \rightarrow Co$

   B. Co^{3+} is produced at the cathode   **D.** Reduction half-reaction: $C + 2\,O^{2-} \rightarrow CO_2 + e^-$

        E. None of the above

24. Which is true regarding the following redox reaction occurring in a spontaneous electrochemical cell?

$$Sn\,(s) + Cu^{2+}\,(aq) \rightarrow Cu\,(s) + Sn^{2+}\,(aq)$$

   A. Anions in the salt bridge flow from the Cu half-cell to the Sn half-cell

   B. Electrons flow from the Cu electrode to the Sn electrode

   C. Cu^{2+} is oxidized at the cathode

   D. Sn is reduced at the anode

   E. None of the above

25. In galvanic cells, reduction occurs at the:

   I. salt bridge   II. cathode   III. anode

   A. I only   **C.** III only

   B. II only   **D.** I and II only

        E. I, II and III

26. For a battery, what is undergoing a reduction in the following oxidation-reduction reaction?

$$Mn_2O_3 + ZnO \rightarrow 2\ MnO_2 + Zn$$

A. Mn_2O_3 **B.** MnO_2 **C.** ZnO   **D.** Zn   **E.** Mn_2O_3 and Zn

27. Which of the metals listed below has the greatest tendency to undergo oxidation, given that the following redox reactions go essentially to completion?

$$Ni\ (s) + Ag^+\ (aq) \rightarrow Ag\ (s) + Ni^{2+}\ (aq)$$

$$Al\ (s) + Cd^{2+}\ (aq) \rightarrow Cd\ (s) + Al^{3+}\ (aq)$$

$$Cd\ (s) + Ni^{2+}\ (aq) \rightarrow Ni\ (s) + Cd^{2+}\ (aq)$$

$$Ag\ (s) + H^+\ (aq) \rightarrow \text{no reaction}$$

A. (H)   **B.** Cd   **C.** Ni   **D.** Ag   **E.** Al

28. What is the purpose of the salt bridge in a voltaic cell?

A. Allows for a balance of charge between the two chambers
B. Allows the Fe^{2+} and the Cu^{2+} to flow freely between the two chambers
C. Allows for the buildup of positively charged ions in one container and negatively charged ions in the other container
D. Prevents any further migration of electrons through the wire
E. None of the above

29. If $\Delta G°$ for a cell is positive, the E° is:

A. neutral   **C.** positive
B. negative   **D.** unable to be determined
                                 E. greater than 1

30. In an electrochemical cell, which of the following statements is FALSE?

A. Oxidation occurs at the anode
B. Reduction occurs at the cathode
C. Electrons flow through the salt bridge to complete the cell
D. A spontaneous electrochemical cell is called a voltaic cell
E. All of the above are true

31. Which of the following statements about electrochemistry is NOT true?

    A. The study of how protons are transferred from one chemical compound to another

    B. The use of electrical current to produce an oxidation-reduction reaction

    C. The use of a set of oxidation-reduction reactions to produce electrical current

    D. The study of how electrical energy and chemical reactions are related

    E. The study of how electrons are transferred from one chemical compound to another

32. In an operating photovoltaic cell, electrons move through the external circuit to the negatively charged p-type silicon wafer. How can the electrons move to the negatively charged silicon wafer if electrons are negatively charged?

    A. The p-type silicon wafer is positively charged

    B. The energy of the sunlight moves electrons in a nonspontaneous direction

    C. Advancements in photovoltaic technology have solved this technological impediment

    D. An electric current occurs because the energy from the sun reverses the charge of the electrons

    E. The p-type silicon wafer is negatively charged

33. What is the term for the value assigned to an atom in a substance that indicates whether the atom is electron-poor or electron-rich compared to a free atom?

    A. Reduction number   **C.** Cathode number

    B. Oxidation number   **D.** Anode number

                                       E. None of the above

34. What is the term for the relative ability of a substance to undergo reduction?

    A. Oxidation potential   **C.** Anode potential

    B. Reduction potential   **D.** Cathode potential

                                       E. None of the above

35. What is the primary difference between a fuel cell and a battery?

    A. Fuel cells oxidize to supply electricity, while batteries reduce to supply electricity

    B. Batteries supply electricity, while fuel cells supply heat

    C. Batteries can be recharged, while fuel cells cannot

    D. Fuel cells do not run down because they can be refueled, while batteries run down and need to be recharged

    E. Fuel cells do not use metals as oxidants and reductants, while batteries have a static reservoir of oxidants and reductants

36. In balancing the equation for a disproportionation reaction, using the oxidation number method, the substance undergoing disproportionation is:

    A. initially written twice on the reactant side of the equation

    B. initially written twice on the product side of the equation

    C. initially written on both the reactant and product sides of the equation

    D. assigned an oxidation number of zero

    E. assigned an oxidation number of +1

37. What is the term for an electrochemical cell in which electrical energy is generated from a spontaneous redox reaction?

    A. Voltaic cell   **C.** Dry cell

    B. Photoelectric cell   **D.** Electrolytic cell

                                 E. Wet cell

38. Which observation describes the solution near the anode when an aqueous solution of sodium sulfate is electrolyzed, and gas is evolved at the anode?

    A. Colored   **B.** Acidic   **C.** Basic   **D.** Frothy   **E.** Viscous

39. Which of the statements listed below is true regarding the following redox reaction occurring in a nonspontaneous electrochemical cell?

$$Cd\ (s) + Zn(NO_3)_2\ (aq) + electricity \rightarrow Zn\ (s) + Cd(NO_3)_2\ (aq)$$

    A. Oxidation half-reaction: $Zn^{2+} + 2\ e^- \rightarrow Zn$ **C.** Cd metal is produced at the anode

    B. Reduction half-reaction: $Cd \rightarrow Cd^{2+} + 2\ e^-$ **D.** Zn metal is produced at the cathode

                                            E. None of the above

40. Which statement is true for electrolysis?

    A. A spontaneous redox reaction produces electricity

    B. A nonspontaneous redox reaction is forced to occur by applying an electric current

    C. Only pure, drinkable water is produced

    D. There is a cell that reverses the flow of ions

    E. None of the above

===

Practice Set 3: Questions 41–60

===

41. Which relationship explains an element's electronegativity and its ability to act as an oxidizing and reducing agent?

    A. Atoms with large electronegativity tend to act as strong oxidizing and strong reducing agents

    B. Atoms with large electronegativity tend to act as weak oxidizing and strong reducing agents

    C. Atoms with large electronegativity tend to act as strong oxidizing and weak reducing agents

    D. Atoms with large electronegativity tend to act as weak oxidizing and weak reducing agents

    E. None of the above

42. Which substance is undergoing reduction for a battery if the following two oxidation-reduction reactions take place?

        Reaction I: $Zn + 2\,OH^- \rightarrow ZnO + H_2O + 2\,e^-$

        Reaction II: $2\,MnO_2 + H_2O + 2\,e^- \rightarrow Mn_2O_3 + 2\,OH^-$

    A. OH^- **B.** H_2O **C.** Zn **D.** ZnO **E.** MnO_2

43. What is the term for an electrochemical cell in which a nonspontaneous redox reaction occurs by forcing electricity through the cell?

    I. voltaic cell   II. dry cell   III. electrolytic cell

    A. I only   **C.** III only

    B. II only   **D.** I and II only

                                                    E. I and III only

44. Which of the statements is true regarding the following redox reaction occurring in a galvanic cell?

    $Cl_2\,(g) + 2\,Br^-\,(aq) \rightarrow Br_2\,(l) + 2\,Cl^-\,(aq)$?

    A. Cations in the salt bridge flow from the Cl_2 half-cell to the Br_2 half-cell

    B. Electrons flow from the anode to the cathode

    C. Cl_2 is reduced at the anode

    D. Br^- is oxidized at the cathode

    E. None of the above

45. Which of the statements below is true regarding the following redox reaction occurring in a nonspontaneous electrolytic cell?

Electricity

$$3\ C\ (s) + 4\ AlCl_3\ (l) \quad \rightarrow \quad 4\ Al\ (l) + 3\ CCl_4\ (g)$$

A. Cl^- is produced at the anode

B. Al metal is produced at the cathode

C. Oxidation half-reaction is $Al^{3+} + 3\ e^- \rightarrow Al$

D. Reduction half-reaction is $C + 4\ Cl^- \rightarrow CCl_4 + 4\ e^-$

E. None of the above

46. Which of the following materials is most likely to undergo oxidation?

    I. Cl^- II. Na   III. Na^+

A. I only

B. II only

C. III only

D. I and II only

E. I, II and III

47. Which of the statements is NOT true regarding the redox reaction occurring in a spontaneous electrochemical cell?

$$Zn\ (s) + Cd^{2+}\ (aq) \rightarrow Cd\ (s) + Zn^{2+}\ (aq)$$

A. Anions in the salt bridge flow from the Zn half-cell to the Cd half-cell

B. Electrons flow from the Zn electrode to the Cd electrode

C. Cd^{2+} is reduced at the cathode

D. Zn is oxidized at the anode

E. Anions in the salt bridge flow from the Cd half-cell to the Zn half-cell

48. Using the provided electrochemical series, determine which is capable of oxidizing Cu (s) to Cu^{2+} (aq) when added to Cu (s) in solution.

A. Al^{3+} (aq)

B. Ag^+ (aq)

C. K^+ (aq)

D. Pb^{2+} (aq)

E. Li^+ (aq)

Equilibrium	E°
$Li^+\ (aq) + e^- \leftrightarrow Li\ (s)$	–3.03 volts
$K^+\ (aq) + e^- \leftrightarrow K\ (s)$	–2.92
$*Ca^{2+}\ (aq) + 2\ e^- \leftrightarrow Ca\ (s)$	–2.87
$*Na^+\ (aq) + e^- \leftrightarrow Na\ (s)$	–2.71
$Mg^{2+}\ (aq) + 2\ e^- \leftrightarrow Mg\ (s)$	–2.37
$Al^{3+}\ (aq) + 3\ e^- \leftrightarrow Al\ (s)$	–1.66
$Zn^{2+}\ (aq) + 2\ e^- \leftrightarrow Zn\ (s)$	–0.76
$Fe^{2+}\ (aq) + 2\ e^- \leftrightarrow Fe\ (s)$	–0.44
$Pb^{2+}\ (aq) + 2\ e^- \leftrightarrow Pb\ (s)$	–0.13
$2\ H^+\ (aq) + 2\ e^- \leftrightarrow H_2\ (g)$	0.0
$Cu^{2+}\ (aq) + 2\ e^- \leftrightarrow Cu\ (s)$	+0.34
$Ag^+\ (aq) + e^- \leftrightarrow Ag\ (s)$	+0.80
$Au^{3+}\ (aq) + 3\ e^- \leftrightarrow Au\ (s)$	+1.50

49. Why is the anode of a battery indicated with a negative (–) sign?

A. Electrons move to the anode to react with NH_4Cl in the battery

B. It indicates the electrode where the chemicals are reduced

C. Electrons are attracted to the negative electrode

D. The cathode is the source of negatively charged electrons

E. The electrode is the source of negatively charged electrons

50. What is the term for the conversion of chemical energy to electrical energy from redox reactions?

A. Redox chemistry
B. Electrochemistry

C. Cell chemistry
D. Battery chemistry
E. None of the above

51. In electrolysis, E° tends to be:

A. zero
B. neutral

C. positive
D. greater than 1
E. negative

52. A major source of chlorine gas is the electrolysis of concentrated saltwater, NaCl (*aq*). What is the sign of the electrode where the chlorine gas is formed?

A. Neither, since chlorine gas is a neutral molecule and there is no electrode attraction
B. Negative, since the chlorine gas needs to deposit electrons to form chloride ions
C. Positive, since the chloride ions lose electrons to form chlorine molecules
D. Both, since the chloride ions from NaCl (*aq*), are attracted to the positive electrode to form chlorine molecules, while the produced chlorine gas molecules move to deposit electrons at the negative electrode
E. Positive, since the chloride ions gain electrons to form chlorine molecules

53. Which statement is correct for an electrolytic cell that has two electrodes?

A. Oxidation occurs at the anode, which is negatively charged
B. Oxidation occurs at the anode, which is positively charged
C. Oxidation occurs at the cathode, which is positively charged
D. Oxidation occurs at the cathode, which is negatively charged
E. Oxidation occurs at the dynode, which is uncharged

54. Which of the statements listed below is true regarding the following redox reaction occurring in a nonspontaneous electrochemical cell?

$$Cd\ (s) + Zn(NO_3)_2\ (aq) + electricity \rightarrow Zn\ (s) + Cd(NO_3)_2\ (aq)$$

A. Oxidation half-reaction: $Zn^{2+} + 2\ e^- \rightarrow Zn$
B. Reduction half-reaction: $Cd \rightarrow Cd^{2+} + 2\ e^-$

C. Cd metal dissolves at the anode
D. Zn metal dissolves at the cathode
E. None of the above

55. Which battery system is based on half-reactions involving zinc metal and manganese dioxide?

A. Alkaline batteries
B. Fuel cells

C. Lead-acid storage batteries
D. Dry-cell batteries
E. None of the above

56. Which of the following equations is a disproportionation reaction?

A. $2 H_2O \rightarrow 2 H_2 + O_2$

B. $H_2SO_3 \rightarrow H_2O + SO_2$

C. $HNO_2 \rightarrow NO + HNO_3$

D. $Mg + H_2SO_4 \rightarrow MgSO_4 + H_2$

E. None of the above

57. How is electrolysis different from the chemical process inside a battery?

A. Pure compounds cannot be generated *via* electrolysis

B. Electrolysis only uses electrons from a cathode

C. Electrolysis does not use electrons

D. They are the same process in reverse

E. Chemical changes do not occur in electrolysis

58. Which fact about fuel cells is FALSE?

A. Fuel cell automobiles are powered by water and only emit hydrogen

B. Fuel cells are based on the tendency of some elements to gain electrons from other elements

C. Fuel cell automobiles are quiet

D. Fuel cell automobiles are environmentally friendly

E. All of the above

59. Which process occurs when copper is refined using the electrolysis technique?

A. Impure copper goes into solution at the anode, and pure copper plates out on the cathode

B. Impure copper goes into solution at the cathode, and pure copper plates out on the anode

C. Pure copper goes into solution from the anode and forms a precipitate at the bottom of the tank

D. Pure copper goes into solution from the cathode and forms a precipitate at the bottom of the tank

E. Pure copper on the bottom of the tank goes into solution and plates out on the cathode

60. Which of the following statements describes electrolysis?

A. A chemical reaction which results when electrical energy is passed through a metallic liquid

B. A chemical reaction which results when electrical energy is passed through a liquid electrolyte

C. The splitting of atomic nuclei by electrical energy

D. The splitting of atoms by electrical energy

E. The passage of electrical energy through a split-field armature

==

Practice Set 4: Questions 61–80

==

61. In a battery, which of the following species in the two oxidation-reduction reactions is undergoing oxidation?

Reaction I: $Zn + 2 OH^- \rightarrow ZnO + H_2O + 2 e^-$

Reaction II: $2 MnO_2 + H_2O + 2 e^- \rightarrow Mn_2O_3 + 2 OH^-$

A. H_2O **B.** MnO_2 **C.** ZnO **D.** Zn **E.** OH^-

62. What is the term for an electrochemical cell in which the anode and cathode reactions do not take place in aqueous solutions?

A. Voltaic cell

B. Electrolytic cell

C. Dry cell

D. Galvanic cell

E. None of the above

63. How many electrons are needed to balance the charge for the following half-reaction in an acidic solution?

$C_2H_6O \rightarrow HC_2H_3O_2$

A. 6 electrons to the right side

B. 4 electrons to the right side

C. 3 electrons to the left side

D. 2 electrons to the left side

E. 8 electrons to the right side

64. Which of the statements is true regarding the following redox reaction occurring in an electrolytic cell?

$$3 C (s) + 4 AlCl_3 (l) \xrightarrow{\text{Electricity}} 4 Al (l) + 3 CCl_4 (g)$$

A. Al^{3+} is produced at the cathode

B. CCl_4 gas is produced at the anode

C. Reduction half-reaction: $C + 4 Cl^- \rightarrow CCl_4 + 4 e^-$

D. Oxidation half-reaction: $Al^{3+} + 3 e^- \rightarrow Al$

E. None of the above

65. What is the term for an electrochemical cell that spontaneously produces electrical energy?

I. half-cell   II. electrolytic cell   III. dry cell

A. I and II only

B. II only

C. III only

D. II and III only

E. None of the above

---

66. The anode of a battery is indicated with a negative (–) sign because the anode is:

A. where electrons are adsorbed

B. positive

C. negative

D. where electrons are generated

E. determined by convention

67. Which is the strongest reducing agent for the following half-reaction potentials?

$$Sn^{4+} (aq) + 2 e^- \rightarrow Sn^{2+} (aq) \qquad E° = -0.13 \text{ V}$$

$$Ag^+ (aq) + e^- \rightarrow Ag (s) \qquad E° = +0.81 \text{ V}$$

$$Cr^{3+} (aq) + 3 e^- \rightarrow Cr (s) \qquad E° = -0.75 \text{ V}$$

$$Fe^{2+} (aq) + 2 e^- \rightarrow Fe (s) \qquad E° = -0.43 \text{ V}$$

A. Cr (s)

B. Fe^{2+} (aq)

C. Sn^{2+} (aq)

D. Ag (s)

E. Sn^{2+} (aq) and Ag (s)

68. Which is true for the redox reaction occurring in a spontaneous electrochemical cell?

$$Zn (s) + Cd^{2+} (aq) \rightarrow Cd (s) + Zn^{2+} (aq)$$

A. Electrons flow from the Cd electrode to the Zn electrode

B. Anions in the salt bridge flow from the Cd half-cell to the Zn half-cell

C. Zn is reduced at the anode

D. Cd^{2+} is oxidized at the cathode

E. None of the above

69. Based upon the reduction potential: $Zn^{2+} + 2 e^- \rightarrow Zn (s)$; $E° = -0.76$ V, does a reaction take place when Zinc (s) is added to aqueous HCl, under standard conditions?

A. Yes, because the reduction potential for H^+ is negative

B. Yes, because the reduction potential for H^+ is zero

C. No, because the oxidation potential for Cl^- is positive

D. No, because the reduction potential for Cl^- is negative

E. Yes, because the reduction potential for H^+ is positive

70. If 1 amp of current passes a cathode for 10 minutes, how much Zn (s) forms in the following reaction? (Use the molecular mass of Zn = 65 g/mole)

$$Zn^{2+} + 2 e^- \rightarrow Zn (s)$$

A. 0.10 g

B. 10.0 g

C. 0.65 g

D. 2.20 g

E. 0.20 g

71. In the oxidation–reduction reaction Mg (*s*) + Cu^{2+} (*aq*) → Mg^{2+} (*aq*) + Cu (*s*), which atom/ion is reduced and which atom/ion is oxidized?

    A. The Cu^{2+} ion is reduced (gains electrons) to form Cu metal, while Mg metal is oxidized (loses electrons) to form Mg^{2+}

    B. Since Mg is transformed from a solid to an aqueous solution and Cu is transformed from an aqueous solution to a solid, no oxidation–reduction reaction occurs

    C. The Cu^{2+} ion is oxidized (i.e., gains electrons) from Cu metal, while Mg metal is reduced (i.e., loses electrons) to form Mg^{2+}

    D. The Mg^{2+} ion is reduced (i.e., gains electrons) from Cu metal, while Cu^{2+} is oxidized (i.e., loses electrons) to the Mg^{2+}

    E. None of the above

72. Electrolysis is an example of a(n):

    A. acid-base reaction   **C.** physical change

    B. exothermic reaction   **D.** chemical change

                                      E. state function

73. What is the term for a process characterized by the loss of electrons?

    A. Redox   **C.** Electrochemistry

    B. Reduction   **D.** Oxidation

                                      E. None of the above

74. Which of the following statements is true about a galvanic cell?

    A. The standard reduction potential for the anode reaction is always positive.

    B. The standard reduction potential for the anode reaction is always negative.

    C. The standard reduction potential for the cathode reaction is always positive.

    D. E° for the cell is always positive.

    E. E° for the cell is always negative.

75. Which of the following statements is true?

    A. Galvanic cells were invented by Thomas Edison

    B. Galvanic cells generate electrical energy rather than consume it

    C. Electrolysis cells generate alternating current when their terminals are reversed

    D. Electrolysis was discovered by Lewis Latimer

    E. The laws of electrolysis were discovered by Granville Woods

76. Which statement is correct for an electrolysis cell that has two electrodes?

    A. Reduction occurs at the anode, which is positively charged

    B. Reduction occurs at the anode, which is negatively charged

    C. Reduction occurs at the cathode, which is positively charged

    D. Reduction occurs at the cathode, which is negatively charged

    E. Reduction occurs at the dynode, which is uncharged

77. How long would it take to deposit 4.00 grams of Cu from a $CuSO_4$ solution if a current of 2.5 A is applied? (Use the molecular mass of Cu = 63.55 g/mol and the conversion of 1 mole $e^- = 9.65 \times 10^4$ C and 1 C = A·s)

    A. 1.02 hours   **C.** 4.46 hours

    B. 1.36 hours   **D.** 6.38 hours

                                 E. 7.72 hours

78. Which battery system is completely rechargeable?

    A. Alkaline batteries   **C.** Lead-acid storage batteries

    B. Fuel cells   **D.** Dry-cell batteries

                                 E. None of the above

79. In which type of cell does the following reaction occur when electrons are forced into a system by applying an external voltage?

$$Fe^{2+} + 2e^- \rightarrow Fe\ (s) \qquad\qquad E° = -0.44\ V$$

    A. Concentration cell   **C.** Electrochemical cell

    B. Battery   **D.** Galvanic cell

                                 E. Electrolytic cell

80. What is the term for a reaction that represents separate oxidation or reduction processes?

    A. Reduction reaction   **C.** Oxidation reaction

    B. Redox reaction   **D.** Half-reaction

                                 E. None of the above

Diagnostic Tests

Diagnostic Test 1

This Diagnostic Test is designed to assess your proficiency on each topic and NOT to mimic the test. Use your test results and identify areas of strength and weakness to adjust your study plan and enhance your fundamental knowledge. The length of the Diagnostic Tests is optimal for a single study session.

#	Answer:					Review	#	Answer:					Review
1:	A	B	C	D	E	___	31:	A	B	C	D	E	___
2:	A	B	C	D	E	___	32:	A	B	C	D	E	___
3:	A	B	C	D	E	___	33:	A	B	C	D	E	___
4:	A	B	C	D	E	___	34:	A	B	C	D	E	___
5:	A	B	C	D	E	___	35:	A	B	C	D	E	___
6:	A	B	C	D	E	___	36:	A	B	C	D	E	___
7:	A	B	C	D	E	___	37:	A	B	C	D	E	___
8:	A	B	C	D	E	___	38:	A	B	C	D	E	___
9:	A	B	C	D	E	___	39:	A	B	C	D	E	___
10:	A	B	C	D	E	___	40:	A	B	C	D	E	___
11:	A	B	C	D	E	___	41:	A	B	C	D	E	___
12:	A	B	C	D	E	___	42:	A	B	C	D	E	___
13:	A	B	C	D	E	___	43:	A	B	C	D	E	___
14:	A	B	C	D	E	___	44:	A	B	C	D	E	___
15:	A	B	C	D	E	___	45:	A	B	C	D	E	___
16:	A	B	C	D	E	___	46:	A	B	C	D	E	___
17:	A	B	C	D	E	___	47:	A	B	C	D	E	___
18:	A	B	C	D	E	___	48:	A	B	C	D	E	___
19:	A	B	C	D	E	___	49:	A	B	C	D	E	___
20:	A	B	C	D	E	___	50:	A	B	C	D	E	___
21:	A	B	C	D	E	___	51:	A	B	C	D	E	___
22:	A	B	C	D	E	___	52:	A	B	C	D	E	___
23:	A	B	C	D	E	___	53:	A	B	C	D	E	___
24:	A	B	C	D	E	___	54:	A	B	C	D	E	___
25:	A	B	C	D	E	___	55:	A	B	C	D	E	___
26:	A	B	C	D	E	___	56:	A	B	C	D	E	___
27:	A	B	C	D	E	___	57:	A	B	C	D	E	___
28:	A	B	C	D	E	___	58:	A	B	C	D	E	___
29:	A	B	C	D	E	___	59:	A	B	C	D	E	___
30:	A	B	C	D	E	___	60:	A	B	C	D	E	___

This page is intentionally left blank

1. Which of the following is/are a characteristic of metals?

      I. conduction of heat

      II. high density

      III. malleable

A. I only   **C.** III only

B. II only   **D.** I and II only   **E.** I, II and III

2. What is the number of valence electrons in a molecule of SOF_2?

A. 18   **B.** 20   **C.** 8   **D.** 19   **E.** 26

3. What is the name given to the transition of a compound from the gas phase directly to the solid phase?

A. deposition   **C.** freezing

B. sublimation   **D.** condensation   **E.** evaporation

4. Which are the spectator ions in the following balanced reaction?

$$2\ AgNO_3\ (aq) + K_2SO_4\ (aq) \rightarrow 2\ KNO_3\ (aq) + Ag_2SO_4\ (s)$$

A. silver ion and sulfate ion   **C.** potassium ion and sulfate ion

B. potassium ion and nitrate ion   **D.** silver ion and nitrate ion

                                         E. hydrogen ion and hydroxide ion

5. Calculate the mass of a sample of gold if it requires 488 J to be heated from 21.8 °C to 34.4 °C? (Use the specific heat of gold $c = 0.130$ J/g·°C)

A. 597 g   **B.** 383 g   **C.** 472 g   **D.** 163 g   **E.** 298 g

6. The reaction shown is an example of which step in a free radical reaction?

      $\cdot OH + H_2 \rightarrow H_2O + H\cdot$

      I. initiation

      II. propagation

      III. termination

A. I only   **C.** III only

B. II only   **D.** II and III only   **E.** None of the above

7. Which substance shown below does NOT produce ions when dissolved in water?

A. manganese (II) nitrate   **C.** CH_2O

B. CsCN   **D.** KClO   **E.** KSH

8. What are the products from the complete neutralization of sulfuric acid with aqueous sodium hydroxide?

A. Na_2SO_3 (*aq*) and H_2O (*l*)   **C.** $NaHSO_3$ (*aq*) and H_2O (*l*)

B. $NaHSO_4$ (*aq*) and H_2O (*l*)   **D.** Na_2S (*aq*) and H_2O (*l*)

      E. Na_2SO_4 (*aq*) and H_2O (*l*)

9. A galvanic cell consists of an Ag (*s*)|Ag^+ (*aq*) half-cell and a Zn (*s*)|Zn^{2+} (*aq*) half-cell connected by a salt bridge. If oxidation occurs in the zinc half-cell, which of the following represents the cell in standard notation?

A. Ag^+ (*aq*)|Ag (*s*) || Zn (*s*)|Zn^{2+} (*aq*)   **C.** Zn (*s*)|Zn^{2+} (*aq*) || Ag^+ (*aq*)|Ag (*s*)

B. Zn^{2+} (*aq*)|Zn (*s*) || Ag (*s*)|Ag^+ (*aq*)   **D.** Zn (*s*)|Zn^{2+} (*aq*) || Ag (*s*)|Ag^+ (*aq*)

      E. Ag (*s*)|Ag^+ (*aq*) || Zn (*s*)|Zn^{2+} (*aq*)

10. What is the maximum number of electrons in the n = 3 shell?

A. 32   **B.** 18   **C.** 10   **D.** 8   **E.** 12

11. What is the molecular geometry of Cl_2CO?

A. tetrahedral   **C.** trigonal pyramidal

B. trigonal planar   **D.** linear

      E. bent

12. Which of the following statements best describes a gas?

A. Definite shape, but indefinite volume   **C.** Indefinite shape and volume

B. Indefinite shape, but definite volume   **D.** Definite shape and volume

      E. Definite shape, but indefinite mass

13. Which of the following is NOT an electrolyte?

A. NaBr   **B.** Ne   **C.** KOH   **D.** HCl   **E.** HI

14. A process or reaction that releases heat into the surroundings is:

A. isothermal   **C.** endothermic

B. exothermic   **D.** conservative   **E.** exergonic

15. As the temperature of a reaction increases, what happens to the reaction rate and rate constant?

A. Increases, but the rate constant remains the same

B. Increases, in proportion with the rate constant

C. Remains constant, but the rate constant increases

D. Remains constant, as does the rate constant

E. Is not affected

16. Would a precipitate form if 1 liter of 0.04 M $Mg(NO_3)_2$ was mixed with 3 liters of 0.08 M K_2SO_4? (Assume complete dissociation of solutions and use the K_{sp} of $MgSO_4 = 4 \times 10^{-5}$)

    A. Yes

    B. No, because the K_{sp} for $MgSO_4$ is not exceeded

    C. No, because the solution does not contain $MgSO_4$

    D. No, because the K_{sp} for $MgSO_4$ is exceeded

    E. Not enough data is provided

17. Which of the following is a strong base?

    A. $Ba(OH)_2$ **C.** NH_3

    B. CH_3CH_2COOH **D.** CH_3CH_2OH **E.** NaCl

18. Uranium exists in nature in the form of several isotopes that have different:

    A. numbers of electrons   **C.** atomic numbers

    B. numbers of protons   **D.** charges   **E.** numbers of neutrons

19. Which of the following pairs is NOT correctly matched?

Formula	Molecular polarity
A. SiF_4	nonpolar
B. H_2O	polar
C. HCN	nonpolar
D. H_2CO	polar
E. CH_2Cl_2	polar

20. If the pressure of a gas sample is doubled, according to Gay–Lussac's law, the sample's:

    A. temperature is doubled   **C.** volume is doubled

    B. temperature decreases by a factor of 2   **D.** volume decreases by a factor of 2

                                                    E. volume decreases by a factor of 4

21. What is the coefficient (n) of O_2 gas for the balanced equation?

    $nP\ (s) + nO_2\ (g) \rightarrow nP_2O_5\ (s)$

    A. 1   **B.** 2   **C.** 4   **D.** 5   **E.** none of the above

22. Consider the enthalpy of the following balanced reaction:

$$C_3H_8 + 5\ O_2 \rightarrow 3\ CO_2 + 4\ H_2O = +488\ kcal$$

The reaction is [   ] and ΔH is [   ].

A. exothermic … negative

B. endothermic … negative

C. exothermic … positive

D. endothermic … zero

E. exothermic … zero

23. Which of the following causes the reaction to shift towards yielding more carbon dioxide gas in the following balanced endothermic reaction?

$$CaCO_3\ (s) \leftrightarrow CaO\ (s) + CO_2\ (g)$$

A. Increasing the pressure and decreasing the temperature of the system

B. Increasing the pressure of the system

C. Decreasing the temperature of the reaction

D. Increasing the pressure and increasing the temperature of the system

E. Increasing the temperature of the reaction

24. Assuming the same solute and solvent, which of the following solutions is the least concentrated?

A. 2.8 g solute in 2 mL solution

B. 2.8 g solute in 5 mL solution

C. 25 g solute in 50 mL solution

D. 40 g solute in 160 mL solution

E. 50 g solute in 180 mL solution

25. What is the pH of a solution with a $[H_3O^+] = 1.4 \times 10^{-3}$?

A. 1.67   B. 2.86   C. 3.42   D. 4.68   E. 11.68

26. In acidic media, how many electrons are needed to balance the following half-reaction?

$$NO_3^- \rightarrow NH_4^+$$

A. 8 e⁻ on the reactant side

B. 4 e⁻ on the reactant side

C. 3 e⁻ on the product side

D. 2 e⁻ on the reactant side

E. 8 e⁻ on the product side

27. Which of the following elements is a metalloid?

A. antimony (Sb)

B. uranium (U)

C. iodine (I)

D. zinc (Zn)

E. selenium (Se)

28. Which of the following is the strongest intermolecular force between molecules?

A. hydrophobic

B. van der Waals

C. polar

D. coordinate covalent bonds

E. hydrogen bonding

29. As the strength of the attractive intermolecular force increases, [   ] decreases.

A. melting point
B. vapor pressure of a liquid

C. viscosity
D. normal boiling temperature
E. density

30. What is the oxidation number of Cl in $HClO_4$?

A. −1   B. +5   C. +4   D. +8   E. +7

31. What is the ΔH for the following reaction?

$$CH_4\,(g) + 2\,O_2\,(g) \rightarrow CO_2\,(g) + 2\,H_2O\,(l)$$

(Use the overall $\Delta H = -806$ kJ/mol and ΔH for $2\,H_2O\,(g) \rightarrow 2\,H_2O\,(l) = -86$ kJ/mol)

A. −892 kJ/mol
B. −720 kJ/mol

C. 720 kJ/mol
D. 892 kJ/mol   E. −978 kJ/mol

32. Which of the changes listed below shifts the equilibrium to the left for the following reversible reaction?

$$PbI_2\,(s) \leftrightarrow Pb^{2+}\,(aq) + 2\,I^-\,(aq)$$

A. Add $Pb(NO_3)_2$ (s)
B. Add NaCl (s)

C. Decrease $[Pb^{2+}]$
D. Decrease $[I^-]$ E. Add HNO_3

33. If the chemical formula of the solute is known, which additional information is necessary to determine the molarity of a solution?

A. Mass of the solute dissolved and the volume of the solvent added
B. Molar mass of both the solute and the solvent used
C. Volume of the solvent used
D. Mass of the solute dissolved
E. Mass of the solute dissolved and the final volume of the solution

34. Which of the following is a general property of an acidic solution?

A. pH greater than 7
B. Turns litmus paper blue

C. Feels slippery
D. Tastes sour   E. Tastes bitter

35. An element can be determined by:

A. protons plus neutrons
B. neutrons only

C. protons plus electrons
D. atomic number   E. atomic mass

36. Which of the following is the strongest form of an intramolecular attraction?

A. ion-dipole interaction

B. dipole-dipole interaction

C. dipole-induced dipole interaction

D. covalent bond

E. induced dipole-induced dipole interaction

37. Which is an assumption of the kinetic molecular theory of gases?

A. Nonelastic collisions

B. Constant interaction of molecules

C. Nonrandom collisions

D. Gas particles take up space

E. Elastic collisions

38. Which of the following reaction is NOT classified correctly?

A. $BaCl_2 + H_2SO_4 \rightarrow BaSO_4 + 2\,HCl$ (single-replacement)

B. $F_2 + 2\,NaCl \rightarrow Cl_2 + 2\,NaF$ (single-replacement)

C. $Fe + CuSO_4 \rightarrow Cu + FeSO_4$ (single-replacement)

D. $2\,NO_2 + H_2O_2 \rightarrow 2\,HNO_3$ (synthesis)

E. All are classified correctly

39. Which statement is true about the formation of the stable ionic compound NaCl?

A. NaCl is stable because it forms from the combination of isolated gaseous ions

B. The first ionization energy of sodium contributes favorably to the overall formation of NaCl

C. The net absorption of 147 kJ/mol of energy for the formation of the Na^+ (g) and Cl^- (g) drives the formation of the stable ionic compound

D. The release of energy as NaCl (s) forms leads to an overall increase in the potential energy

E. The lattice energy provides the necessary stabilization force for the formation of NaCl

40. What is the order of B in a reactant with the rate law of $k[A]^2$?

A. 0th order

B. 1st order

C. 2nd order

D. ½th order

E. Requires more information

41. Which of the following illustrates the *like dissolves like* rule for a solid solute in a liquid solvent?

    I.   A nonpolar compound is soluble in a polar solvent

    II. A polar compound is soluble in a polar solvent

    III. An ionic compound is soluble in a polar solvent

A. I only

B. II only

C. I and II only

D. II and III only

E. I, II and III

42. If an unknown solution is a good conductor of electricity, which of the following statements is true?

    A. The solution is highly reactive
    B. The solution is slightly reactive

    C. The solution is highly ionized
    D. The solution is slightly ionized
    E. None of the above

43. Which of the following is the balanced chemical reaction for the electrolysis of brine (i.e., concentrated saltwater) as a major source of chlorine gas?

    A. $2\ NaCl\ (aq) + H_2O \rightarrow 2\ NaH\ (aq) + Cl_2O^-\ (aq)$
    B. $2\ NaCl\ (aq) + 2\ H_2O \rightarrow 2\ Na\ (s) + O_2\ (g) + 2H_2\ (g) + Cl_2\ (g)$
    C. $2\ NaCl\ (aq) + H_2O \rightarrow Na_2O\ (aq) + 2\ HCl\ (aq)$
    D. $2\ NaOH\ (aq) + Cl_2\ (g) + 2\ H_2\ (g) \rightarrow 2\ NaCl\ (aq) + 2\ H_2O$
    E. $2\ NaCl\ (aq) + 2\ H_2O \rightarrow 2\ NaOH\ (aq) + Cl_2\ (g) + H_2\ (g)$

44. Which element has the electron configuration of $1s^2 2s^2 2p^6 3s^2$ in a neutral state?

    A. Na
    B. Si

    C. Ca
    D. Mg
    E. None of the above

45. The valence shell is the:

    A. innermost shell that is complete with electrons
    B. last partially filled orbital of an atom

    C. shell of electrons in the least reactive atom
    D. outermost shell of electrons around an atom
    E. same as the orbital configuration

46. What is the new internal pressure of a given mass of nitrogen gas in a 350-ml vessel at 24 °C and 1.2 atm, when heated to 64 °C and compressed to 300 ml?

    A. $(1.2) \cdot (0.300) \cdot (333) / (0.350) \cdot (297)$
    B. $(0.350) \cdot (333) / (1.2) \cdot (0.300) \cdot (297)$

    C. $(1.2) \cdot (0.350) \cdot (297) / (0.300) \cdot (337)$
    D. $(1.2) \cdot (0.350) \cdot (0.300) / (333) \cdot (278)$
    E. $(1.2) \cdot (0.350) \cdot (337) / (0.300) \cdot (297)$

47. Which response represents the balanced oxidation half-reaction for the following reaction?

    $HCl\ (aq) + Fe\ (s) \rightarrow FeCl_3\ (aq) + H_2\ (g)$

    A. $2\ Fe \rightarrow 2\ Fe^{3+} + 6\ e^-$
    B. $3\ Fe \rightarrow Fe^{3+} + 3\ e^-$

    C. $Fe + 3\ e^- \rightarrow Fe^{3+}$
    D. $6\ e^- + 6\ H^+ \rightarrow 3\ H_2$
    E. None of the above

48. Which component of an insulated vessel design minimizes heat radiation?

    A. Tight-fitting screw-on lid
    B. Heavy-duty aluminum material

    C. Double-walled material
    D. Reflective interior coating
    E. Heavy-duty plastic casing

49. Which of the following changes the value of the equilibrium constant?

    A. Changing the initial concentrations of reactants

    B. Changing the initial concentrations of products

    C. Changing temperature

    D. Adding a catalyst at the onset of the reaction

    E. Adding other substances that do not react with any of the species involved in the equilibrium

50. What is the molar concentration of Ca^{2+} (*aq*) in a solution prepared by mixing 20 mL of a 0.03 M $CaCl_2$ (*aq*) solution with a 15 mL of a 0.06 M $CaSO_4$ (*aq*) solution?

    A. 1.84 M   **B.** 0.012 M   **C.** 0.043 M   **D.** 1.22 M   **E.** 0.086 M

51. Which of the following compounds is the strongest weak acid?

    A. H_2CO_3 ($K_a = 4.3 \times 10^{-7}$)   **C.** HCN ($K_a = 6.2 \times 10^{-10}$)

    B. HNO_2 ($K_a = 4.0 \times 10^{-4}$)   **D.** HClO ($K_a = 3.5 \times 10^{-8}$)   **E.** HF ($K_a = 7.2 \times 10^{-4}$)

52. Which of the following particles would NOT be deflected by charged plates?

    A. alpha particles   **C.** hydrogen atoms

    B. protons   **D.** cathode rays

                             E. all are deflected by charged plates

53. Which of the general statements regarding covalent bond characteristics is NOT correct?

    A. Double and triple bonds are shorter than single bonds

    B. Triple bonds are possible when 3 or more electrons are needed to complete an octet

    C. Triple bonds are stronger than double bonds

    D. Double bonds are stronger than single bonds

    E. Double bonding can occur with Group VIIA elements

54. What are the conditions for a real gas to behave most like an ideal gas?

    A. Low temperature, high-pressure   **C.** High temperature, high-pressure

    B. Low temperature, low-pressure   **D.** High temperature, low-pressure

                                 E. None of the above

55. If *A* of element Y equals 14, then 28 grams of element Y represents approximately:

    A. 28 moles of atoms   **C.** 2 atoms

    B. ½ mole of atoms   **D.** ½ of an atom   **E.** 2 moles of atoms

56. How many grams of Ag can be heated from 21.4 °C to 34 °C by the heat released when 26 g of Au cools from 97.2 °C to 29.4 °C? (Use the specific heat c of Ag = 0.240 J/g·°C and the specific heat c of Au = 0.130 J/g·°C)

A. 75.8 g

C. 4.4×10^3 g

B. 2.7×10^3 g

D. 22.3 g

E. 66.7 g

57. Which of the following influences the rate of a chemical reaction?

    I. Collision orientation

    II. Collision kinetic energy

    III. Collision frequency

A. II only

C. II and III only

B. III only

D. I and II only

E. I, II and III

58. What is the %v/v concentration of red wine that contains 25 mL of ethyl alcohol in 225 mL?

A. 0.110 %v/v

C. 0.011 %v/v

B. 0.055 %v/v

D. 5.5 %v/v

E. 11.1 %v/v

59. Which statement most specifically describes the following reaction?

$$HCl\ (aq) + KOH\ (aq) \rightarrow KCl\ (aq) + H_2O\ (l)$$

A. HCl and potassium hydroxide solutions produce potassium chloride and H_2O

B. HCl and potassium hydroxide solutions produce potassium chloride solution and H_2O

C. Aqueous HCl and potassium hydroxide produce aqueous potassium chloride and H_2O

D. HCl and potassium hydroxide produce potassium chloride and H_2O

E. An acid plus a base produce H_2O and a salt

60. A galvanic cell is constructed with the following two elements and their ions:

    $Mg\ (s) \rightarrow Mg^{2+} + 2\ e^-$ $E° = 2.35$ V

    $Pb\ (s) \rightarrow Pb^{2+} + 2\ e^-$ $E° = 0.13$ V

What is the $E°$ for the net reaction of the oxidation of Mg (s) and the reduction of Pb (s)?

A. –2.48 V   **B.** 2.2 V   **C.** –2.22 V   **D.** 0.31 V   **E.** 2.22 V

> Use the answer key to check your answers. Review the explanations in detail, focusing on questions you didn't answer correctly. Note the topic of those questions. Complete this BEFORE taking the next Diagnostic Test.

Notes for active learning

Diagnostic Test 2

> This Diagnostic Test is designed to assess your proficiency on each topic and NOT to mimic the test. Use your test results and identify areas of strength and weakness to adjust your study plan and enhance your fundamental knowledge. The length of the Diagnostic Tests is optimal for a single study session.

#	Answer:					Review	#	Answer:					Review
1:	A	B	C	D	E	___	31:	A	B	C	D	E	___
2:	A	B	C	D	E	___	32:	A	B	C	D	E	___
3:	A	B	C	D	E	___	33:	A	B	C	D	E	___
4:	A	B	C	D	E	___	34:	A	B	C	D	E	___
5:	A	B	C	D	E	___	35:	A	B	C	D	E	___
6:	A	B	C	D	E	___	36:	A	B	C	D	E	___
7:	A	B	C	D	E	___	37:	A	B	C	D	E	___
8:	A	B	C	D	E	___	38:	A	B	C	D	E	___
9:	A	B	C	D	E	___	39:	A	B	C	D	E	___
10:	A	B	C	D	E	___	40:	A	B	C	D	E	___
11:	A	B	C	D	E	___	41:	A	B	C	D	E	___
12:	A	B	C	D	E	___	42:	A	B	C	D	E	___
13:	A	B	C	D	E	___	43:	A	B	C	D	E	___
14:	A	B	C	D	E	___	44:	A	B	C	D	E	___
15:	A	B	C	D	E	___	45:	A	B	C	D	E	___
16:	A	B	C	D	E	___	46:	A	B	C	D	E	___
17:	A	B	C	D	E	___	47:	A	B	C	D	E	___
18:	A	B	C	D	E	___	48:	A	B	C	D	E	___
19:	A	B	C	D	E	___	49:	A	B	C	D	E	___
20:	A	B	C	D	E	___	50:	A	B	C	D	E	___
21:	A	B	C	D	E	___	51:	A	B	C	D	E	___
22:	A	B	C	D	E	___	52:	A	B	C	D	E	___
23:	A	B	C	D	E	___	53:	A	B	C	D	E	___
24:	A	B	C	D	E	___	54:	A	B	C	D	E	___
25:	A	B	C	D	E	___	55:	A	B	C	D	E	___
26:	A	B	C	D	E	___	56:	A	B	C	D	E	___
27:	A	B	C	D	E	___	57:	A	B	C	D	E	___
28:	A	B	C	D	E	___	58:	A	B	C	D	E	___
29:	A	B	C	D	E	___	59:	A	B	C	D	E	___
30:	A	B	C	D	E	___	60:	A	B	C	D	E	___

This page is intentionally left blank

1. Which characteristics describe a neutron's mass, charge, and location, respectively?

    A. Approximate mass 1 amu; charge 0; inside nucleus

    B. Approximate mass 5×10^{-4} amu; charge 0; inside nucleus

    C. Approximate mass 1 amu; charge +1; inside nucleus

    D. Approximate mass 1 amu; charge –1; inside nucleus

    E. Approximate mass 5×10^{-4} amu; charge –1; outside nucleus

2. Which bond is formed by the equal sharing of electrons?

    A. covalent

    B. dipole

    C. ionic

    D. London

    E. van der Waals

3. At constant pressure, what effect does decreasing the temperature of a liquid by 20 °C have on the magnitude of its vapor pressure?

    A. Inversely proportional

    B. Requires more information

    C. Increase

    D. No effect

    E. Decrease

4. Which compound contains the lowest number of moles?

    A. 15 g CH_4

    B. 15 g Si

    C. 15 g CO

    D. 15 g N_2

    E. 15 g AlH_3

5. Which is/are general guidelines for balancing an equation?

    I. Balance polyatomic ions as a single unit

    II. Begin balancing with the most complex formula

    III. Write correct formulas for reactants and products

    A. I only

    B. II only

    C. III only

    D. II and III only

    E. I, II and III

6. In the reaction energy diagrams below, the reaction on the right is [ ] and ΔG is [ ] than for the reaction on the left. (Assume the reactions are drawn to scale).

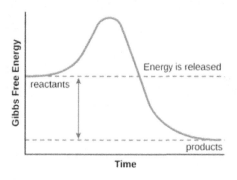 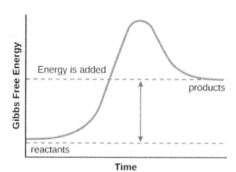

A. exergonic … less than

B. endergonic … greater than

C. exergonic … greater than

D. exergonic … the same as

E. endergonic … less than

7. Which of the following methods does NOT extract colloidal particles?

A. distillation

B. extraction

C. evaporation

D. dialysis

E. simple filtration

8. Which of the following indicators is/are colorless in an acidic solution and pink in a basic solution?

    I. phenolphthalein   II. bromothymol blue   III. methyl red

A. I only

B. II only

C. III only

D. I and II only

E. I, II and III

9. How many grams of chromium would be electroplated by passing a constant current of 4.8 amps through a solution containing chromium (III) sulfate for 50.0 minutes? (Use the molecular mass for Cr = 52.00 g/mol and 1 Faraday = 96,500 C/mol)

A. 2.27×10^3 g

B. 21.82 g

C. 0.095 g

D. 8.43×10^{-4} g

E. 2.58 g

10. Which electron configuration represents an excited state of potassium that has an atomic number of 19?

A. $1s^2 2s^2 2p^6 3s^2 3p^6$

B. $1s^2 2s^2 2p^6 3s^2 3p^6 4s^2$

C. $1s^2 2s^2 2p^6 3s^2 3p^5 4s^2$

D. $1s^2 2s^2 2p^6 3s^2 3p^7$

E. $1s^2 2s^2 2p^6 3s^2 3p^2 4s^6$

11. Which formula for an ionic compound is NOT correct as written?

    A. NH_4ClO_4 **B.** MgS   **C.** KOH   **D.** $Ca(SO_4)_2$ **E.** KF

12. According to Boyle's law, what happens to a gas as the volume increases?

    A. The temperature increases   **C.** The pressure increases

    B. The temperature decreases   **D.** The pressure decreases

                                                      E. None of the above

13. By definition, which of the following is true for a strong electrolyte?

    A. Must be highly soluble in water

    B. Contains metal and nonmetal atoms

    C. Is an ionic compound

    D. Dissociates almost completely into its ions in solution

    E. None of the above

14. What describes a reaction with no change in entropy if ΔG is negative?

    A. Spontaneous and endothermic   **C.** Nonspontaneous and exothermic

    B. Spontaneous and exothermic   **D.** Nonspontaneous and endothermic

                                                        E. Nonspontaneous and endergonic

15. For the following reaction of $A + B \rightarrow C$, which proceeds at the fastest rate at STP? (Assume that each graph is drawn to proportion)

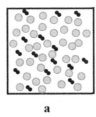 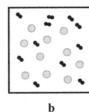 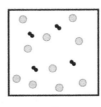

          a   **b**   **c**

    A. a   **C.** c

    B. b   **D.** All proceed at the same rate

                                                      E. Requires more information

16. Which ions are the spectator ions in the reaction below?

$$K_2SO_4\ (aq) + Ba(NO_3)_2\ (aq) \rightarrow BaSO_4\ (s) + 2\ KNO_3\ (aq)$$

    A. K^+ and SO_4^{2-} **C.** Ba^{2+} and K^+

    B. Ba^{2+} and NO_3^- **D.** Ba^{2+} and SO_4^{2-}

                                                     E. K^+ and NO_3^-

17. The neutralization of $Cr(OH)_3$ with H_2SO_4 produces which product?

A. $Cr_2(SO_4)_3$
B. SO_2

C. ^-OH
D. H_3O^+
E. H_2SO_4

18. Which of the following is/are a general characteristic of a nonmetal?

I. low density II. heat conduction III. brittle

A. I only
B. II only

C. III only
D. I and III only
E. I, II and III

19. Which of the following statements about electronegativity is NOT correct?

A. Electronegativity increases from left to right within a row
B. Electronegativity increases from bottom to top within a group
C. Fluorine is the most electronegative atom of the elements
D. Francium is the least electronegative element
E. Metals are generally more electronegative than nonmetals

20. When NaCl dissolves in water, what is the force of attraction between Na^+ and H_2O?

A. van der Waals
B. hydrogen bonding

C. ion-ion
D. dipole-dipole
E. ion-dipole

21. Which of the following has the greatest mass?

A. 34 protons, 34 neutrons and 39 electrons
B. 34 protons, 35 neutrons and 37 electrons

C. 35 protons, 34 neutrons and 37 electrons
D. 34 protons, 35 neutrons and 34 electrons
E. 35 protons, 35 neutrons and 33 electrons

22. What is the final temperature after 340 J of heat energy is removed from 30.0 g of H_2O at 19.8 °C? (Use specific heat c of H_2O = 4.184 J/g·°C)

A. 28.4 °C
B. 23.2 °C

C. 12.4 °C
D. 9.7 °C
E. 17.1 °C

23. Which statement describes how a catalyst affects the rate of a reaction?

   A. increases ΔG
   B. increases E_{act}

   C. decreases ΔH
   D. increases ΔH
   E. decreases E_{act}

24. What is the concentration in mass-volume percent for 14.6 g of CaF_2 in 260 mL of solution?

   A. 5.62%   B. 9.95%   C. 0.180%   D. 2.73%   E. 0.360%

25. Which of the following is/are an example of an Arrhenius acid?

   I. HCl (aq) II. H_2SO_4 (aq) III. HNO_3 (aq)

   A. I only
   B. II only

   C. III only
   D. I and II only
   E. I, II and III

26. Which of the statements is true for the following two half-reactions occurring in a voltaic cell?

   Reaction I: $Cr_2O_7^{2-}$ (aq) + 14 H^+ (aq) + 6 e^- → 2 Cr^{3+} (aq) + 7 H_2O (l)

   Reaction II: 6 I^- (aq) → 3 I_2 (s) + 6 e^-

   A. Reaction II is oxidation and occurs at the cathode
   B. Reaction II is oxidation and occurs at the anode
   C. Reaction I is reduction and occurs at the anode
   D. Reaction I is oxidation and occurs at the anode
   E. Reaction I is reduction and occurs at the anode and cathode

27. The l quantum number refers to the electron's:

   A. angular momentum
   B. magnetic orientation

   C. spin
   D. energy level
   E. shell size

28. Which statement is true about the valence shell?

   A. The valence shell determines the electron dot structure
   B. The electron dot structure is made up of each of the valence shells
   C. The valence shell is the innermost shell
   D. The valence shell is usually the most unreactive shell
   E. None of the above

29. What is the relationship between temperature and volume of a fixed amount of gas at constant pressure?

A. Equal
B. Indirectly proportional

C. Directly proportional
D. Decreased by a factor of 2
E. None of the above

30. What is the empirical formula of the compound with the mass percent of 71.7% Cl, 24.3% C and 4.0% H?

A. ClC_2H_5 B. Cl_2CH_2 C. $ClCH_3$ D. $ClCH_2$ E. CCl_4

31. What is the term given for a reaction when the bonds formed during the reaction are stronger than the bonds broken?

A. endergonic
B. exergonic

C. endothermic
D. exothermic
E. spontaneous

32. Which statement about chemical equilibrium is NOT true?

A. At equilibrium, the forward reaction rate equals the reverse reaction rate
B. The same equilibrium state can be attained starting either from the reactant or product side of the equation
C. Chemical equilibrium can only be attained by starting with reagents from the reactant side of the equation
D. At equilibrium, the concentration of reactants and products are constant over time
E. At equilibrium, the concentration of reactants and products may be different

33. What is the solubility of CaF_2 when it is added to a 0.03 M solution of NaF?

A. Low because $[F^-]$ present in solution inhibits dissociation of CaF_2
B. High because of the common ion effect
C. Unaffected by the presence of NaF
D. Low because less water is available for solvation due to the presence of NaF
E. High because $[F^-]$ present in solution facilitates dissociation of CaF_2

34. Which of the following properties is NOT characteristic of a base?

A. Has a slippery feel
B. Is neutralized by an acid

C. Has a bitter taste
D. Produces H^+ in water
E. All of the above

35. The element calcium belongs to which family?

A. representative elements
B. transition metals

C. alkali metals
D. lanthanides
E. alkaline earth metals

36. Which of the following statements is true?

A. The buildup of electron density between two atoms repels each nucleus, making the nuclei less stable

B. Bond energy is the minimum energy required to bring about the pairing of electrons in a covalent bond

C. As the distance between the nuclei decreases for covalent bond formation; there is a corresponding decreased probability of finding both electrons near either nucleus

D. One mole of hydrogen atoms is more stable than one mole of hydrogen molecules

E. Two electrons in a single covalent bond are paired according to the Pauli exclusion principle

37. Which substance would be expected to have the highest boiling point?

A. Nonvolatile liquid

B. Nonpolar liquid with van der Waals interactions

C. Polar liquid with hydrogen bonding

D. Nonpolar liquid

E. Nonpolar liquid with dipole-induced dipole interactions

38. Ethane (C_2H_6) gives off carbon dioxide and water when burning. What is the coefficient of oxygen in the balanced equation for the reaction?

$$\underline{}\,C_2H_6\,(g) + \underline{}\,O_2\,(g) \xrightarrow{\text{Spark}} \underline{}\,CO_2\,(g) + \underline{}\,H_2O\,(g)$$

A. 5   **B.** 7   **C.** 10   **D.** 14   **E.** none of the above

39. An ideal gas fills a closed rigid container. As the number of moles of gas in the chamber is increased at a constant temperature:

A. volume increases

B. pressure remains constant

C. pressure decreases

D. the effect on pressure cannot be determined

E. pressure increases

40. Which of the following drives this exothermic reaction towards products?

$$CH_4\,(g) + 2\,O_2\,(g) \leftrightarrow CO_2\,(g) + 2\,H_2O\,(g)$$

A. Addition of CO_2

B. Removal of CH_4

C. Increase in temperature

D. Decrease in temperature

E. Addition of H_2O

41. What mass of H_2O is needed to prepare 160 grams of 14.0% (m/m) $KHCO_3$ solution?

A. 22.4 g   **B.** 138 g   **C.** 1.9 g   **D.** 11.4 g   **E.** 14.0 g

42. A 40.0 mL sample of aqueous sulfuric acid was titrated with 0.4 M NaOH (*aq*) until neutralized. What is the molarity of the sulfuric acid solution if the residue was dried and has a mass of 840 mg?

A. 0.034 M   **B.** 0.148 M   **C.** 0.623 M   **D.** 0.745 M   **E.** 0.944 M

43. How many electrons are needed to balance the charge for the following half-reaction in an acidic solution?

$$C_2H_6O \rightarrow HC_2H_3O_2$$

A. 6 electrons to the right side
B. 4 electrons to the right side
C. 3 electrons to the left side
D. 2 electrons to the left side
E. 8 electrons to the right side

44. The three isotopes of hydrogen have different numbers of:

A. neutrons
B. electrons
C. protons
D. charges
E. protons and neutrons

45. Which statement describes the lengths of the sulfur-oxygen bonds in SO_3^{2-}?

A. All bonds are of different lengths
B. All bonds are of the same length
C. Two are the same length, while the other is longer
D. Two are the same length, while the other is shorter
E. Depends if the molecule is solid or liquid

46. Which of the following would have the highest boiling point?

A. F_2
B. Br_2
C. Cl_2
D. I_2
E. Not enough information given

47. What substance is the oxidizing agent in the following redox reaction?

$$Co\ (s) + 2\ HCl\ (aq) \rightarrow CoCl_2\ (aq) + H_2\ (g)$$

A. H_2
B. $CoCl_2$
C. HCl
D. Co
E. None of the above

48. Which component of an insulated vessel design minimizes heat loss by convection?

A. Heavy-duty aluminum construction
B. Reflective interior coating
C. Tight-fitting, screw-on lid
D. Double-walled construction
E. Heavy-duty plastic casing

49. Dinitrogen tetraoxide decomposes to produce nitrogen dioxide according to the reaction below. Calculate the equilibrium constant (K_{eq}) at 100 °C. (Use the equilibrium concentrations [N_2O_4] = 0.800 and [NO_2] = 0.400)

$$N_2O_4\ (g) \leftrightarrow 2\ NO_2\ (g)$$

A. $K_{eq} = 0.725$
B. $K_{eq} = 2.50$

C. $K_{eq} = 0.200$
D. $K_{eq} = 0.500$
E. $K_{eq} = 0.250$

50. Which solid compound is most soluble in water?

A. CuS
B. AgCl

C. PbBr$_2$
D. NiCO$_3$
E. Ba(OH)$_2$

51. What is the pH of an aqueous solution if [H^+] = 0.000001 M?

A. 1   **B.** 4   **C.** 6   **D.** 7   **E.** 9

52. Which of the following statements is/are true?

I. There are ten *d* orbitals in the *d* subshell
II. The third energy shell (n=3) has *d* orbitals
III. The *p* subshell can accommodate a maximum of 6 electrons

A. II only
B. III only

C. I and III only
D. II and III only
E. I, II and III

53. Which of these compounds has a trigonal pyramidal molecular geometry?

A. SO$_3$ **B.** XeO$_3$ **C.** AlF$_3$ **D.** BF$_3$ **E.** CH$_4$

54. What is the partial pressure due to CO$_2$ of a mixture of gases at 700 torr that contains 40% CO$_2$, 40% O$_2$ and 20% H$_2$ by pressure?

A. 760 torr   **B.** 280 torr   **C.** 420 torr   **D.** 560 torr   **E.** 700 torr

55. What is the mass of 8.50 moles of glucose (C$_6$H$_{12}$O$_6$)?

A. 1.54×10^6 g
B. 21 g

C. 964 g
D. 1,532 g
E. 6.34×10^3 g

56. Which statement describes a spontaneous reaction?

A. Releases heat to the surroundings

B. Proceeds without external influence once it has begun

C. Proceeds in both the forward and reverse directions

D. Has the same rate in the forward and reverse direction

E. Increases in disorder

57. Which of the following stresses would shift the equilibrium to the left for the following chemical system at equilibrium?

$$6\ H_2O\ (g) + 2\ N_2\ (g) + heat \leftrightarrow 4\ NH_3\ (g) + 3\ O_2\ (g)$$

A. Increasing the concentration of H_2O

B. Decreasing the concentration of NH_3

C. Increasing the concentration of O_2

D. Increasing the reaction temperature

E. Decreasing the concentration of O_2

58. What is the molarity of a solution formed by dissolving 30.0 g of NaCl in water to make 675 mL of solution? (Use the molecular mass of NaCl = 58.45 g/mol)

A. 0.49 M   **B.** 0.76 M   **C.** 0.32 M   **D.** 0.53 M   **E.** 0.12 M

59. Which of these salts is basic in an aqueous solution?

A. NaF

B. $CrCl_3$

C. KBr

D. NH_4ClO_4

E. None are basic

60. Which of the statements is NOT true regarding the following redox reaction occurring in a nonspontaneous electrochemical cell?

$$\text{Electricity}$$
$$3\ C\ (s) + 4\ AlCl_3\ (l) \quad \rightarrow \quad 4\ Al\ (l) + 3\ CCl_4\ (g)$$

A. Al metal is produced at the cathode

B. Oxidation half-reaction: $C + 4\ Cl^- \rightarrow CCl_4 + 4\ e^-$

C. Reduction half-reaction: $Al^{3+} + 3\ e^- \rightarrow Al$

D. $CCl_4\ (g)$ is produced at the cathode

E. $CCl_4\ (g)$ is produced at the anode

> Use the answer key to check your answers. Review the explanations in detail, focusing on questions you didn't answer correctly. Note the topic of those questions. Complete this BEFORE taking the next Diagnostic Test.

Notes for active learning

Notes for active learning

Diagnostic Test 3

This Diagnostic Test is designed to assess your proficiency on each topic and NOT to mimic the test. Use your test results and identify areas of strength and weakness to adjust your study plan and enhance your fundamental knowledge. The length of the Diagnostic Tests is optimal for a single study session.

#	Answer:					Review	#	Answer:					Review
1:	A	B	C	D	E	___	31:	A	B	C	D	E	___
2:	A	B	C	D	E	___	32:	A	B	C	D	E	___
3:	A	B	C	D	E	___	33:	A	B	C	D	E	___
4:	A	B	C	D	E	___	34:	A	B	C	D	E	___
5:	A	B	C	D	E	___	35:	A	B	C	D	E	___
6:	A	B	C	D	E	___	36:	A	B	C	D	E	___
7:	A	B	C	D	E	___	37:	A	B	C	D	E	___
8:	A	B	C	D	E	___	38:	A	B	C	D	E	___
9:	A	B	C	D	E	___	39:	A	B	C	D	E	___
10:	A	B	C	D	E	___	40:	A	B	C	D	E	___
11:	A	B	C	D	E	___	41:	A	B	C	D	E	___
12:	A	B	C	D	E	___	42:	A	B	C	D	E	___
13:	A	B	C	D	E	___	43:	A	B	C	D	E	___
14:	A	B	C	D	E	___	44:	A	B	C	D	E	___
15:	A	B	C	D	E	___	45:	A	B	C	D	E	___
16:	A	B	C	D	E	___	46:	A	B	C	D	E	___
17:	A	B	C	D	E	___	47:	A	B	C	D	E	___
18:	A	B	C	D	E	___	48:	A	B	C	D	E	___
19:	A	B	C	D	E	___	49:	A	B	C	D	E	___
20:	A	B	C	D	E	___	50:	A	B	C	D	E	___
21:	A	B	C	D	E	___	51:	A	B	C	D	E	___
22:	A	B	C	D	E	___	52:	A	B	C	D	E	___
23:	A	B	C	D	E	___	53:	A	B	C	D	E	___
24:	A	B	C	D	E	___	54:	A	B	C	D	E	___
25:	A	B	C	D	E	___	55:	A	B	C	D	E	___
26:	A	B	C	D	E	___	56:	A	B	C	D	E	___
27:	A	B	C	D	E	___	57:	A	B	C	D	E	___
28:	A	B	C	D	E	___	58:	A	B	C	D	E	___
29:	A	B	C	D	E	___	59:	A	B	C	D	E	___
30:	A	B	C	D	E	___	60:	A	B	C	D	E	___

This page is intentionally left blank

1. Which of the following is the effect of adding one proton to the nucleus of an atom?

   A. No change in the atomic number and decrease in the atomic mass
   B. The increase of its atomic number by one unit, but no change in atomic mass
   C. The increase of its atomic mass by one unit, but no change in atomic number
   D. The increase of the atomic number and the mass number by one unit
   E. No change in the atomic mass and decrease of the atomic number

2. How many covalent bonds can neutral sulfur (S) form?

   A. 0   B. 1   C. 2   D. 3   E. 4

3. Fifteen grams of O_2 are placed in a 15-liter container at 32 °C. Compared to an equal mass of H_2 placed in an identical container at 32 °C, what is the pressure of O_2?

   A. Less than the pressure of the H_2
   B. Double the pressure of the H_2
   C. Greater than the pressure of the H_2
   D. Equal to the pressure of the H_2
   E. Equal to the square root of the pressure of the H_2

4. What is the percent mass composition of each element in H_2SO_4?

   A. 48% oxygen, 50% sulfur and 2% hydrogen
   B. 65% oxygen, 33% sulfur and 2% hydrogen
   C. 72% oxygen, 24% sulfur and 4% hydrogen
   D. 84% oxygen, 14% sulfur and 2% hydrogen
   E. 52% oxygen, 44% sulfur and 8% hydrogen

5. Which of the following conditions result in a negative ΔG for a reaction?

   A. ΔH is negative and ΔS is positive
   B. ΔH is negative and ΔS is negative
   C. ΔS is positive
   D. ΔH is negative
   E. ΔH is negative and ΔS is zero

6. Which change causes the greatest increase in the reaction rate, if $k[X] \cdot [Y]^2$?

   A. Quadrupling [X]
   B. Tripling [Y]
   C. Decreasing temperature at constant concentrations
   D. Doubling [Y]
   E. Doubling [X]

7. What mass of KOH is needed to produce 26.0 mL of 0.65 M solution?

   A. 2.210 g
   B. 0.898 g
   C. 1.48 g
   D. 0.443 g
   E. 3.43 g

8. What volume of 0.045 M NaOH is needed to titrate 40.0 ml of 0.10 N H_3PO_4?

    A. 49.4ml   **B.** 8.2 ml   **C.** 88.9 ml   **D.** 26.6 ml   **E.** 39.9 ml

9. What is the term for a device that allows ions to travel between two half-cells to maintain an ionic charge balance in each compartment?

    A. Salt bridge

    B. Reduction half-cell

    C. Oxidation half-cell

    D. Electrochemical cell

    E. None of the above

10. Which subshell has a principal quantum number of 4 and an angular momentum quantum number of 2?

    A. 4d **B.** 4f **C.** 2p **D.** 4s **E.** 3s

11. Which of the following describes a polyatomic ion?

    A. Develops a charge as a result of the combination of two or more types of atoms

    B. Does not bond with other ions

    C. Has a negative charge of less than −1

    D. Contains a metal and a nonmetal

    E. Remains neutral from the combination of two or more types of atoms

12. Which of the following pairs of compounds contain the same intermolecular forces?

    A. H_2S and CH_4

    B. NH_3 and CH_4

    C. CH_3CH_3 and H_2O

    D. CH_3CH_2OH and H_2O

    E. CCl_4 and CH_3OH

13. What is the net ionic equation for the following reaction?

$$Ca(OH)_2 + 2\ HCl \rightarrow 2\ H_2O + CaCl_2$$

    A. $^-OH + H^+ \rightarrow H_2O$

    B. $2\ ^-OH + 2\ HCl \rightarrow 2\ H_2O$

    C. $Ca^{2+} + 2\ Cl^- \rightarrow CaCl_2$

    D. $Ca(OH)_2 + 2\ H^+ \rightarrow Ca^{2+} + H_2O$

    E. None of the above

14. Which of the following statement regarding the symbol ΔH is NOT correct?

    A. It is referred to as a change in entropy

    B. It is referred to as a change in enthalpy

    C. It is referred to as heat of reaction

    D. It represents the difference between the energy used in breaking bonds and the energy released in forming bonds during the chemical reaction

    E. It has a negative value for an exothermic reaction

15. In writing an equilibrium constant expression, which of the following is NOT correct?

A. Reactant concentrations are placed in the denominator of the expression

B. Concentrations of pure solids and pure liquids, when placed in the equilibrium expression, are never raised to any power

C. Concentrations are expressed as molarities

D. Product concentrations are placed in the numerator of the expression

E. None of the above

16. What is the concentration of ^-OH if the concentration of H_3O^+ is 1×10^{-4}M? (Use the expression $K_w = [H_3O^+] \times [^-OH] = 1 \times 10^{-14}$)

A. 1×10^5

B. 1×10^{10}

C. 1×10^{-4}

D. 1×10^{-14}

E. 1×10^{-10}

17. What is the molarity of a sulfuric acid solution if 40.0 mL of H_2SO_4 is required to neutralize 0.90 g of sodium hydrogen carbonate? (Use the molecular mass of $NaHCO_3 = 84.01$ g/mol)

$$H_2SO_4 \ (aq) + 2 \ NaHCO_3 \ (aq) \rightarrow Na_2SO_4 \ (aq) + 2 \ H_2O \ (l) + 2 \ CO_2 \ (g)$$

A. 0.334 M   **B.** 0.500 M   **C.** 0.138 M   **D.** 3.33 M   **E.** 0.667 M

18. What is the maximum number of electrons that can be placed into the *f* subshell, *d* subshell, and *p* subshell, respectively?

A. 14, 10 and 6

B. 12, 10 and 6

C. 10, 14 and 6

D. 2, 12 and 20

E. 16, 8 and 2

19. What is the molecular geometry of a CH_3^+ molecule?

A. bent

B. trigonal pyramidal

C. tetrahedral

D. linear

E. trigonal planar

20. For the following unbalanced reaction, if 14.5 L of nitrogen are reacted to form NH_3 at STP, how many liters of hydrogen are required to completely consume the nitrogen?

$$N_2 + H_2 \rightarrow NH_2$$

A. 4.8 L   **B.** 9.6 L   **C.** 14.5 L   **D.** 22.6 L   **E.** 43.5 L

21. Which substance listed below is the strongest oxidizing agent given the following spontaneous redox reaction?

$$Mg\ (s) + Sn^{2+}\ (aq) \rightarrow Mg^{2+}\ (aq) + Sn\ (s)$$

A. Mg^{2+} **B.** Sn   **C.** Mg   **D.** Sn^{2+} **E.** None of the above

22. How do potential energies of reactants and products compare in an endothermic reaction?

A. The potential energy of the reactants equals the potential energy of the products

B. Initially, the potential energy of the reactants is higher, but the potential energy of the products is higher as the reaction proceeds

C. The potential energy of the products is higher than the potential energy of the reactants

D. The potential energy of the reactants is higher than the potential energy of the products

E. Cannot be concluded

23. What is the rate law, given the data from the table below, for the following reaction?

$$2\ H_2 + 2\ NO \rightarrow N_2 + 2\ H_2O$$

Reaction	[H$_2$]	[NO]	Rate (M/s)
1	0.10 M	0.10 M	1.23×10^{-3}
2	0.10 M	0.20 M	2.46×10^{-3}
3	0.20 M	0.10 M	4.92×10^{-3}
4	0.10 M	0.30 M	1.11×10^{-2}

A. rate = $k[H_2] \times [NO]^2$

B. rate = $k[H_2]^2 \times [NO]^3$

C. rate = $k[H_2]^3 \times [NO]$

D. rate = $k[H_2]^3 \times [NO]^2$

E. rate = $k[H_2]^2 \times [NO]$

24. The reason a water solution of sucrose (i.e., table sugar) does not conduct electricity is because sucrose is a:

A. semiconductor

B. non-electrolyte

C. strong electrolyte

D. weak electrolyte

E. semi-electrolyte

25. What is the pH of a solution that has a hydronium ion concentration of 3.82×10^{-9} M?

A. 7.20   **B.** 3.96   **C.** 5.74   **D.** 2.60   **E.** 8.42

26. Which is NOT true for the following redox reaction occurring in a spontaneous electrochemical cell?

$$Sn\ (s) + Cu^{2+}\ (aq) \rightarrow Cu\ (s) + Sn^{2+}\ (aq)$$

A. Electrons flow from the Sn electrode to the Cu electrode

B. Anions in the salt bridge flow from the Cu half-cell to the Sn half-cell

C. Sn is oxidized at the cathode

D. Cu^{2+} is reduced at the anode

E. Cu^{2+} is reduced at the cathode

27. Which statement about a neutron is FALSE?

A. It is a nucleon

B. It is often associated with protons

C. It is more difficult to detect than a proton or an electron

D. It is much more massive than an electron

E. It has a charge equivalent but opposite to an electron

28. Two atoms are held together by a chemical bond because bonding electrons:

A. are attracted to the atomic nuclei

B. form an electrostatic cloud with the nuclei on the exterior

C. have the same quantum numbers

D. bonding electrons attract each other

E. form an electrostatic cloud that contains both nuclei

29. A balloon originally had a volume of 4.8 L at 46 °C and a pressure of 720 torrs. At constant pressure, to what temperature must the balloon be cooled to reduce its volume to 3.8 L?

A. –41.3 °C

B. 273 °C

C. 16 °C

D. –20.5 °C

E. 68 °C

30. From the periodic table, how many atoms of cobalt equal a mass of 58.9 g?

A. 58.9

B. 117.8

C. 1

D. 6.02×10^{23}

E. 29.5

31. The energy change occurring in a chemical reaction at constant pressure is:

A. ΔE **B.** ΔG **C.** ΔS **D.** ΔH **E.** ΔP

32. What is the composition of the equilibrium mixture in terms of moles of each substance present when 0.68 mol of NO and 0.42 mol of Br_2 are placed in a container and allowed to react according to the following reaction until equilibrium producing 0.54 mol of NOBr?

$$2\ NO\ (g) + Br_2\ (g) \rightarrow 2\ NOBr\ (g)$$

A. 0.56 mol NO and 0.30 mol Br_2 C. 0.56 mol NO and 0.42 mol Br_2

B. 0.47 mol NO and 0.075 mol Br_2 D. 0.33 mol NO and 0.60 mol Br_2

                                             E. 0.14 mol NO and 0.15 mol Br_2

33. If the solubility of CO_2 in a bottle of champagne is 1.60 g per liter at 1.00 atmosphere, what is the solubility of carbon dioxide at 10.0 atmospheres?

A. 3.2 g/L   B. 1.6 g/L   C. 0.80 g/L   D. 0.16 g/L   E. 16.0 g/L

34. When a base is added to a buffered solution, the buffer:

A. accepts H^+ C. releases H_2O

B. releases H^+ D. releases ^-OH

                                             E. accepts H_3O^+

35. How many neutrons are in a neutral atom of ^{40}Ar?

A. 18   B. 22   C. 38   D. 40   E. 60

36. Which of the statements concerning covalent double bonds is correct?

A. They involve the sharing of 2 electron pairs

B. They occur only between atoms containing 4 valence electrons

C. They are found only in molecules containing polyatomic ions

D. They are found only in molecules containing carbon

E. They occur only between atoms containing 8 valence electrons

37. According to Gay–Lussac's law, what happens to gas as temperature increases?

A. The pressure increases   C. The volume increases

B. The volume decreases   D. The pressure decreases

                                             E. None of the above

38. Octane (C_8H_{18}) is a major component of gasoline. What is the coefficient of oxygen in the balanced equation for the combustion of octane?

$$\_\_C_8H_{18}\ (g) + \_\_O_2\ (g) \xrightarrow{\text{Spark}} \_\_CO_2\ (g) + \_\_H_2O\ (g)$$

A. 16   **B.** 25   **C.** 36   **D.** 48   **E.** none of the above

39. What kind of system is represented when an investigator compresses gas within a hermetically sealed system by pushing down on the inside of a piston?

A. isolated
B. closed

C. open
D. endergonic
E. endothermic

40. Which of the following occurs when the concentration of reactants increases?

A. The amount of product decreases
B. The rate of the reaction increases

C. The heat of the reaction increases
D. The energy of activation decreases
E. The rate of the reaction decreases

41. What is the molarity of 100.0 mL of a sucrose solution that contains 12.0 g of $C_{12}H_{22}O_{11}$ dissolved in it? (Use the molecular mass of $C_{12}H_{22}O_{11}$ = 342.0 g/mol)

A. 0.351 M   **B.** 3.33 M   **C.** 0.007 M   **D.** 0.071 M   **E.** 0.333 M

42. If 30.0 mL of 0.600 M KOH is titrated with 0.350 M HNO_3, what volume of nitric acid is required to neutralize the base?

$$HNO_3\ (aq) + KOH\ (aq) \rightarrow KNO_3\ (aq) + H_2O\ (l)$$

A. 10.0 mL   **B.** 20.0 mL   **C.** 30.0 mL   **D.** 50.0 mL   **E.** 60.0 mL

43. Rust tends to form when iron comes in contact with water. According to the reaction, what is happening in the cathode region?

$$4\ Fe + 3\ O_2 \rightarrow 2\ Fe_2O_3$$

A. Oxygen is being reduced
B. Iron is being reduced

C. Iron is being oxidized
D. Oxygen is being oxidized
E. None of the above

44. Which statement(s) correctly describe(s) the alkali metals (group IA)?

  I. Form strong ionic bonds with nonmetals

  II. Little or no reaction with water

  III. More reactive than group IIA elements

A. I only

B. II only

C. III only

D. I and II only

E. I and III only

45. An alkaline earth element is expected to have [   ] ionization energy and [   ] electron affinity.

A. low; high

B. low; low

C. high; low

D. high; high

E. none of the above

46. What is the term for a change of state from a gas to a liquid?

A. vaporizing

B. melting

C. freezing

D. condensation

E. sublimation

47. How many moles of Mg are in a 4.80 g sample? (Use the atomic mass of Mg = 24.31 g/mol)

A. 0.07 moles

B. 0.20 moles

C. 1.17 moles

D. 0.17 moles

E. 1.23×10^{23} moles

48. Which of the following quantities are necessary for calculating the amount of heat energy required to change liquid H_2O at 80 °C to steam at 120 °C?

A. Specific heat of ice and specific heat of H_2O

B. Heat of fusion for H_2O and heat of condensation for H_2O

C. Specific heat of steam and heat of fusion for H_2O

D. Specific heat of H_2O, specific heat of steam, and heat of vaporization for H_2O

E. Specific heat of H_2O and specific heat of steam

49. Which is NOT a difference between a first-order and second-order elementary reaction?

A. When concentrations of reactants are doubled, the rate of a first-order reaction doubles, while the rate of a second-order elementary reaction quadruples

B. The rate of a first-order reaction is greater than the rate of a second-order reaction because collisions are not required

C. A first-order reaction is unimolecular, while a second-order reaction is bimolecular

D. The half-life of a first-order reaction is independent of the starting concentration of the reactant, while the half-life of a second-order reaction depends on the starting concentration of the reactant

E. The rate of a first-order reaction is less than the rate of a second-order reaction because collisions are required

50. Which of the following compounds is NOT soluble in water?

A. NH_4F **B.** $FeCl_3$ **C.** NaOH   **D.** CH_3OH **E.** CuS

51. Which substance ionizes completely when dissolved in water to form an aqueous solution?

A. $C_6H_{12}O_6$ (*s*)   **C.** $NaClO_4$ (*s*)

B. NH_3 (*l*)   **D.** HNO_3 (*aq*)

                             E. $Ba(OH)_2$ (*s*)

52. Compared to the atomic radius of calcium (Ca), the atomic radius of gallium (Ga) is:

A. smaller because increased nuclear charge causes electrons to be held more tightly

B. larger because its additional electrons increase the atomic volume

C. smaller because gallium loses more electrons, thereby decreasing its size

D. larger because increased electron charge requires that the same force be distributed over a greater number of electrons

E. larger because decreased electron charge requires that the same force be distributed over a smaller number of electrons

53. Two molecules, X and Y, are not miscible and have very different physical properties. Molecule X boils at 70 °C and freezes at –20 °C, while molecule Y boils at 45 °C and freezes at –90 °C. Which molecule is likely to have the largest dipole?

A. Molecule X   **C.** Both have similar dipoles

B. Molecule Y   **D.** Molecule Y does not have a dipole

                          E. Not enough information to determine

54. What is the boiling point of the solution where 80 g of glucose ($C_6H_{12}O_6$) is dissolved in 500 g of H_2O? (Use the molecular mass of glucose = 180 g/mol and K = 0.52)

A. 372.08 K

B. 273.92 K

C. 278.82 K

D. 373.46 K

E. 393.28 K

55. Which of the following statements is true regarding a chemical reaction?

A. The number of products is equal to the number of reactants

B. The number of atoms is equal on both sides of the reaction

C. The number of atoms in a reaction varies when the conditions change during the reaction

D. The number of molecules is equal on both sides of the reaction

E. The sum of the coefficients is equal on both sides of the reaction

56. Given the data, what is the standard enthalpy change for the following reaction?

$4\ PCl_3\ (l) \rightarrow P_4\ (s) + 6\ Cl_2\ (g)$

$P_4\ (s) + 6\ Cl_2\ (g) \rightarrow 4\ PCl_3\ (l)$ $\Delta H° = -1,274$ kJ/mol

A. 637 kJ/mol

B. −637 kJ/mol

C. −1,917 kJ/mol

D. 1,274 kJ/mol

E. −2,548 kJ/mol

57. What is the equilibrium reaction if the solubility product expression for slightly soluble iron (III) chromate is given as $K_{sp} = [Fe^{3+}]^2 \times [CrO_4{}^{2-}]^3$?

A. $Fe_2(CrO_4)_3\ (s) \leftrightarrow 3\ Fe^{3+}\ (aq) + 2\ CrO_4{}^{2-}\ (aq)$

B. $Fe_3(CrO_4)_2\ (s) \leftrightarrow 2\ Fe^{3+}\ (aq) + 3\ CrO_4{}^{2-}\ (aq)$

C. $Fe_2(CrO_4)_3\ (s) \leftrightarrow Fe^{3+}\ (aq) + CrO_4{}^{2-}\ (aq)$

D. $Fe_2(CrO_4)_3\ (s) \leftrightarrow 2\ Fe^{3+}\ (aq) + 3\ CrO_4{}^{2-}\ (aq)$

E. $Fe_3(CrO_4)_2\ (s) \leftrightarrow 3\ Fe^{3+}\ (aq) + 2\ CrO_4{}^{2-}\ (aq)$

58. What volume of 0.125 M hydrochloric acid reacts completely with 0.130 g of sodium carbonate? (Use the molar mass of Na_2CO_3 = 106.0 g/mol)

$Na_2CO_3\ (s) + 2\ HCl\ (aq) \rightarrow CO_2\ (g) + H_2O\ (l) + 2\ NaCl\ (aq)$

A. 19.7 mL

B. 36.2 mL

C. 9.25 mL

D. 12.8 mL

E. 2.65 mL

59. Which of the following is an acid-base neutralization reaction?

   A. H^+ (*aq*) with O_2 (*g*) to form H_2O (*l*)

   B. Na^+ (*aq*) with ^-OH (*aq*) to form NaOH (*aq*)

   C. H_2 (*g*) with O_2 (*g*) to form H_2O (*l*)

   D. H_2 (*aq*) with ^-OH (*aq*) to form H_2O (*l*)

   E. H^+ (*aq*) with ^-OH (*aq*) to form H_2O (*l*)

60. How many e^- are gained or lost in the following half-reaction: $Br_2 \rightarrow 2\ Br^-$?

   A. 2 e^- are gained   **C.** ½ e^- is gained

   B. ½ e^- is lost   **D.** 2 e^- are lost

                                E. 4 e^- are gained

Use the answer key to check your answers. Review the explanations in detail, focusing on questions you didn't answer correctly. Note the topic of those questions.

Notes for active learning

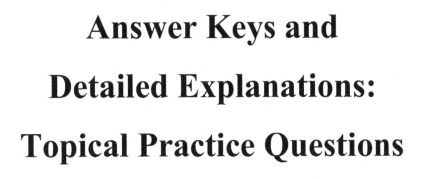

Answer Keys and
Detailed Explanations:
Topical Practice Questions

Topical Practice Questions – Answer Keys

Electronic and Atomic Structure; Periodic Table

1: B	21: C	41: B	61: D	81: C
2: C	22: D	42: A	62: B	82: A
3: E	23: E	43: D	63: C	83: C
4: B	24: C	44: A	64: B	84: B
5: C	25: D	45: D	65: E	85: A
6: B	26: D	46: A	66: B	86: E
7: C	27: C	47: E	67: D	87: C
8: D	28: D	48: D	68: A	88: E
9: B	29: E	49: B	69: B	89: B
10: A	30: B	50: A	70: C	90: B
11: B	31: A	51: E	71: D	91: E
12: E	32: D	52: E	72: E	92: A
13: A	33: B	53: D	73: A	93: E
14: C	34: C	54: A	74: E	94: D
15: E	35: A	55: D	75: B	95: D
16: C	36: C	56: D	76: D	96: A
17: D	37: B	57: A	77: C	97: B
18: C	38: B	58: A	78: E	98: E
19: C	39: C	59: A	79: E	99: A
20: E	40: D	60: C	80: C	100: C
				101: D
				102: B
				103: A
				104: A

Chemical Bonding

1: D	21: D	41: E	61: E	81: C
2: A	22: C	42: A	62: D	82: B
3: C	23: B	43: C	63: A	83: C
4: E	24: D	44: D	64: E	84: E
5: B	25: D	45: C	65: C	85: C
6: B	26: E	46: A	66: A	86: B
7: E	27: A	47: E	67: B	87: B
8: A	28: C	48: A	68: D	88: C
9: D	29: A	49: D	69: B	
10: E	30: D	50: B	70: B	
11: B	31: B	51: D	71: E	
12: E	32: E	52: D	72: B	
13: E	33: E	53: D	73: C	
14: A	34: A	54: C	74: C	
15: D	35: D	55: D	75: B	
16: C	36: D	56: A	76: C	
17: C	37: B	57: C	77: E	
18: A	38: A	58: C	78: B	
19: D	39: B	59: E	79: D	
20: C	40: A	60: C	80: D	

Phases and Phase Equilibria

1: A	11: B	21: E	31: B	41: C	51: B	61: D	71: D	81: A	91: B
2: C	12: C	22: A	32: D	42: E	52: C	62: E	72: B	82: C	92: E
3: C	13: A	23: E	33: C	43: B	53: A	63: B	73: E	83: B	93: A
4: A	14: D	24: D	34: B	44: E	54: D	64: E	74: C	84: E	94: E
5: A	15: C	25: D	35: C	45: E	55: C	65: D	75: A	85: D	95: C
6: B	16: A	26: A	36: E	46: A	56: B	66: E	76: B	86: D	96: C
7: B	17: C	27: D	37: A	47: C	57: B	67: A	77: C	87: D	97: E
8: B	18: C	28: B	38: C	48: C	58: B	68: C	78: E	88: A	98: E
9: E	19: C	29: E	39: B	49: C	59: A	69: E	79: D	89: C	99: B
10: D	20: E	30: A	40: C	50: B	60: C	70: A	80: D	90: E	100: C

Stoichiometry

1: C	21: D	41: B	61: E	81: B	101: B	121: A
2: C	22: A	42: E	62: D	82: C	102: B	122: A
3: C	23: B	43: D	63: E	83: B	103: A	123: A
4: E	24: C	44: E	64: B	84: E	104: E	124: E
5: C	25: C	45: D	65: E	85: D	105: C	125: D
6: B	26: E	46: E	66: D	86: C	106: B	126: E
7: A	27: A	47: C	67: B	87: D	107: D	127: C
8: C	28: E	48: E	68: C	88: C	108: A	128: B
9: B	29: E	49: C	69: C	89: C	109: C	129: E
10: C	30: C	50: E	70: B	90: B	110: E	130: E
11: C	31: A	51: D	71: A	91: D	111: B	131: B
12: D	32: C	52: E	72: D	92: B	112: A	132: B
13: C	33: E	53: C	73: B	93: B	113: A	133: A
14: D	34: A	54: B	74: C	94: A	114: A	134: E
15: D	35: A	55: D	75: D	95: C	115: C	135: A
16: E	36: C	56: B	76: A	96: A	116: B	136: B
17: B	37: D	57: C	77: E	97: C	117: C	137: A
18: C	38: A	58: D	78: C	98: B	118: B	138: A
19: E	39: B	59: E	79: D	99: A	119: E	139: E
20: D	40: A	60: D	80: B	100: D	120: A	140: B

Thermochemistry

1: A	11: C	21: A	31: B	41: A	51: C	61: C	71: D
2: B	12: A	22: C	32: C	42: C	52: E	62: C	72: C
3: E	13: B	23: A	33: D	43: B	53: C	63: C	73: D
4: E	14: D	24: B	34: B	44: E	54: C	64: B	74: E
5: A	15: C	25: C	35: D	45: A	55: D	65: A	75: C
6: D	16: B	26: C	36: A	46: D	56: E	66: B	76: E
7: E	17: A	27: A	37: B	47: C	57: C	67: D	77: B
8: A	18: B	28: D	38: D	48: C	58: D	68: C	78: D
9: C	19: C	29: A	39: E	49: B	59: E	69: E	79: A
10: E	20: C	30: E	40: E	50: B	60: E	70: B	80: A

Kinetics and Equilibrium

1: A	21: E	41: A	61: E	81: E	101: C
2: B	22: A	42: E	62: B	82: A	102: A
3: A	23: E	43: B	63: D	83: C	103: E
4: A	24: B	44: E	64: A	84: C	104: D
5: C	25: C	45: B	65: D	85: A	105: D
6: D	26: D	46: C	66: C	86: B	106: C
7: B	27: B	47: B	67: D	87: C	107: B
8: A	28: A	48: D	68: D	88: C	108: D
9: E	29: E	49: C	69: B	89: A	109: E
10: D	30: A	50: E	70: A	90: D	110: D
11: D	31: C	51: C	71: B	91: D	111: D
12: B	32: B	52: B	72: B	92: E	112: C
13: D	33: A	53: B	73: B	93: C	113: B
14: B	34: A	54: C	74: D	94: A	114: B
15: B	35: B	55: A	75: D	95: D	115: C
16: D	36: E	56: D	76: B	96: E	116: E
17: B	37: A	57: E	77: D	97: A	117: A
18: E	38: A	58: E	78: C	98: E	118: A
19: A	39: D	59: E	79: B	99: E	119: A
20: C	40: E	60: C	80: E	100: D	120: C

Solution Chemistry

1: E	21: B	41: E	61: D	81: D	101: A
2: E	22: A	42: B	62: A	82: A	102: A
3: A	23: C	43: B	63: A	83: E	103: D
4: A	24: E	44: A	64: A	84: E	104: B
5: D	25: A	45: D	65: C	85: B	105: B
6: A	26: C	46: B	66: D	86: E	106: D
7: B	27: A	47: A	67: C	87: E	107: D
8: D	28: D	48: E	68: B	88: B	
9: A	29: D	49: D	69: E	89: D	
10: C	30: A	50: C	70: C	90: C	
11: C	31: E	51: B	71: E	91: C	
12: A	32: B	52: E	72: A	92: D	
13: C	33: B	53: B	73: E	93: A	
14: C	34: B	54: A	74: A	94: E	
15: B	35: D	55: A	75: A	95: E	
16: E	36: A	56: B	76: C	96: C	
17: B	37: A	57: B	77: C	97: A	
18: A	38: C	58: B	78: C	98: E	
19: A	39: D	59: D	79: E	99: C	
20: D	40: B	60: E	80: D	100: D	

Acids and Bases

1: A	21: E	41: A	61: A	81: B	101: C	121: D
2: B	22: A	42: A	62: C	82: C	102: D	122: C
3: A	23: D	43: D	63: D	83: A	103: A	123: C
4: C	24: D	44: E	64: A	84: B	104: C	124: D
5: C	25: B	45: C	65: A	85: B	105: E	125: B
6: C	26: B	46: E	66: E	86: D	106: D	126: E
7: E	27: B	47: A	67: A	87: E	107: E	127: B
8: E	28: D	48: D	68: A	88: E	108: A	128: B
9: B	29: C	49: B	69: B	89: D	109: D	129: B
10: B	30: C	50: B	70: E	90: E	110: E	130: D
11: B	31: E	51: C	71: A	91: D	111: A	131: E
12: E	32: C	52: A	72: D	92: E	112: B	132: A
13: C	33: A	53: A	73: C	93: C	113: C	133: D
14: B	34: D	54: D	74: E	94: E	114: D	134: C
15: C	35: E	55: A	75: E	95: C	115: B	135: E
16: A	36: C	56: E	76: A	96: E	116: A	136: A
17: D	37: B	57: A	77: C	97: D	117: D	137: B
18: C	38: B	58: C	78: A	98: C	118: E	138: E
19: D	39: B	59: B	79: B	99: D	119: C	139: A
20: C	40: E	60: C	80: C	100: A	120: C	140: B

Electrochemistry

1: D	11: A	21: A	31: A	41: C	51: E	61: D	71: A
2: C	12: E	22: D	32: B	42: E	52: C	62: C	72: D
3: E	13: A	23: A	33: B	43: C	53: B	63: B	73: D
4: B	14: B	24: A	34: B	44: B	54: C	64: B	74: D
5: C	15: C	25: B	35: D	45: B	55: D	65: C	75: B
6: E	16: E	26: C	36: A	46: B	56: C	66: D	76: D
7: C	17: B	27: E	37: A	47: A	57: D	67: A	77: B
8: D	18: C	28: A	38: B	48: B	58: A	68: B	78: C
9: D	19: C	29: B	39: D	49: E	59: A	69: B	79: E
10: E	20: E	30: C	40: B	50: B	60: B	70: E	80: D

Electronic and Atomic Structure; Periodic Table – Detailed Explanations

===

Practice Set 1: Questions 1–20

===

1. B is correct.

Ionization energy (IE) is the amount of energy required to remove the most loosely bound electron of an isolated gaseous atom to form a cation. This is an endothermic process.

Ionization energy is expressed as:

$$X + energy \rightarrow X^+ + e^-$$

where X is an atom (or molecule) capable of being ionized (i.e., having an electron removed), X^+ is that atom or molecule after an electron is removed, and e^- is the removed electron.

The principal quantum number (n) describes the size of the orbital and the energy of an electron, and the most probable distance of the electron from the nucleus. It refers to the size of the orbital and the energy level of an electron.

The elements with larger shell sizes (n is large) listed at the bottom of the periodic table have low ionization energies. This is due to the shielding (by the inner shell electrons) from the positive charge of the nucleus. The greater the distance between the electrons and the nucleus, the less energy is needed to remove the outer valence electrons.

2. C is correct.

Accepting electrons to form anions is a characteristic of non-metals to obtain the electron configuration of the noble gases (i.e., complete octet).

Except for helium (which has a complete octet with 2 electrons, $1s^2$), the noble gases have complete octets with ns^2 and np^6 orbitals.

Donating electrons to form cations (e.g., Ca^{2+}, Fr^+, Na^+) is a characteristic of metals to obtain the electron configuration of the noble gases (i.e., complete octet).

3. E is correct.

Metalloids are semimetallic elements (i.e., between metals and nonmetals). The metalloids are boron (B), silicon (Si), germanium (Ge), arsenic (As), antimony (Sb), and tellurium (Te). Some literature reports polonium (Po) and astatine (At) as metalloids.

Seventeen elements are generally classified as nonmetals. Eleven are gases: hydrogen (H), helium (He), nitrogen (N), oxygen (O), fluorine (F), neon (Ne), chlorine (Cl), argon (Ar), krypton (Kr), xenon (Xe) and radon (Rn). One nonmetal is a liquid – bromine (Br) – and five are solids: carbon (C), phosphorus (P), sulfur (S), selenium (Se), and iodine (I).

---

4. B is correct.

An element is a pure chemical substance consisting of a single type of atom, distinguished by its atomic number (Z) (i.e., the number of protons it contains). 118 elements have been identified, of which the first 94 occur naturally on Earth, with the remaining 24 being synthetic elements.

The properties of the elements on the periodic table repeat at regular intervals, creating "groups" or "families" of elements. Each column on the periodic table is a group, and elements within each group have similar physical and chemical characteristics due to the orbital location of their outermost electron. These groups only exist because the elements of the periodic table are listed by increasing atomic numbers.

5. C is correct.

For $n = 2$ shell, it has 2 orbitals: s, p

Each orbital can hold two electrons.

A maximum number of electrons in each shell:

 The s subshell has 1 spherical orbital and can accommodate 2 electrons.

 The p subshell has 3 dumbbell-shaped orbitals and can accommodate 6 electrons.

Maximum number of electrons in $n = 2$ shell is:

 2 (for s) + 6 (for p) = 8 electrons

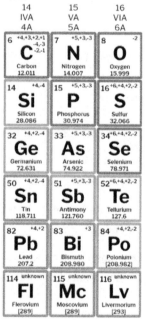

6. B is correct.

Groups IVA, VA and VIA each contain at least one metal and one nonmetal.

Group IVA has three metals (tin, lead, and flerovium) and one nonmetal (carbon).

Group VA has two metals (bismuth and moscovium) and two nonmetals (nitrogen and phosphorous).

Group VIA has one metal (livermorium) and three nonmetals (oxygen, sulfur, and selenium).

All three groups are part of the p-block of the periodic table.

7. C is correct.

The majority of elements on the periodic table (over 100 elements) are metals. Currently, there are 84 metal elements on the Periodic Table.

Seventeen elements are generally classified as nonmetals. Eleven are gases: hydrogen (H), helium (He), nitrogen (N), oxygen (O), fluorine (F), neon (Ne), chlorine (Cl), argon (Ar), krypton (Kr), xenon (Xe) and radon (Rn). One nonmetal is a liquid – bromine (Br) – and five are solids: carbon (C), phosphorus (P), sulfur (S), selenium (Se), and iodine (I).

Therefore, with the ratio of 84:17, there are about five times more metals than nonmetals.

8. D is correct.

English chemist John Dalton is known for his Atomic Theory, which states that *elements are made of small particles called atoms, which cannot be created or destroyed*.

9. B is correct.

Isotopes are variants of a particular element, which differ in the number of neutrons. Isotopes of the element have the same number of protons and occupy the same position on the periodic table.

The number of protons within the atom's nucleus is the atomic number (Z) and is equal to the number of electrons in the neutral (non-ionized) atom. Each atomic number identifies a specific element, but not the isotope; an atom of a given element may have a wide range in its number of neutrons.

The number of protons and neutrons (i.e., nucleons) in the nucleus is the atom's mass number (A), and each isotope of an element has a different mass number.

The atomic mass unit (amu) was designed using ^{12}C isotope as the reference.

1 amu = 1/12 mass of a ^{12}C atom.

Masses of other elements are measured against this standard.

If the mass of an atom 55.91 amu, the atom's mass is 55.91 × (1/12 mass of ^{12}C).

10. A is correct.

The three coordinates that come from Schrodinger's wave equations are the principal (n), angular (l) and magnetic (m) quantum numbers.

These quantum numbers describe the size, shape, and orientation in space of the orbitals on an atom.

The principal quantum number (n) describes the size of the orbital and the energy of an electron, and the most probable distance of the electron from the nucleus. It refers to the size of the orbital and the energy level of an electron.

The angular momentum quantum number (l) describes the shape of the orbital within the subshells.

The magnetic quantum number (m) determines the number of orbitals and their orientation within a subshell. Consequently, its value depends on the orbital angular momentum quantum number (l).

Given a certain l, m is an interval ranging from $-l$ to $+l$ (i.e., it can be zero, a negative integer, or a positive integer).

The s is the spin quantum number (e.g., $+\frac{1}{2}$ or $-\frac{1}{2}$).

11. B is correct.

Electron shells represent the orbit that electrons allow around an atom's nucleus. Each shell comprises one or more subshells, which are named using lowercase letters (*s, p, d, f*).

The first shell has one subshell (1*s*); the second shell has two subshells (2*s*, 2*p*); the third shell has three subshells (3*s*, 3*p*, 3*d*), etc.

An *s* subshell holds 2 electrons, and each subsequent subshell in the series can hold 4 more (*p* holds 6, *d* holds 10, *f* holds 14).

The shell number (i.e., principal quantum number) before the *s* (i.e., 4 in this example) does not affect how many electrons can occupy the subshell.

Subshell name	Subshell max electrons	Shell max electrons
1s	2	**2**
2s	2	2 + 6 = 8
2p	6	
3s	2	2 + 6 + 10 = **18**
3p	6	
3d	10	
4s	2	2 + 6 + + 10 + 14 = **32**
4p	6	
4d	10	
4f	14	

12. E is correct.

The specific characteristic line spectra for atoms result from photons being emitted when excited electrons drop to lower energy levels.

13. A is correct.

In general, the size of neutral atoms increases down a group (i.e., increasing shell size) and decreases from left to right across the periodic table.

Negative ions (anions) are *much larger* than their neutral element, while positive ions (cations) are *much smaller*. The examples are isoelectronic because of the same number of electrons.

Atomic numbers:

Br = 35; K = 19; Ar = 18; Ca = 20 and Cl = 17

The general trend for the atomic radius is to decrease from left to right and increase from top to bottom in the periodic table. When the ion gains or loses an electron to create a new charged ion, its radius would change slightly, but the general trend of radius still applies.

The ions K^+, Ca^{2+}, Cl^-, and Ar have identical numbers of electrons.

However, Br is located below Cl (larger principal quantum number *n*), and its atomic number is almost twice the others. This indicates that Br has more electrons, and its radius must be significantly larger than the other atoms.

14. C is correct.

The ground state configuration of sulfur is $[Ne]3s^23p^4$.

According to Hund's rule, *p* orbitals are filled separately and pair electrons by +½ or –½ spin.

The first three *p* electrons fill separate orbitals, and then the fourth electron pairs with two remaining unpaired electrons.

15. E is correct.

There are two ways to obtain the proper answer to this problem:

1. Using an atomic number.

Calculate the atomic number by adding the electrons:

$2 + 2 + 6 + 2 + 6 + 1 = 19$

Find element number 19 in the periodic table.

Check the group where it is located to see other elements that belong to the same group.

Element number 19 is potassium (K), so the element that belongs to the same group (IA) is lithium (Li).

2. Using subshells.

Identify the outermost subshell and use it to identify its group in the periodic table:

In this problem, the outermost subshell is $4s^1$.

Relationship between outermost subshell and group:

s^1 = Group IA

s^2 = Group IIA

p^1 = Group IIIA

p^2 = Group IVA

…

p^6 = Group VIII A

d = transition element

f = lanthanide/actinide element

16. C is correct.

The number of valence electrons for an element can be determined by its group (i.e., vertical column) on the periodic table.

Except for the transition metals (i.e., groups 3-12), the group number identifies how many valence electrons are associated with a particular element: elements of the same group have the same number of valence electrons.

Atoms are most stable when they contain 8 electrons (i.e., complete octet) in the valence shell.

17. D is correct.

The principal quantum number (*n*) describes the size of the orbital and the energy of an electron, and the most probable distance of the electron from the nucleus. It refers to the size of the orbital and the energy level of an electron.

The elements with larger shell sizes (*n* is large) listed at the bottom of the periodic table have low ionization energies. This is due to the shielding (by the inner shell electrons) from the positive charge of the nucleus. The greater the distance between the electrons and the nucleus, the less energy is needed to remove the outer valence electrons.

Ionization energy decreases with increasing shell size (i.e., *n* value) and generally increases to the right across a period (i.e., row) in the periodic table.

Moving down a column corresponds to increasing shell size with electrons further from the nucleus and decreasing nuclear attraction.

18. C is correct.

The *f* subshell has 7 orbitals.

Each orbital can hold two electrons.

The capacity of an *f* subshell is 7 orbitals × 2 electrons/orbital = 14 electrons.

19. C is correct.

The term "electron affinity" does not use the word energy as a reference; it is one of the measurable energies, just like ionization energy.

20. E is correct.

The attraction of the nucleus on the outermost electrons determines the ionization energy, which increases towards the right and increases up on the periodic table.

===

Practice Set 2: Questions 21–40

===

21. C is correct.

The mass number (A) is the sum of protons and neutrons in an atom.

The mass number is an approximation of the atomic weight of the element as amu (grams per mole).

The problem only specifies the atomic mass (A) of Cl: 35 amu.

The atomic number (Z) is not given, but the information is available in the periodic table (atomic number = 17).

> number of neutrons = atomic weight – atomic number
>
> number of neutrons = 35 – 17
>
> number of neutrons = 18

22. D is correct.

An element is a pure chemical substance consisting of a single type of atom, distinguished by its atomic number (Z) (i.e., the number of protons it contains). One hundred eighteen elements have been identified, of which the first 94 occur naturally on Earth, with the remaining 24 being synthetic elements.

The properties of the elements on the periodic table repeat at regular intervals, creating "groups" or "families" of elements. Each column on the periodic table is a group, and elements within each group have similar physical and chemical characteristics due to the orbital location of their outermost electron. These groups only exist because the elements of the periodic table are listed by increasing atomic numbers.

23. E is correct.

Metals are elements that form positive ions by losing electrons during chemical reactions. Thus metals are electropositive elements.

Metals are characterized by bright luster, hardness, ability to resonate sound, and excellent heat and electricity conductors.

Metals, except mercury, are solids under normal conditions. Potassium has the lowest melting point of the solid metals at 146 °F.

24. C is correct.

Electron affinity is defined as the energy liberated when an electron is added to a gaseous neutral atom converting it to an anion.

25. D is correct.

Metalloids are semimetallic elements (i.e., between metals and nonmetals). The metalloids are boron (B), silicon (Si), germanium (Ge), arsenic (As), antimony (Sb), and tellurium (Te). Some literature reports polonium (Po) and astatine (At) as metalloids.

Metalloids have properties between metals and nonmetals. They typically have a metallic appearance but are only fair conductors of electricity (as opposed to metals that are excellent conductors), making them useable in the semiconductor industry.

Metalloids tend to be brittle, and chemically they behave more like nonmetals.

26. D is correct.

Halogens (Group VIIA) include fluorine (F), chlorine (Cl), bromine (Br), iodine (I), and astatine (At).

Halogens gain one electron to become a –1 anion, and the resulting ion has a complete octet of valence electrons.

27. C is correct.

Dalton's Atomic Theory, developed in the early 1800s, states that atoms of a given element are identical in mass and properties.

The masses of atoms of a particular element may be not identical, although atoms of an element must have the same number of protons; they can have different numbers of neutrons (i.e., isotopes).

28. D is correct.

Elements are defined by the number of protons (i.e., atomic number).

The isotopes are neutral atoms: # electrons = # protons.

Isotopes are variants of a particular element that differ in the number of neutrons. Isotopes of the element have the same number of protons and occupy the same position on the periodic table.

The number of protons is denoted by the subscript on the left.

The number of protons and neutrons is denoted by the superscript on the left.

The charge of an ion is denoted by the superscript on the right.

The number of protons within the atom's nucleus is the atomic number (Z) and is equal to the number of electrons in the neutral (non-ionized) atom. Each atomic number identifies a specific element, but not the isotope; an atom of a given element may have a wide range in its number of neutrons.

The number of protons and neutrons (i.e., nucleons) in the nucleus is the atom's mass number (A), and each isotope of an element has a different mass number.

29. E is correct.

A cathode-ray particle is a different name for an electron. Those particles (i.e., electrons) are attracted to the positively charged cathode, implying that they are negatively charged.

30. B is correct.

The three coordinates that come from Schrodinger's wave equations are the principal (n), angular (l), and magnetic (m) quantum numbers. These quantum numbers describe the size, shape, and orientation in the space of the orbitals on an atom.

The principal quantum number (n) describes the size of the orbital, the energy of an electron, and the most probable distance of the electron from the nucleus. It refers to the size of the orbital and the energy level of an electron.

The angular momentum quantum number (l) describes the shape of the orbital of the subshells.

Carbon has an atomic number of 6 and an electron configuration of $1s^2$, $2s^2$, $2p^2$.

Therefore, electrons are in the second shell of $n = 2$, and two subshells are in the outermost shell of $l = 1$.

The values of l are 0 and 1 whereby, only the largest value of l ($l = 1$) is reported.

31. A is correct.

The alkaline earth metals (group IIA), in the ground state, have a filled s subshell with 2 electrons.

32. D is correct.

The 3rd shell consists of s, p and d subshells.

Each orbital can hold two electrons.

       The s subshell has 1 spherical orbital and can accommodate 2 electrons

       The p subshell has 3 dumbbell-shaped orbitals and can accommodate 6 electrons

       The d subshell has 5 lobe-shaped orbitals and can accommodate 10 electrons

1s
2s 2p
3s 3p 3d
4s 4p 4d 4f
5s 5p 5d 5f ...
6s 6p 6d ... ...

Order of filling orbitals

The $n = 3$ shell can accommodate 18 electrons.

The element with the electron configuration terminating in $3p^4$ is sulfur (i.e., 16 electrons).

33. B is correct.

Ions typically form with the same electron configuration as noble gases. Except for helium (which has a complete octet with 2 electrons, $1s^2$), the noble gases have complete octets with ns^2 and np^6 orbitals.

Halogens (Group VIIA) include fluorine (F), chlorine (Cl), bromine (Br), iodine (I), and astatine (At).

Halogens gain one electron to become -1 anion, and the resulting ion has a complete octet of valence electrons.

34. C is correct.

A continuous spectrum refers to a broad, uninterrupted spectrum of radiant energy.

The visible spectrum refers to the light that humans can see and includes the colors of the rainbow.

The ultraviolet spectrum refers to electromagnetic radiation with a wavelength shorter than visible light but longer than X-rays.

The radiant energy spectrum refers to electromagnetic (EM) waves of all wavelengths. However, the frequency bands in an EM signal may be sharply defined with interruptions or broad.

35. A is correct.

In general, the size of neutral atoms increases down a group (i.e., increasing shell size) and decreases from left to right across the periodic table.

Positive ions (cations) are *much smaller* than the neutral element (due to the greater effective nuclear charge). In comparison, negative ions (anions) are *much larger* (due to the smaller effective nuclear charge and repulsion of valence electrons).

36. C is correct.

In the ground state, the 3*p* orbitals fill before the 3*d* orbitals.

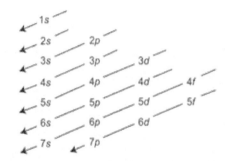

The lowest energy orbital fills before an orbital of a higher energy level.

Aufbau principle determines the order of energy levels in subshells:

From the table above, the orbitals increase in energy from:

$$1s < 2s < 2p < 3s < 3p < 4s < 3d < 4p < 5s < 4d < 5p < 6s < 4f < 5d < 6p < 7s < 5f < 6d < 7p$$

37. B is correct.

Ionization energy is defined as the energy needed to remove an electron from a neutral atom of the element in the gas phase.

The principal quantum number (*n*) describes the size of the orbital and the energy of an electron, and the most probable distance of the electron from the nucleus. It refers to the size of the orbital and the energy level of an electron.

The elements with larger shell sizes (*n* is large) listed at the bottom of the periodic table have low ionization energies. This is due to the shielding (by the inner shell electrons) from the positive charge of the nucleus. The greater the distance between the electrons and the nucleus, the less energy is needed to remove the outer valence electrons.

Ionization energy decreases with increasing shell size (i.e., *n* value) and generally increases to the right across a period (i.e., row) in the periodic table.

Neon (Ne) has an atomic number of 10 and a shell size of $n = 2$.

Rubidium (Rb) has an atomic number of 37 and a shell size of $n = 5$.

Potassium (K) has an atomic number of 19 and a shell size of $n = 4$.

Calcium (Ca) has an atomic number of 20 and a shell size is $n = 4$.

Magnesium (Mg) has an atomic number of 12 and a shell size of $n = 3$.

38. B is correct.

Seventeen elements are generally classified as nonmetals. Eleven are gases: hydrogen (H), helium (He), nitrogen (N), oxygen (O), fluorine (F), neon (Ne), chlorine (Cl), argon (Ar), krypton (Kr), xenon (Xe) and radon (Rn). One nonmetal is a liquid – bromine (Br) – and five are solids: carbon (C), phosphorus (P), sulfur (S), selenium (Se), and iodine (I).

Alkali metals (group IA) include lithium (Li), potassium (K), sodium (Na), rubidium (Rb), cesium (Cs) and francium (Fr).

Alkali metals lose one electron to become +1 cations, and the resulting ion has a complete octet of valence electrons.

Alkaline earth metals (group IIA) include beryllium (Be), magnesium (Mg), calcium (Ca), strontium (Sr), barium (Ba), and radium (Ra). Alkaline earth metals lose two electrons to become +2 cations, and the resulting ion has a complete octet of valence electrons.

39. C is correct.

The atom has 47 protons, 47 electrons, and 60 neutrons.

Because the periodic table is arranged by atomic number (Z), the fastest way to identify an element is to determine its atomic number. The atomic number is equal to the number of protons or electrons, which means that this atom's atomic number is 47. Use this information to locate element #47 in the table, which is Ag (silver).

Check the atomic mass (A), equal to the atomic number + number of neutrons.

For this atom, the mass is $60 + 47 = 107$.

The mass of Ag on the periodic table is listed as 107.87, which is the average mass of Ag isotopes.

Usually, isotopes of an element have similar masses (within 1-3 amu).

40. D is correct.

The three coordinates that come from Schrodinger's wave equations are the principal (n), angular (l), and magnetic (m) quantum numbers.

These quantum numbers describe the size, shape, and orientation in the space of the orbitals in an atom.

The principal quantum number (n) describes the size of the orbital, the energy of an electron, and the most probable distance of the electron from the nucleus.

==

Practice Set 3: Questions 41–60

==

41. B is correct.

The atomic number (Z) is the sum of protons in an atom which determines the chemical properties of an element and its location on the periodic table.

The mass number (A) is the sum of protons and neutrons in an atom.

The mass number is an approximation of the atomic weight of the element as amu (grams per mole).

Atomic mass – atomic number = number of neutrons

$9 - 4 = 5$ neutrons

42. A is correct.

Electrons are the negatively-charged particles (charge –1) located in the electron cloud, orbiting around the nucleus.

Electrons are extremely tiny particles, much smaller than protons and neutrons, with a mass of about 5×10^{-4} amu.

43. D is correct.

A compound consists of two or more different atoms which associate via chemical bonds.

Calcium chloride ($CaCl_2$) is an ionic compound of calcium and chloride.

Dichloromethane has the molecular formula of CH_2Cl_2

Dichlorocalcium exists as a hydrate with the molecular formula of $CaCl_2 \cdot (H_2O)_2$.

Carbon chloride (i.e., carbon tetrachloride) has the molecular formula of CCl_4.

Dicalcium chloride is not the proper IUPAC name for calcium chloride ($CaCl_2$).

44. A is correct.

Congeners are chemical substances related by origin, structure, or function. In regards to the periodic table, congeners are the elements of the same group which share similar properties (e.g., copper, silver, and gold are congeners of Group 11).

Stereoisomers, diastereomers, and epimers are terms commonly used in organic chemistry.

Stereoisomers: are chiral molecules (attached to 4 different substituents and are non-superimposable mirror images. They have the same molecular formula and the same sequence of bonded atoms but are oriented differently in 3-D space (e.g., *R* / *S* enantiomers).

Diastereomers are chiral molecules that are not mirror images. The most common form is a chiral molecule with more than 1 chiral center. Additionally, *cis* / *trans* (*Z* / *E*) geometric isomers are diastereomers.

Epimers: diastereomers that differ in absolute configuration at only one chiral center.

Anomers: is a type of stereoisomer used in carbohydrate chemistry to describe the orientation of the glycosidic bond of adjacent saccharides (e.g., α and β linkage of sugars). A refers to the hydroxyl group – of the anomeric carbon – pointing downward while β points upward.

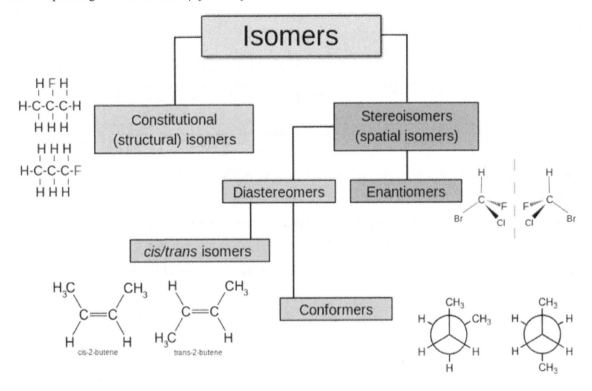

Summary of isomers

45. D is correct.

A group (or family) is a vertical column, and elements within each group share similar properties.

A period is a horizontal row in the periodic table of elements.

46. A is correct.

Metals, except mercury, are solids under normal conditions.

Potassium has the lowest melting point of the solid metals at 146 °F.

The relatively low melting temperature for potassium is due to its fourth shell ($n = 4$), which means its valence electrons are further from the nucleus; therefore, there is less attraction between its electrons and protons.

47. E is correct.

Alkaline earth metals (group IIA) include beryllium (Be), magnesium (Mg), calcium (Ca), strontium (Sr), barium (Ba), and radium (Ra). Alkaline earth metals lose two electrons to become +2 cations, and the resulting ion has a complete octet of valence electrons.

Transition metals (or transition elements) are defined as elements that have a partially-filled *d* or *f* subshell in a common oxidative state.

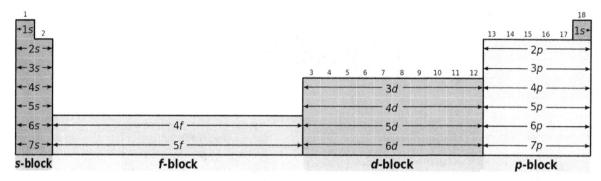

Transition metals occur in groups (vertical columns) 3–12 of the period table. They occur in periods (horizontal rows) 4–7. This group of elements includes silver, iron, and copper.

The *f*-block lanthanides (i.e., rare earth metals) and actinides (i.e., radioactive elements) are considered transition metals and are known as *inner transition metals*.

Noble gases (group VIIIA) include helium (He), neon (Ne), argon (Ar), krypton (Kr), xenon (Xe), radon (Rn), and oganesson (Og).

Alkali metals (group IA) include lithium (Li), potassium (K), sodium (Na), rubidium (Rb), cesium (Cs) and francium (Fr). Alkali metals lose one electron to become +1 cations, and the resulting ion has a complete octet of valence electrons.

Halogens (Group VIIA) include fluorine (F), chlorine (Cl), bromine (Br), iodine (I), and astatine (At). Halogens gain one electron to become a –1 anion, and the resulting ion has a complete octet of valence electrons.

48. D is correct.

The metalloids have some properties of metals and some properties of nonmetals.

Metalloids are semimetallic elements (i.e., between metals and nonmetals). The metalloids are boron (B), silicon (Si), germanium (Ge), arsenic (As), antimony (Sb), and tellurium (Te). Some literature reports polonium (Po) and astatine (At) as metalloids.

They have properties between metals and nonmetals. They typically have a metallic appearance but are only fair conductors of electricity (as opposed to metals that are excellent conductors), making them useable in the semiconductor industry.

Metalloids tend to be brittle, and chemically they behave more like nonmetals.

However, the elements in the IIIB group are transition metals, not metalloids.

49. B is correct.

Halogens (Group VIIA) include fluorine (F), chlorine (Cl), bromine (Br), iodine (I), and astatine (At).

Halogens gain one electron to form a –1 anion, and the resulting ion has a complete octet of valence electrons.

50. A is correct.

Isotopes are variants of a particular element that differ in the number of neutrons. Isotopes of the element have the same number of protons and occupy the same position on the periodic table.

The number of protons within the atom's nucleus is the atomic number (Z) and is equal to the number of electrons in the neutral (non-ionized) atom. Each atomic number identifies a specific element, but not the isotope; an atom of a given element may have a wide range in its number of neutrons.

The number of protons and neutrons (i.e., nucleons) in the nucleus is the atom's mass number (A), and each isotope of an element has a different mass number.

From the periodic table, the atomic mass of a natural sample of Si is 28.1, which is less than the mass of ^{29}Si or ^{30}Si. Therefore, ^{28}Si is the most abundant isotope.

51. E is correct.

The initial explanation was that the ray was present in the gas, and the cathode activated it.

The ray was observed even when gas was not present, so the conclusion was that the ray must have been coming from the cathode itself.

52. E is correct.

The choices correctly describe the spin quantum number (*s*).

The three coordinates that come from Schrodinger's wave equations are the principal (*n*), angular (*l*), and magnetic (*m*) quantum numbers. These quantum numbers describe the size, shape, and orientation in the space of the orbitals of an atom.

The principal quantum number (*n*) describes the size of the orbital and the energy of an electron, and the most probable distance of the electron from the nucleus. It refers to the size of the orbital and the energy level of an electron.

The angular momentum quantum number (*l*) describes the shape of the orbitals of the subshells.

The magnetic quantum number (*m*) determines the number of orbitals and their orientation within a subshell. Consequently, its value depends on the orbital angular momentum quantum number (*l*).

Given a certain *l*, *m* is an interval ranging from –*l* to +*l* (i.e., it can be zero, a negative integer, or a positive integer).

The *s* is the spin quantum number (e.g., +½ or –½).

53. D is correct.

The three coordinates that come from Schrodinger's wave equations are the principal (*n*), angular (*l*), and magnetic (*m*) quantum numbers. These quantum numbers describe the size, shape, and orientation in the space of the orbitals on an atom.

The principal quantum number (*n*) describes the size of the orbital and the energy of an electron, and the most probable distance of the electron from the nucleus. It refers to the size of the orbital and the energy level of an electron.

The angular momentum quantum number (*l*) describes the shape of the orbital of the subshells.

The magnetic quantum number (*m*) determines the number of orbitals and their orientation within a subshell. Consequently, its value depends on the orbital angular momentum quantum number (*l*). Given a certain *l*, *m* is an interval ranging from –*l* to +*l* (i.e., it can be zero, a negative integer, or a positive integer).

The fourth quantum number is *s*, which is the spin quantum number (e.g., +½ or –½).

Electrons cannot be precisely located in space at any point in time, and orbitals describe probability regions for finding the electrons.

The values needed to locate an electron are *n*, *m* and *l*. The spin can be either +½ or –½, so four values are needed to describe a single electron.

54. A is correct.

The lowest energy orbital fills before an orbital of a higher energy level.

Aufbau principle to determine the order of energy levels in subshells:

From the graphic shown, the orbitals increase in energy from:

$1s < 2s < 2p < 3s < 3p < 4s < 3d < 4p < 5s < 4d < 5p < 6s < 4f < 5d < 6p < 7s < 5f < 6d < 7p$

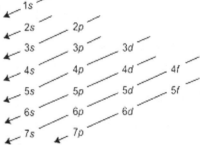

55. D is correct.

In general, the size of neutral atoms increases down a group (i.e., increasing shell size) and decreases from left to right across the periodic table.

Positive ions (cations) are *much smaller* than the neutral element (due to greater effective nuclear charge), while negative ions (anions) are *much larger* (due to smaller effective nuclear charge and repulsion of valence electrons).

56. D is correct.

Boron's atomic number is 5; therefore, it contains 5 electrons.

Use the Aufbau principle to determine the order of filling orbitals.

Remember that each electron shell (principal quantum number *n*) starts with a new *s* orbital.

57. A is correct.

Identify an element using the periodic table is its atomic number.

The atomic number is equal to the number of protons or electrons.

The number of electrons can be determined by adding the electrons in the provided electron configuration: 2 + 2 + 6 + 2 + 6 + 2 + 10 + 6 + 2 + 10 + 2 = 50.

Element #50 in the periodic table is tin (Sn).

58. A is correct.

Ionization energy (IE) is the amount of energy required to remove the most loosely bound electron of an isolated gaseous atom to form a cation. This is an endothermic process.

Ionization energy is expressed as $X + energy \rightarrow X^+ + e^-$

where X is an atom (or molecule) capable of being ionized (i.e., having an electron removed), X^+ is that atom or molecule after an electron is removed, and e^- is the removed electron.

The principal quantum number (n) describes the size of the orbital and the energy of an electron, and the most probable distance of the electron from the nucleus. It refers to the size of the orbital and the energy level of an electron.

The elements with larger shell sizes (n is large) listed at the bottom of the periodic table have low ionization energies. This is due to the shielding (by the inner shell electrons) from the positive charge of the nucleus. The greater the distance between the electrons and the nucleus, the less energy is needed to remove the outer valence electrons.

Ionization energy decreases with increasing shell size (i.e., n value) and generally increases to the right across a period (i.e., row) in the periodic table.

Argon (Ar) has an atomic number of 18 and a shell size of $n = 3$.

Strontium (Sr) has an atomic number of 38 and a shell size of $n = 5$.

Bromine (Br) has an atomic number of 35 and a shell size of $n = 4$.

Indium (In) has an atomic number of 49 and a shell size of $n = 5$.

Tin (Sn) has an atomic number of 50 and a shell size of $n = 5$.

59. A is correct.

Electronegativity is defined as the ability of an atom to attract electrons when it bonds with another atom. The most common use of electronegativity pertains to polarity along the sigma (single) bond.

The trend for increasing electronegativity within the periodic table is up and toward the right. The most electronegative atom is fluorine (F), while the least electronegative atom is francium (Fr).

The greater the difference in electronegativity between two atoms, the more polar a is the bond these atoms form.

The atom with the higher electronegativity is the partial (delta) negative end of the dipole.

60. C is correct.

Seventeen elements are generally classified as nonmetals. Eleven are gases: hydrogen (H), helium (He), nitrogen (N), oxygen (O), fluorine (F), neon (Ne), chlorine (Cl), argon (Ar), krypton (Kr), xenon (Xe) and radon (Rn). One nonmetal is a liquid – bromine (Br) – and five are solids: carbon (C), phosphorus (P), sulfur (S), selenium (Se), and iodine (I).

Metals, except mercury, are solids under normal conditions. Potassium has the lowest melting point of the solid metals at 146 °F.

===

Practice Set 4: Questions 61–80

===

61. D is correct.

When an electron absorbs energy, it moves temporarily to a higher energy level. It then drops back to its initial state (or _ground state_) while emitting the excess energy. This emission can be observed as visible spectrum lines. The protons do not move between energy levels, so they cannot absorb or emit energy.

62. B is correct.

The three coordinates that come from Schrodinger's wave equations are the principal (n), angular (l), and magnetic (m) quantum numbers. These quantum numbers describe the size, shape, and orientation in the space of the orbitals on an atom.

The principal quantum number (n) describes the size of the orbital, the energy of an electron, and the most probable distance of the electron from the nucleus.

The angular momentum quantum number (l) describes the shape of the orbital of the subshells.

The magnetic quantum number (m) determines the number of orbitals and their orientation within a subshell. Consequently, its value depends on the orbital angular momentum quantum number (l). Given a certain l, m is an interval ranging from $-l$ to $+l$ (i.e., it can be zero, a negative integer, or a positive integer).

l must be less than n, while m_l must be less than or equal to l.

63. C is correct.

The mass number (A) is the total number of nucleons (i.e., protons and neutrons) in an atom.

The atomic number (Z) is the number of protons in an atom.

The number of neutrons in an atom can be calculated by subtracting the atomic number (Z) from the mass number (A).

Mass number – atomic number = number of neutrons

64. B is correct.

A compound consists of two or more different types of atoms that associate via chemical bonds.

An element is a pure chemical substance that consists of a single type of atom, defined by its atomic number (Z), which is the number of protons.

One hundred eighteen elements have been identified, of which the first 94 occur naturally on Earth.

65. E is correct.

Alkali metals (group IA) include lithium (Li), potassium (K), sodium (Na), rubidium (Rb), cesium (Cs) and francium (Fr).

Alkaline earth metals (group IIA) include beryllium (Be), magnesium (Mg), calcium (Ca), strontium (Sr), barium (Ba), and radium (Ra).

Halogens (Group VIIA) include fluorine (F), chlorine (Cl), bromine (Br), iodine (I), and astatine (At). Halogens gain one electron to become –1 anion, and the resulting ion has a complete octet of valence electrons.

Noble gases (group VIIIA) include helium (He), neon (Ne), argon (Ar), krypton (Kr), xenon (Xe), radon (Rn), and oganesson (Og). Except for helium (which has a complete octet with 2 electrons, $1s^2$), the noble gases have complete octets with ns^2 and np^6 orbitals.

Representative elements on the periodic table are groups IA and IIA (on the left) and groups IIIA – VIIIA (on the right).

Polonium (Po) element 84 is highly radioactive, with no stable isotopes, and is classified as either a metalloid or a metal.

66. B is correct.

Metalloids are semimetallic elements (i.e., between metals and nonmetals). The metalloids are boron (B), silicon (Si), germanium (Ge), arsenic (As), antimony (Sb), and tellurium (Te). Some literature reports polonium (Po) and astatine (At) as metalloids.

They have properties between metals and nonmetals. They typically have a metallic appearance but are only fair conductors of electricity (as opposed to metals that are excellent conductors), making them useable in the semiconductor industry.

Metalloids tend to be brittle, and chemically they behave more like nonmetals.

Alkali metals (group IA) include lithium (Li), potassium (K), sodium (Na), rubidium (Rb), cesium (Cs) and francium (Fr).

Lewis dot structures showing valence electrons for elements

Alkaline earth metals (group IIA) include beryllium (Be), magnesium (Mg), calcium (Ca), strontium (Sr), barium (Ba), and radium (Ra).

Seventeen elements are generally classified as nonmetals. Eleven are gases: hydrogen (H), helium (He), nitrogen (N), oxygen (O), fluorine (F), neon (Ne), chlorine (Cl), argon (Ar), krypton (Kr), xenon (Xe) and radon (Rn). One nonmetal is a liquid – bromine (Br) – and five are solids: carbon (C), phosphorus (P), sulfur (S), selenium (Se), and iodine (I).

Nonmetals tend to be highly volatile (i.e., easily vaporized), have low elasticity, and are good insulators of heat and electricity. Nonmetals tend to have high ionization energy and electronegativity and share (or gain) an electron when bonding with other elements.

Halogens (Group VIIA) include fluorine (F), chlorine (Cl), bromine (Br), iodine (I), and astatine (At). Halogens gain one electron to become a –1 anion, and the resulting ion has a complete octet of valence electrons.

67. D is correct.

In general, the size of neutral atoms increases down a group (i.e., increasing shell size) and decreases from left to right across the periodic table.

Positive ions (cations) are *much smaller* than the neutral element (due to the greater effective nuclear charge). In comparison, negative ions (anions) are *much larger* (due to the smaller effective nuclear charge and repulsion of valence electrons).

Sulfur (S, atomic number = 16) is smaller than aluminum (Al, atomic number = 13) due to the increase in the number of protons (effective nuclear charge) from left to right across a period (i.e., horizontal rows).

Al^{3+} has the same electronic configuration as Ne ($1s^22s^22p^6$) compared to Al ($1s^22s^22p^63s^23p^1$).

68. A is correct.

The valence shell is the outermost shell (i.e., highest principal quantum number n) of an atom.

Valence electrons are those electrons of the outermost electron shell that can participate in a chemical bond.

The number of valence electrons for an element can be determined by its group (i.e., vertical column) on the periodic table. Except for the transition metals (i.e., groups 3-12), the group number identifies how many valence electrons are associated with a particular element: elements of the same group have the same number of valence electrons.

69. B is correct.

The semimetallic elements are arsenic (As), antimony (Sb), bismuth (Bi), and graphite (a crystalline form of carbon). Arsenic and antimony are considered metalloids (along with boron, silicon, germanium, and tellurium), but the terms semimetal and metalloid are not synonymous.

Semimetals, in contrast to metalloids, can be chemical compounds.

70. C is correct.

Noble gases (group VIIIA) include helium (He), neon (Ne), argon (Ar), krypton (Kr), xenon (Xe), radon (Rn), and oganesson (Og).

Except for helium (which has a complete octet with 2 electrons, $1s^2$), the noble gases have complete octets with ns^2 and np^6 orbitals.

The metalloids are boron (B), silicon (Si), germanium (Ge), arsenic (As), antimony (Sb), and tellurium (Te). Some literature reports polonium (Po) and astatine (At) as metalloids.

71. D is correct.

Isotopes are variants of a particular element that differ in the number of neutrons. Isotopes of the element have the same number of protons and occupy the same position on the periodic table.

The experimental results should depend on the mass of the gas molecules.

Deuterium (D or ^{2}H) is known as *heavy hydrogen*. It is one of two stable isotopes of hydrogen. The deuterium nucleus contains one proton and one neutron, compared to H, which has 1 proton and 0 neutrons. The mass of deuterium is 2.0141 daltons, compared to 1.0078 daltons for hydrogen.

Based on the difference of mass between the isotopes, the density, rate of gas effusion, and atomic vibrations would be different.

72. E is correct.

Elements are defined by the number of protons (i.e., atomic number).

The isotopes are neutral atoms: # electrons = # protons.

Isotopes are variants of a particular element that differ in the number of neutrons. Isotopes of the element have the same number of protons and occupy the same position on the periodic table.

The superscript on the left denotes the number of protons and neutrons.

Since naturally occurring lithium has a mass of 6.9 g/mol and both protons and neutrons have a mass of approximately 1 g/mol, ^{7}lithium is the predominant isotope.

73. A is correct.

The charge on 1 electron is negative.

One mole of electrons:

Avogadro's number × e$^-$.

$(6.02 \times 10^{23}) \times (1.60 \times 10^{-19}) = 96{,}485$ coulombs (values are rounded).

74. E is correct.

Each orbital can hold two electrons.

The *f* subshell has 7 orbitals and can accommodate 14 electrons.

The *d* subshell has 5 lobed orbitals and can accommodate 10 electrons.

The $n = 3$ shell contains only *s*, *p* and *d* subshells.

75. B is correct.

Ionization energy (IE) is the amount of energy required to remove the most loosely bound electron of an isolated gaseous atom to form a cation. This is an endothermic process.

Ionization energy is expressed as:

$$X + energy \rightarrow X^+ + e^-$$

where X is an atom (or molecule) capable of being ionized (i.e., having an electron removed), X^+ is that atom or molecule after an electron is removed, and e$^-$ is the removed electron.

The principal quantum number (n) describes the size of the orbital and the energy of an electron, and the most probable distance of the electron from the nucleus. It refers to the size of the orbital and the energy level of an electron.

The elements with larger shell sizes (n is large) listed at the bottom of the periodic table have low ionization energies. This is due to the shielding (by the inner shell electrons) from the positive charge of the nucleus. The greater the distance between the electrons and the nucleus, the less energy is needed to remove the outer valence electrons.

Ionization energy decreases with increasing shell size (i.e., n value) and generally increases to the right across a period (i.e., row) in the periodic table.

Chlorine (Cl) has an atomic number of 10 and a shell size of $n = 3$.

Francium (Fr) has an atomic number of 87 and a shell size of $n = 7$.

Gallium (Ga) has an atomic number of 31 and a shell size of $n = 4$.

Iodine (I) has an atomic number of 53 and a shell size of $n = 5$.

Cesium (Cs) has an atomic number of 55 and a shell size of $n = 6$.

76. D is correct.

Electrons are electrostatically (i.e., negative and positive charge) attracted to the nucleus and an atom's electrons generally occupy outer shells only if other electrons have already filled the more inner shells. However, there are exceptions to this rule, with some atoms having two or even three incomplete outer shells.

The Aufbau (German for building up) principle is based on the Madelung rule to fill the subshells based on the lowest energy levels.

Order of filling electron's orbitals

77. C is correct.

In Bohr's atom model, electrons can jump to higher energy levels, gain energy, or drop to lower energy levels, releasing energy. When electric current flows through an element in the gas phase, glowing light is produced.

By directing this light through a prism, a pattern of lines known as the *atomic spectra* can be seen.

These lines are produced by an excited electron dropping to lower energy levels. Since the energy levels in each element are different, each element has a unique set of lines it produces, which is why the spectrum is called the "atomic fingerprint" of the element.

78. E is correct.

Obtain the atomic number of Mn from the periodic table.

Mn is a transition metal, and it is located in Group VIIB/7; its atomic number is 25.

Use the Aufbau principle to fill up the orbitals of Mn: $1s^2 2s^2 2p^6 3s^2 3p^6 4s^2 3d^5$

The transition metals occur in groups 3–12 (vertical columns) of the periodic table. They occur in periods 4–7 (horizontal rows).

Transition metals are defined as elements that have a partially-filled d or f subshell in a common oxidative state. This group of elements includes silver, iron, and copper.

The transition metals are elements whose atom has an incomplete d sub-shell, giving rise to cations with an incomplete d sub-shell. By this definition, elements in groups 3–11 (or 12 by some literature) are transition metals.

The transition elements have characteristics that are not found in other elements, resulting from the partially filled d shell. These include the formation of compounds whose color is due to d–d electronic transitions and compounds in many oxidation states due to the relatively low reactivity of unpaired d electrons.

The transition elements form many paramagnetic (i.e., attracted to an externally applied magnetic field) compounds due to the presence of unpaired d electrons. By exception to their unique traits, a few compounds of main group elements are paramagnetic (e.g., nitric oxide and oxygen).

79. E is correct.

Electronegativity is defined as the ability of an atom to attract electrons when it bonds with another atom. The most common use of electronegativity pertains to polarity along the sigma (single) bond.

The trend for increasing electronegativity within the periodic table is up and toward the right. The most electro-negative atom is fluorine (F), while the least electronegative atom is francium (Fr).

80. C is correct.

Alkali metals (group IA) include lithium (Li), potassium (K), sodium (Na), rubidium (Rb), cesium (Cs) and francium (Fr). Alkali metals lose one electron to become +1 cations, and the resulting ion has a complete octet of valence electrons.

Alkaline earth metals (group IIA) include beryllium (Be), magnesium (Mg), calcium (Ca), strontium (Sr), barium (Ba), and radium (Ra). Alkaline earth metals lose two electrons to become +2 cations, and the resulting ion has a complete octet of valence electrons.

===

Practice Set 5: Questions 81–104

===

81. C is correct.

The mass number (A) is the sum of protons and neutrons in an atom.

The mass number is an approximation of the atomic weight of the element as amu (grams per mole).

The mass number is already provided by the problem: ^{79}Br means the mass number is 79.

The atomic number of ^{79}Br can be obtained from the periodic table. Br is in group VIIA/17, and its atomic number is 35.

82. A is correct.

The mass number (A) is the total number of nucleons (i.e., protons and neutrons) in an atom.

The number of protons and neutrons is denoted by the superscript on the left.

The atomic number (Z) is the number of protons in an atom.

The number of neutrons in an atom can be calculated by subtracting the atomic number (Z) from the mass number (A).

83. C is correct.

An element is a pure chemical substance consisting of a single type of atom, distinguished by its atomic number (Z) (i.e., the number of protons it contains). One hundred eighteen elements have been identified, of which the first 94 occur naturally on Earth, with the remaining 24 being synthetic elements.

The properties of the elements on the periodic table repeat at regular intervals, creating "groups" or "families" of elements. Each column on the periodic table is a group, and elements within each group have similar physical and chemical characteristics due to the orbital location of their outermost electron. These groups only exist because the elements of the periodic table are listed by increasing atomic numbers.

84. B is correct.

Alkali metals (group IA) include lithium (Li), potassium (K), sodium (Na), rubidium (Rb), cesium (Cs) and francium (Fr).

Alkali metals lose one electron to become +1 cations, and the resulting ion has a complete octet of valence electrons.

The alkali metals have low electronegativity and react violently with water (e.g., metallic sodium with water).

85. A is correct.

Seventeen elements are generally classified as nonmetals. Eleven are gases: hydrogen (H), helium (He), nitrogen (N), oxygen (O), fluorine (F), neon (Ne), chlorine (Cl), argon (Ar), krypton (Kr), xenon (Xe) and radon (Rn). One nonmetal is a liquid – bromine (Br) – and five are solids: carbon (C), phosphorus (P), sulfur (S), selenium (Se), and iodine (I).

Nonmetals tend to be highly volatile (i.e., easily vaporized), have low elasticity, and are good insulators of heat and electricity.

Nonmetals tend to have high ionization energy and electronegativity and share (or gain) an electron when bonding with other elements.

86. E is correct.

Group VIA (16) has three nonmetals: oxygen, sulfur, and selenium.

Metalloids are semimetallic elements (i.e., between metals and nonmetals). The metalloids are boron (B), silicon (Si), germanium (Ge), arsenic (As), antimony (Sb), and tellurium (Te). Some literature reports polonium (Po) and astatine (At) as metalloids.

87. C is correct.

Transition metals (or transition elements) are defined as partially-filled *d* or *f* subshell elements in a common oxidative state.

Transition metals occur in groups (vertical columns) 3–12 of the period table. They occur in periods (horizontal rows) 4–7. This group of elements includes silver, iron, and copper.

The *f*-block lanthanides (i.e., rare earth metals) and actinides (i.e., radioactive elements) are considered transition metals and are known as inner transition metals.

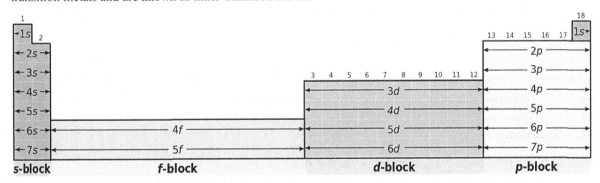

Transition elements have characteristics that are not found in other elements, which result from the partially filled *d* shell. These include the formation of compounds whose color is due to *d* electronic transitions, the formation of compounds in many oxidation states due to the relatively low reactivity of unpaired *d* electrons. The incomplete *d* sub-shell can give rise to cations with an incomplete *d* sub-shell.

Transition elements form many paramagnetic (i.e., attracted to an externally applied magnetic field) compounds due to the presence of unpaired *d* and *f* electrons. A few compounds of main group elements are paramagnetic (e.g., nitric oxide and oxygen).

88. E is correct.

An element is a pure chemical substance that consists of a single type of atom, distinguished by its atomic number (Z) for the number of protons.

One hundred eighteen elements have been identified, of which the first 94 occur naturally on Earth.

89. B is correct.

Isotopes are variants of an element that differ in the number of neutrons.

Isotopes of the element have the same number of protons and occupy the same position on the periodic table.

Alpha decay results in the loss of two protons.

Beta-decay is a type of radioactive decay in which a neutron is transformed into a proton, or a proton is transformed into a neutron.

Since isotopes of the same element have the same number of protons (Z), the number of protons lost by α decay must equal the number gained by β decay.

Therefore, twice as many β decays as α decays occurs for a ratio of 1:2 for α to β decay.

90. B is correct.

Elements are defined by the number of protons (i.e., atomic number).

The isotopes are neutral atoms: # electrons = # protons.

Isotopes are variants of a particular element that differ in the number of neutrons. Isotopes of the element have the same number of protons and occupy the same position on the periodic table.

Cu has an atomic weight of 63.5 grams.

The other isotope of Cu must be heavier than the more common ^{63}Cu, and the atomic weight is closer to 65.

91. E is correct.

A cathode-ray particle is another name for an electron. Those particles (i.e., electrons) are attracted to the positively charged cathode, implying that they are negatively charged.

92. A is correct.

Pauli exclusion principle is the quantum principle that states that two identical electrons cannot have the same four quantum numbers: the principal quantum number (n), the angular momentum quantum number (l), the magnetic quantum number (m_ℓ), and the spin quantum number (m_s). For two electrons in the same orbital (n, m_ℓ, and l), the spin quantum number (m_s) must be different, and the electrons must have opposite half-integer spins (i.e., + ½ and –½).

Hund's rule describes that the electrons enter each orbital of a given type singly and with identical spins before any pairing of electrons of the opposite spin occurs within those orbitals.

Heisenberg's uncertainty principle states that it is impossible to accurately determine both the momentum and the position of an electron simultaneously.

93. E is correct.

The atom's second electron is $n = 2$; it has 2 orbitals: *s, p*

Each orbital can hold two electrons.

Maximum number of electrons in $n = 2$ shell is:

The *s* subshell has 1 spherical orbital and can accommodate 2 electrons.

The *p* subshell has 3 dumbbell-shaped orbitals and can accommodate 6 electrons.

Total number of electrons:

2 (for *s*) + 6 (for *p*) = 8 electrons

94. D is correct.

Each orbital can hold two electrons.

There are 5 *d* orbitals that can accommodate 10 electrons.

95. D is correct.

The number of orbitals in a subshell is different from the maximum number of electrons in the subshell.

The *f* subshell has 7 orbitals.

96. A is correct.

Only photons are released as electrons move from higher to lower energy orbitals.

Alpha particles are helium nuclei, and beta particles are electrons.

Gamma rays (i.e., gamma radiation) are ionizing radiation produced from the decay of an atomic nucleus from a high energy state to a lower energy state. They are biologically hazardous.

97. B is correct.

According to the Bohr model and quantum theory, the energy levels for electrons occur at quantified and calculable energies. For hydrogen, the energy of each level is found by:

$$E = -13.6 \text{ eV}/n^2$$

The negative sign by convention indicates that this is the amount of energy needed to ionize the electron completely.

When promoted by absorbing energy, a single electron can be at higher energy levels.

98. E is correct.

Ions often have the same electronic configuration (i.e., isoelectronic) as neutral atoms of a different element (e.g., F⁻ is isoelectronic with Ne). However, this is not valid for elements with electrons in excited states.

An ion can have an electron configuration that is consistent with the rules of quantum numbers.

99. A is correct.

Consider the last 2 orbitals: $3s^2 3p^6$.

This means that the atom's valence electrons are $2 + 6 = 8$ electrons.

Atoms with 8 valence electrons (i.e., complete octet) are very stable – chemically inert.

100. C is correct.

Ionization energy (IE) is the amount of energy required to remove an isolated gaseous atom's most loosely bound electron to form a cation. This is an endothermic process.

Ionization energy is expressed as:

$$X + \text{energy} \rightarrow X^+ + e^-$$

where X is an atom (or molecule) capable of being ionized (i.e., having an electron removed), X^+ is that atom or molecule after an electron is removed, and e^- is the removed electron.

The principal quantum number (n) describes the size of the orbital and the energy of an electron, and the most probable distance of the electron from the nucleus. It refers to the size of the orbital and the energy level of an electron.

Elements with larger shells (n is large) listed at the bottom of the periodic table have low ionization energies. This is due to the shielding (by the inner shell electrons) from the positive charge of the nucleus. The greater the distance between the electrons and the nucleus, the less energy is needed to remove the outer valence electrons.

Halogens have high ionization energy because they have 7 electrons; they only need to gain one more electron to reach stability (i.e., complete octet). Therefore, halogens have a large electron affinity, which measures the tendency for ions to gain electrons. More energy is required to remove the 7th electron (compared to the 1st or 2nd electrons, such as for elements in group IA or IIA).

101. D is correct.

Chlorine completes its octet when it gains an electron, so chlorine atoms form anions (i.e., gain electrons) much more easily than cations (i.e., lose electrons).

102. B is correct.

The mass number (A) is the sum of protons and neutrons in an atom.

The mass number is an approximation of the atomic weight of the element as amu (grams per mole).

103. A is correct.

The atomic mass of H is 1.0 amu, and a single proton has a mass of 1 amu, the most common isotope contains no neutrons. Hydrogen ions are referred to as protons (H+) with zero neutrons in the nucleus in acid-base chemistry.

104. A is correct. Oxygen contains only s and p orbitals and cannot form six bonds due to the lack of d orbitals. Second-period elements lack d orbitals and cannot have more than four bonds to the central atom.

Notes for active learning

Notes for active learning

Chemical Bonding – Detailed Explanations

==

Practice Set 1: Questions 1–20

==

1. D is correct.

The valence shell is the outermost shell (i.e., highest principal quantum number n) of an atom.

Valence electrons are those electrons of the outermost electron shell that can participate in a chemical bond.

The number of valence electrons for an element can be determined by its group (i.e., vertical column) on the periodic table. Except for the transition metals (i.e., groups 3-12), the group number identifies how many valence electrons are associated with a particular element: elements of the same group have the same number of valence electrons.

2. A is correct.

Three degenerate p orbitals exist for an atom with an electron configuration in the second shell or higher. The first shell only has access to s orbitals.

The d orbitals become available from $n = 3$ (third shell).

3. C is correct.

The valence shell is the outermost shell (i.e., highest principal quantum number n) of an atom.

Valence electrons are those electrons of the outermost electron shell that can participate in a chemical bond.

The number of valence electrons for an element can be determined by its group (i.e., vertical column) on the periodic table.

Except for the transition metals (i.e., groups 3-12), the group number identifies how many valence electrons are associated with a particular element: elements of the same group have the same number of valence electrons.

To find the number of valence electrons in a sulfite ion, SO_3^{2-}, add the valence electrons of each atom:

Sulfur = 6; Oxygen = $(6 \times 3) = 18$

Total = 24

This ion has a net charge of -2, which indicates that it has 2 extra electrons.

Therefore, thel number of valence electrons would be $24 + 2 = 26$ electrons.

4. E is correct.

London dispersion forces result from the momentary flux of valence electrons and are present in compounds; they are the attractive forces that hold molecules together.

They are the weakest of the intermolecular forces, and their strength increases with increasing size (i.e., surface area contact) and polarity of the molecules involved.

5. B is correct.

Each hydroxyl group (alcohol or ~OH) has oxygen with 2 lone pairs and one attached hydrogen.

Therefore, each hydroxyl group can participate in 3 hydrogen bonds:

5 hydroxyl groups × 3 bonds = 15 hydrogen bonds.

The oxygen of the ether group (C–O–C) in the ring has 2 lone pairs for an additional 2 H–bonds.

6. B is correct.

The valence shell is the outermost shell (i.e., highest principal quantum number *n*) of an atom.

Valence electrons are those electrons of the outermost electron shell that can participate in a chemical bond.

The number of valence electrons for an element can be determined by its group (i.e., vertical column) on the periodic table. Except for the transition metals (i.e., groups 3-12), the group number identifies how many valence electrons are associated with a particular element: elements of the same group have the same number of valence electrons.

Lewis dot structure for ethane

7. E is correct.

In covalent bonds, the electrons can be shared either equally or unequally.

Polar covalent bonded atoms are covalently bonded compounds that involve unequal sharing of electrons due to large electronegativity differences (Pauling units of 0.4 to 1.7) between the atoms.

An example of this is water, where there is a polar covalent bond between oxygen and hydrogen. Water is a polar molecule with the oxygen partial negative while the hydrogens are partial positive.

8. A is correct.

Van der Waals forces involve nonpolar (hydrophobic) molecules, such as hydrocarbons. The van der Waals force is the total of attractive or repulsive forces between molecules, and therefore can be either attractive or repulsive. It can include the force between two permanent dipoles, the force between a permanent dipole and a temporary dipole, or the force between two temporary dipoles.

Hydrogens, bonded directly to F, O or N, participate in hydrogen bonds. The hydrogen is partial positive (i.e., delta plus or ∂+) due to the bond to these electronegative atoms. The lone pair of electrons on the F, O or N interacts with the ∂+ hydrogen to form a hydrogen bond.

Hydrogen bonds are a type of dipole–dipole and are the strongest intermolecular forces (i.e., between molecules), followed by other types of dipole–dipole, dipole–induced dipole, and van der Waals forces (i.e., London dispersion).

9. D is correct.

Representative structures include:

$HNCH_2$: one single and one double bond

NH_3 : three single bonds

HCN : one triple bond and one single bond

10. E is correct.

The valence shell is the outermost shell (i.e., highest principal quantum number n) of an atom.

Valence electrons are those electrons of the outermost electron shell that can participate in a chemical bond.

Sample Lewis dot structures for some elements

The number of valence electrons for an element can be determined by its group (i.e., vertical column) on the periodic table. Except for the transition metals (i.e., groups 3-12), the group number identifies how many valence electrons are associated with an element: elements of the same group have the same number of valence electrons.

11. B is correct.

The nitrite ion has the chemical formula NO_2^- with the negative charge distributed between the two oxygen atoms.

Two resonance structures of the nitrite ion

12. E is correct.

Nitrogen has 5 valence electrons. In ammonia, nitrogen has 3 bonds to hydrogen, and a lone pair remains on the central nitrogen atom.

The ammonium ion (NH_4^+) has 4 hydrogens, and the lone pair of the nitrogen has been used to bond to the H^+ that has added to ammonia.

Formal charge is shown on the ammonium ion

13. E is correct.

In atoms or molecules that are ions, the number of electrons is not equal to the number of protons, which gives the molecule a charge (either positive or negative).

Ionic bonds form when the difference in electronegativity of atoms in a compound is greater than 1.7 Pauling units.

Ionic bonds involve transferring an electron from the electropositive element (along the left-hand column/group) to the electronegative element (along the right-hand column/groups) on the periodic table.

Oppositely charged ions are attracted to each other, and this attraction results in ionic bonds.

In an ionic bond, electron(s) are transferred from a metal to a nonmetal, giving both molecules a full valence shell and causing the molecules to associate with each other closely.

14. A is correct.

Electronegativity is a chemical property that describes an atom's tendency to attract electrons to itself.

The most electronegative atom is F, while the least electronegative atom is Fr. The trend for increasing electronegativity within the periodic table is up and toward the right (i.e., fluorine).

15. D is correct.

Hydrogen bonds are the strongest intermolecular forces (i.e., between molecules), followed by dipole–dipole, dipole–induced dipole, and van der Waals forces (i.e., London dispersion).

London dispersion forces are present in compounds; they are the attractive forces that hold molecules together. They are the weakest of the intermolecular forces, and their strength increases with the increasing size and polarity of the molecules involved.

16. C is correct.

For asymmetrical molecules (e.g., water is bent), use the geometry and difference of electronegativity values between the atoms.

For water, H (2.1 Pauling units) is more electropositive than O (3.5 Pauling units).

For sulfur dioxide, S (2.5 Pauling units) is more electropositive than O (3.5 Pauling units).

The reported dipole moment for water is 1.8 D compared to 1.6 D for sulfur dioxide.

CO_2 does not have any nonbonding electrons on the central carbon atom, and it is symmetrical, so it is non-polar and has zero dipole moment. CCl_4 is symmetrical as a tetrahedron and therefore does not exhibit a net dipole.

CH_4 is symmetrical as a tetrahedron and therefore does not exhibit a net dipole.

17. C is correct.

Cohesion is the property of like molecules sticking together. Hydrogen bonds join water molecules.

Adhesion is the attraction between unlike molecules (e.g., the meniscus observed from water molecules adhering to the graduated cylinder).

Polarity is the differences in electronegativity between bonded molecules. Polarity gives rise to the delta plus (on H) and the delta minus (on O), which permits hydrogen bonds to form between water molecules.

18. A is correct.

With little or no difference in electronegativity between the elements (i.e., Pauling units < 0.4), it is a nonpolar covalent bond, whereby the electrons are shared between the two bonding atoms.

Among the answer choices, the atoms of H, C and O are closest in magnitude for Pauling units for electronegativity.

19. D is correct.

Positively charged nuclei repel each other while each attracts the bonding electrons. These opposing forces reach equilibrium at the bond length.

20. C is correct.

In a carbonate ion (CO_3^{2-}), the carbon atom is bonded to 3 oxygen atoms. Two of those bonds are single covalent bonds, and the oxygen atoms each have an extra (third) lone pair of electrons, which imparts a negative formal charge. The remaining oxygen has a double bond with carbon.

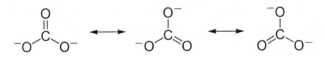

Three resonance structures for the carbonate ion CO_3^{2-}

Practice Set 2: Questions 21–40

21. D is correct.

A dipole is a separation of full (or partial) positive and negative charges due to differences in electronegativity of atoms.

22. C is correct.

Ionic bonds involve transferring an electron from the electropositive element (along the left-hand column/group) to the electronegative atom (along the right-hand column/groups) on the periodic table.
Ca is a group II element with 2 electrons in its valence shell.
I is a group VII element with 7 electrons in its valence shell.
Ca becomes Ca^{2+}, and each of the two electrons is joined to I, which becomes I^-.

23. B is correct.

An atom with 4 valence electrons can make a maximum of 4 bonds. 1 double and 1 triple bond equals 5 bonds, which exceeds the maximum allowable bonds.

24. D is correct.

Water molecules stick to each other (i.e., cohesion) due to the collective action of hydrogen bonds between individual water molecules. These hydrogen bonds are constantly breaking and reforming; many molecules are held together by these bonds.

Water sticks to surfaces (i.e., adhesion) because of water's polarity. On an extremely smooth surface (e.g., glass), the water may form a thin film because the molecular forces between glass and water molecules (adhesive forces) are stronger than the cohesive forces between the water molecules.

25. D is correct.

Hydrogen bonds are the strongest intermolecular forces (i.e., between molecules), followed by dipole–dipole, dipole–induced dipole, and van der Waals forces (i.e., London dispersion).

When ionic compounds are dissolved, each ion is surrounded by more than one water molecule. The combined force ion–dipole interactions of several water molecules are stronger than a single ionic bond.

26. E is correct.

Hydrogens, bonded directly to F, O or N, participate in hydrogen bonds. The hydrogen is partial positive (i.e., delta plus or ∂+) due to the bond to these electronegative atoms. The lone pair of electrons on the F, O or N interacts with the ∂+ hydrogen to form a hydrogen bond.

The two lone pairs of electrons on the oxygen atom can each participate as a hydrogen bond acceptor. The molecule does not have hydrogen bonded directly to an electronegative atom (F, O or N) and cannot be a hydrogen bond donor.

27. A is correct.

Water is a bent molecule with a partial negative charge on the oxygen and a partial positive charge on each hydrogen. Therefore, the water molecule exists as a dipole. The Na^+ is an ion attracted to the partial negative charge on the oxygen in the water molecule.

28. C is correct.

The molecule H_2CO (formaldehyde) is shown below and has one C=O bond and two C–H bonds. The electronegative oxygen pulls electron density away from the carbon atom and creates a net dipole towards the oxygen, resulting in a polar covalent bond.

The electronegativity values of carbon and each hydrogen are similar and result in a covalent bond (i.e., about equal sharing of bonded electrons).

29. A is correct.

The valence shell is the outermost shell (i.e., highest principal quantum number n) of an atom.

Valence electrons are those electrons of the outermost electron shell that can participate in a chemical bond.

The number of valence electrons for an element can be determined by its group (i.e., vertical column) on the periodic table.

Except for the transition metals (i.e., groups 3-12), the group number identifies how many valence electrons are associated with a particular element: elements of the same group have the same number of valence electrons.

30. D is correct.

Group IA elements (e.g., Li, Na, K) tend to lose 1 electron to achieve a complete octet to be cations with a +1 charge.

Group IIA elements (e.g., Mg, Ca) tend to lose 2 electrons to achieve a complete octet to be cations with a +2 charge.

Group VIIA elements (halogens such as F, Cl, Br and I) tend to gain 1 electron to achieve a complete octet to be anions with a –1 charge.

31. B is correct.

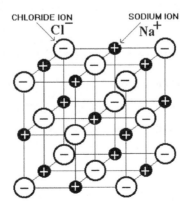

Lattice structure of sodium chloride

32. E is correct.

Each positively charged nuclei attracts the bonding electrons.

33. E is correct.

Electronegativity is a measure of how strongly an element attracts electrons within a bond.

Electronegativity is the relative attraction of the nucleus for bonding electrons. It increases from left to right (i.e., periods) and from bottom to top along a group (similar to the trend for ionization energy). The most electronegative atom is F, while the least electronegative atom is Fr.

The greater the difference in electronegativity between two atoms in a compound, the more polar of a bond these atoms form. The atom with the higher electronegativity is the partial (delta) negative end of the dipole.

34. A is correct.

Hydrogen bonds are the strongest intermolecular forces (i.e., between molecules), followed by dipole–dipole, dipole–induced dipole and van der Waals forces (i.e., London dispersion).

Hydrogens, bonded directly to F, O or N, participate in hydrogen bonds. H–bonding is a polar interaction involving hydrogen forming bonds to the electronegative atoms F, O or N, which accounts for the high boiling points of water. The hydrogen is partial positive (i.e., delta plus or ∂+) due to the bond to these electronegative atoms. The lone pair of electrons on the F, O or N interacts with the ∂+ hydrogen to form a hydrogen bond.

Molecular geometry of H₂S

Polar molecules have high boiling points because of polar interaction. H_2S is a polar molecule but does not form hydrogen bonds; it forms dipole–dipole interactions.

35. D is correct.

With 4 electron pairs, the starting shape is tetrahedral. After removing 2 of the 4 groups surrounding the central atom, the central atom has two groups bound (e.g., H_2O). The molecule has the hybridization (and original angles) from a tetrahedral shape, bent rather than linear.

36. D is correct.

Ionic bonds hold salt crystals. Salts are composed of cations and anions and are electrically neutral. When salts are dissolved in solution, they separate into their constituent ions by breaking of noncovalent interactions.

Water: the hydrogen atoms in the molecule are covalently bonded to the oxygen atom.

Hydrogen peroxide: the molecule has one more oxygen atom than a water molecule and held by covalent bonds.

Ester (RCOOR'): undergo hydrolysis and breaks covalent bonds to separate into its constituent carboxylic acid (RCOOH) and alcohol (ROH).

37. B is correct.

An ionic compound consists of a metal ion and a nonmetal ion.

Ionic bonds are formed between elements with an electronegativity difference greater than 1.7 Pauling units (e.g., a metal atom and a non-metal atom).

(K and I) form ionic bonds because K is metal and I is a nonmetal.

(C and Cl) and (H and O) only have nonmetals, while (Fe and Mg) have only metals.

(Ga and Si) has a metal (Ga) and a metalloid (Si); they *might* form weak ionic bonds.

38. A is correct.

A coordinate bond is a covalent bond (i.e., shared pair of electrons) in which both electrons come from the same atom.

$$H \overset{\cdot\cdot}{\underset{H}{\times N \times}} H \;+\; [H]^{+} \longrightarrow H \overset{H}{\underset{H}{\overset{\cdot\cdot}{\times N \times}}} H$$

The lone pair of the nitrogen is donated to form the fourth N–H bond

39. B is correct.

Dipole moment depends on the overall shape of the molecule, the length of the bond, and whether the electrons are pulled to one side of the bond (or the molecule overall).

For a large dipole moment, one element pulls electrons more strongly (i.e., differences in electronegativity).

Electronegativity is a measure of how strongly an element attracts electrons within a bond.

Electronegativity is the relative attraction of the nucleus for bonding electrons. It increases from left to right (i.e., periods) and from bottom to top along a group (similar to the trend for ionization energy). The most electronegative atom is F, while the least electronegative atom is Fr.

The greater the difference in electronegativity between two atoms in a compound, the more polar a bond these atoms form. The atom with the higher electronegativity is the partial (delta) negative end of the dipole.

40. A is correct.

The valence shell is the outermost shell (i.e., highest principal quantum number n) of an atom.

The noble gas configuration refers to eight electrons in the atom's outermost valence shell, referred to as a complete octet.

Depending on how many electrons it starts with, an atom may have to lose, gain or share an electron to obtain the noble gas configuration.

===

Practice Set 3: Questions 41–60

===

41. E is correct.

The bond between the oxygens is nonpolar, while the bonds between the oxygens and hydrogens are polar (due to the differences in electronegativity).

Line bond structure of H_2O_2 with lone pairs shown

42. A is correct.

The octet rule states that atoms of main-group elements tend to combine so that each atom has eight electrons in its valence shell. This occurs because electron arrangements involving eight valence electrons are extremely stable, as is the case with noble gasses.

Selenium (Se) is in group VI and has 6 valence electrons. By gaining two electrons, it has a complete octet (i.e., stable).

43. C is correct.

When one or more new compounds are formed by rearranging atoms is an example of a chemical reaction.

44. D is correct.

HBr experiences dipole–dipole interactions due to the electronegativity difference resulting in a partial negative charge on bromine and a partial positive charge on hydrogen.

45. C is correct.

An ion is an atom (or a molecule) in which the number of electrons is not equal to the number of protons. Therefore, the atom (or molecule) has a net positive or negative electrical charge.

If a neutral atom loses one or more electrons, it has a net positive charge (i.e., cation).

If a neutral atom gains one or more electrons, it has a net negative charge (i.e., anion).

Aluminum is a group III atom and has proportionally more protons per electron once the cation forms.

All other elements listed are from group I or II.

46. A is correct.

Dipole–dipole (e.g., $CH_3Cl...CH_3Cl$) attraction occurs between neutral molecules, while ion–dipole interaction involves dipole interactions with charged ions (e.g., $CH_3Cl...^-OOCCH_3$).

Hydrogen bonds are the strongest intermolecular forces (i.e., between molecules), followed by dipole–dipole, dipole–induced dipole, and van der Waals forces (i.e., London dispersion).

47. E is correct.

When more than two H_2O molecules are present (e.g., liquid water), more bonds (between 2 and 4) are possible because the oxygen of one water molecule has two lone pairs of electrons, each of which can form a hydrogen bond with hydrogen on another water molecule.

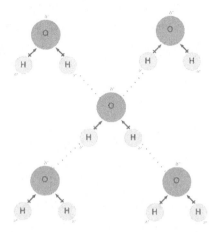

Ice with 4 hydrogen bonds between the water molecules

This bonding can repeat such that every water molecule is H–bonded with up to four other molecules (two through its two lone pairs of O, and two through its two hydrogen atoms).

Like other substances, when liquid water is cooled from room temperature, it becomes increasingly dense. However, at approximately 4 °C (39 °F), water reaches its maximum density, and as it is cooled further, it expands and becomes less dense. This phenomenon is *negative thermal expansion* and is attributed to strong intermolecular interactions that are orientation-dependent.

The density of water is about 1 g/cm^3 and depends on the temperature. When frozen, the density of water is decreased by about 9%. This is due to the decrease in intermolecular vibrations, allowing water molecules to form stable hydrogen bonds with other water molecules around. As these hydrogen bonds form, molecules are locking into positions similar to the hexagonal structure.

Even though hydrogen bonds are shorter in the crystal than in the liquid, this position locking decreases the average coordination number of water molecules as the liquid reaches the solid phase.

48. A is correct.

The phrase "from its elements" in the question stem implies the need to create the formation reaction of the compound from its elements.

Chemical equation:

$$3\ Na^+ + N^{3-} \rightarrow Na_3N$$

Each sodium loses one electron to form Na^+ ions.

Nitrogen gains 3 electrons to form the N^{3-} ion.

Na^+	
Na^+	N^{3-}
Na^+	

Schematic of an ionic compound

49. D is correct.

The valence shell is the outermost shell (i.e., highest principal quantum number n) of an atom.

Valence electrons are those electrons of the outermost electron shell that can participate in a chemical bond.

The number of valence electrons for a neutral element can be determined by its group (i.e., vertical column) on the periodic table. Except for the transition metals (i.e., groups 3-12), the group number identifies how many valence electrons are associated with a particular element: elements of the same group have the same number of valence electrons.

H ·								He:
Li ·	·Be·		·Ḃ·	·Ċ·	·N̈·	:Ö·	:F̈·	:N̈e:
Na·	·Mg·		·Äl·	·Si·	·P̈·	:S̈·	:C̈l·	:Är:
K ·	·Ca·		·Ga·	·Ge·	·Äs·	:Se·	:Br·	:Kr:

Sample Lewis dot structures for some neutral elements

Lewis dot structures show the valence electrons of an individual atom as dots around the symbol of the element. Non-valence electrons (i.e., inner shell electrons) are not represented in Lewis structures.

Mg^{2+} a total of ten electrons but zero valance (outer shell) electrons. The valence shell (i.e., 3s subshell) has zero electrons, and therefore there are no dots shown in the Lewis dot structure: Mg^{2+}

S^{2-} has eight valence electrons (same electronic configuration as Ar).

Ga^+ has two valence electrons (same electronic configuration as Ca).

Ar^+ has seven valence electrons (same electronic configuration as Cl).

F^- has eight valence electrons (same electronic configuration as Ne). Lewis dot structure for F^-.

50. B is correct.

The octet rule states that atoms of main-group elements tend to combine so that each atom has eight electrons in its valence shell.

51. D is correct.

In a chemical reaction, the bonds being formed are different from the ones broken.

52. D is correct.

Hydrogen sulfide (i.e., sewer gas) is H_2S.

Hydrosulfide (bisulfide) ion is HS^-. Sulfide ion is S^{2-}

53. D is correct.

An ionic compound consists of a metal ion and a nonmetal ion.

Ionic bonds are formed between elements with an electronegativity difference greater than 1.7 Pauling units (e.g., a metal atom and a non-metal atom).

The charge on bicarbonate ion (HCO_3) is −1, while the charge on Mg is +2.

The proper formula is $Mg(HCO_3)_2$

54. C is correct.

Lattice: crystalline structure – consists of unit cells

Unit cell: smallest unit of solid crystalline pattern

Covalent unit and *ionic unit* are not valid terms.

55. D is correct.

Formula to calculate dipole moment:

$$\mu = qr$$

where μ is dipole moment (coulomb-meter or C·m), q is a charge (coulomb or C), and r is the radius (meter or m)

Convert the unit of dipole moment from Debye to C·m:

$$0.16 \times 3.34 \times 10^{-30} = 5.34 \times 10^{-31}\,C·m$$

Convert the unit of radius to meter:

$$115\,pm \times 1 \times 10^{-12}\,m/pm = 115 \times 10^{-12}\,m$$

Rearrange the dipole moment equation to solve for q:

$$q = \mu / r$$

$$q = (5.34 \times 10^{-31}\,C·m) / (115 \times 10^{-12}\,m)$$

$$q = 4.64 \times 10^{-21}\,C$$

Express the charge in terms of electron charge (e).

$$(4.64 \times 10^{-21}\,C) / (1.602 \times 10^{-19}\,C/e) = 0.029\,e$$

In this NO molecule, oxygen is the more electronegative atom.

Therefore, the charge experienced by the oxygen atom is negative: –0.029e.

56. A is correct.

The greater the difference in electronegativity between two atoms in a compound, the more polar a bond these atoms form. The atom with the higher electronegativity is the partial (delta) negative end of the dipole.

57. C is correct.

Hybridization of the central atom and the corresponding molecular geometry:

Examples of sp^3 hybridized atoms with 4 substituents (CH4),
3 substituents (NH3) and two substituents (H2O)

Hybridization - Shape

sp – linear

sp^2 – trigonal planar

sp^3 – tetrahedral

sp^3d – trigonal bipyramid

58. C is correct.

Hydrogen bonds are the strongest intermolecular forces (i.e., between molecules), followed by dipole–dipole, dipole–induced dipole, and van der Waals forces (i.e., London dispersion).

Hydrogen is a very electropositive element, and O is a very electronegative element. Therefore, these elements will be attracted to each other, both within the same water molecule and between water molecules.

Ionic or covalent bonding can only form within a molecule and not between adjacent molecules.

59. E is correct.

In a compound, the sum of ionic charges must equal zero.

For the charges of K^+ and CO_3^{2-} to even out, there have to be 2 K^+ ions for every CO_3^{2-} ion.

The formula of this balanced molecule is K_2CO_3.

60. C is correct.

The substantial difference between the two forms of carbon (i.e., graphite and diamonds) is mainly due to their crystal structure, hexagonal for graphite and cubic for diamond.

The conditions to convert graphite into diamond are high pressure and high temperature. That is why creating synthetic diamonds is time-consuming, energy-intensive, and expensive since carbon is forced to change its bonding structure

==

Practice Set 4: Questions 61–88

==

61. E is correct.

Lewis acids are defined as electron-pair acceptors, whereas Lewis bases are electron-pair donors.

Boron has an atomic number of five and has a vacant $2p$ orbital to accept electrons.

62. D is correct.

CH_3SH experiences dipole–dipole interactions due to the electronegativity difference resulting in a partial negative charge on sulfur and a partial positive charge on hydrogen.

CH_3CH_2OH experiences dipole–dipole interactions (due to the electronegativity difference resulting in a partial negative charge on oxygen and a partial positive charge on hydrogen. This example is a type of dipole–dipole known as hydrogen bonding (when hydrogen is bonded directly to fluorine, oxygen, or nitrogen).

63. A is correct.

In carbon–carbon double bonds, there is an overlap of sp^2 orbitals and a p orbital on the adjacent carbon atoms.

The sp^2 orbitals overlap head-to-head as a *sigma* (σ) bond, whereas the p orbitals overlap sideways as a *pi* (π) bond.

Bond lengths and strengths (σ or π) depends on the size and shape of the atomic orbitals and the density of these orbitals to overlap effectively.

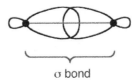

σ bond

Sigma bond formation showing electron density along the internuclear axis

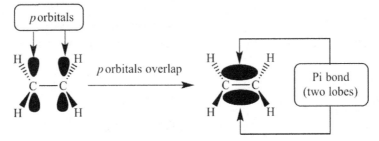

Two pi orbitals showing the pi bond formation during sideways overlap – note the absence of electron density (i.e., node) along the internuclear axis

The σ bonds are stronger than π bonds because head-to-head orbital overlap involves more shared electron density than sideways overlap. The σ bonds formed from two $2s$ orbitals are shorter than those formed from two $2p$ orbitals or two $3s$ orbitals.

Carbon, oxygen, and nitrogen are in the second period ($n = 2$), while sulfur (S), phosphorus (P), and silicon (Si) are in the third period. Therefore, S, P and Si use $3p$ orbitals to form π bonds, while C, N and O use $2p$ orbitals. The $3p$ orbitals are much larger than $2p$ orbitals, and therefore there is a reduced probability for an overlap of the $2p$ orbital of C and the $3p$ orbital of S, P and Si.

B: S, P, and Si can hybridize, but these elements can combine *s* and *p* orbitals and (unlike C, O and N) have *d* orbitals.

C: S, P and Si (ground state electron configurations) have partially occupied *p* orbitals which form bonds.

D: carbon combines with elements below the second row of the periodic table. For example, carbon forms bonds with higher principal quantum number ($n > 2$) halogens (e.g., F, Cl, Br and I).

64. E is correct.

The magnesium atom has a +2 charge, and the sulfur atom has a –2 charge.

65. C is correct.

Hydrogen bonds are the strongest intermolecular forces (i.e., between molecules), followed by dipole–dipole, dipole–induced dipole, and van der Waals forces (i.e., London dispersion).

Larger atoms have more electrons, which means that they can exert stronger induced dipole forces compared to smaller atoms.

Bromine is located under chlorine on the periodic table, which means it has more electrons and stronger induced dipole forces.

66. A is correct.

Hydrogen bonds are the strongest intermolecular forces (i.e., between molecules), followed by dipole–dipole, dipole–induced dipole, and van der Waals forces (i.e., London dispersion).

Hydrogens, bonded directly to F, O or N, participate in hydrogen bonds. The hydrogen is partially positive (i.e., delta plus: $\partial+$) due to the bond to these electronegative atoms. The lone pair of electrons on the F, O or N interacts with the partial positive ($\partial+$) hydrogen to form a hydrogen bond.

When ~NH or ~OH groups are present in a molecule, they form hydrogen bonds with other molecules. Hydrogen bonding is the strongest intermolecular force, and it is the major intermolecular force in this substance.

67. B is correct.

Potassium oxide (K_2O) contains metal and nonmetal, which form a salt.

Salts contain ionic bonds and dissociate in aqueous solutions and therefore are strong electrolytes.

A polyatomic (i.e., molecular) ion is a charged chemical species composed of two or more atoms covalently bonded or composed of a metal complex acting as a single unit. An example is the hydroxide ion consisting of one oxygen atom and one hydrogen atom; hydroxide has a charge of –1.

A polyatomic ion does bond with other ions.

A polyatomic ion has various charges.

A polyatomic ion might contain only metals or nonmetals.

A polyatomic ion is not neutral.

oxidation state	−1	+1	+3	+5	+7
anion name	chloride	hypochlorite	chlorite	chlorate	perchlorate
formula	Cl^-	ClO^-	ClO_2^-	ClO_3^-	ClO_4^-

68. D is correct.

The valence shell is the outermost shell (i.e., highest principal quantum number *n*) of an atom.

Valence electrons are those electrons of the outermost electron shell that can participate in a chemical bond.

69. B is correct.

Nitrogen (NH_3) is neutral with three bonds, negative with two bonds ($^-NH_2$), and positive with 4 bonds ($^+NH_4$)

Line bond structure of methylamine
with the lone pair on nitrogen shown

70. B is correct.

Sulfide ion is S^{2-}.

Hydrogen sulfide (i.e., sewer gas) is H_2S.

Hydrosulfide (bisulfide) ion is HS^-.

Sulfite ion is the conjugate base of bisulfite and has the molecular formula of SO_3^{2-}

Sulfur (S) is a chemical element with an atomic number of 16.

Sulfate ion is a polyatomic anion (i.e., two or more atoms covalently bonded or metal complex) with the molecular formula of SO_4^{2-}

Sulfur acid (HSO_3^-) results from the combination of sulfur dioxide (SO_2) and H_2O.

71. E is correct.

Ionic bonds involve electrostatic interactions between oppositely charged atoms.

Electrons from the metallic atom are transferred to the nonmetallic atom, giving both atoms a full valence shell.

72. B is correct.

If the charge of O is −1, the proper formula of its compound with K (+1) should be KO.

73. C is correct.

The lattice energy of a crystalline solid is usually defined as the energy of formation of the crystal from infinitely-separated ions (i.e., an exothermic process and hence the negative sign).

Some older textbooks define lattice energy with the opposite sign. The older notation refers to the energy required to convert the crystal into infinitely separated gaseous ions in a vacuum (i.e., an endothermic process and hence the positive sign).

Lattice energy decreases downward for a group.

Lattice energy increases with charge.

Because NaCl and LiF contain charges of +1 and −1, they would have smaller lattice energy than the higher-charged ions on other options.

Na is lower than Li, and Cl is lower than F in their respective groups, so that NaCl would have lower lattice energy than LiF.

The bond between ions of opposite charge is strongest when the ions are small.

The force of attraction between oppositely charged particles is directly proportional to the product of the charges on the two objects (q_1 and q_2) and inversely proportional to the square of the distance between the objects (r^2):

$$F = q_1 \times q_2 / r^2$$

Therefore, the strength of the bond between the ions of opposite charge in an ionic compound depends on the charges on the ions and the distance between the centers of the ions when they pack to form a crystal.

For NaCl, the lattice energy is the energy released by the reaction:

Na^+ (g) + Cl^- (g) → NaCl (s), lattice energy = −786 kJ/mol

Other examples:

Al_2O_3, lattice energy = −15,916 kJ/mol

KF, lattice energy = −821 kJ/mol

LiF, lattice energy = −1,036 kJ/mol

NaOH, lattice energy = −900 kJ/mol

74. C is correct.

Formula to calculate dipole moment:

$$\mu = qr$$

where μ is dipole moment (coulomb per meter or C/m), q is a charge (coulomb or C), and r is the radius (meters or m)

Convert the unit of radius to meters:

$$154 \text{ pm} \times 1 \times 10^{-12} \text{ m/pm} = 154 \times 10^{-12} \text{ m}$$

Convert the charge to coulombs:

$$0.167 \ e \times 1.602 \times 10^{-19} \text{ C/e}^- = 2.68 \times 10^{-20} \text{ C}$$

Use these values to calculate the dipole moment:

$$\mu = qr$$

$$\mu = 2.68 \times 10^{-20} \text{ C} \times 154 \times 10^{-12} \text{ m}$$

$$\mu = 4.13 \times 10^{-30} \text{ C·m}$$

Finally, convert the dipole moment to Debye:

$$4.13 \times 10^{-30} \text{ C·m} / 3.34 \times 10^{-30} \text{ C·m·D}^{-1} = 1.24 \text{ D}$$

75. B is correct.

The greater the difference in electronegativity between two atoms in a compound, the more polar of a bond these atoms form. The atom with the higher electronegativity is the partial (delta) negative end of the dipole.

Although each C–Cl bond is very polar, the dipole moments of each of the four bonds in CCl_4 (carbon tetrachloride) cancel because the molecule is a symmetric tetrahedron.

76. C is correct.

The pK_a of the carboxylic acid is about 5 and is deprotonated (exists as an anion) at a pH of 10.

The pK_a of the alcohol is about 15 and is protonated (neutral) at a pH of 10.

77. E is correct.

Hybridization of the central atom and the corresponding electron geometry.

Hybridization	Electron groups	Bonding groups	Lone pairs	Approx. bond angles	Electron geometry	Molecular geometry
sp	2	2	0	180°	Linear	Linear
sp^2	3	3	0	120°	Trigonal planar	Trigonal planar
	3	2	1	<120°	Trigonal planar	Bent
sp^3	4	4	0	109.5°	Tetrahedral	Tetrahedral
	4	3	1	<109.5°	Tetrahedral	Trigonal pyramidal
	4	2	2	<<109.5°	Tetrahedral	Bent
sp^3d	5	5	0	120° (equatorial) 90° (axial)	Trigonal bipyramidal	Trigonal bipyramidal
	5	4	1	<120° (equatorial) <90° (axial)	Trigonal bipyramidal	Seesaw
	5	3	2	<90°	Trigonal bipyramidal	T-shaped
	5	2	3	180°	Trigonal bipyramidal	Linear
sp^3d^2	6	6	0	90°	Octahedral	Octahedral
	6	5	1	<90°	Octahedral	Square pyramidal
	6	4	2	90°	Octahedral	Square planar

78. B is correct.

The valence shell is the outermost shell (i.e., highest principal quantum number n) of an atom. The octet rule states that atoms of main-group elements tend to combine so that each atom has eight electrons in its valence shell. This occurs because electron arrangements involving eight valence electrons are extremely stable, as is the case with noble gasses.

79. D is correct.

Solubility depends on the solvent's ability to overcome the intermolecular forces in a solute.

80. D is correct.

Electronegativity is defined as the ability of an atom to attract electrons when it bonds with another atom. The most electronegative atom is F, while the least electronegative atom is Fr. The trend for increasing electronegativity within the periodic table is up and toward the right (i.e., fluorine).

Chlorine has the highest electronegativity of the elements listed. Electronegativity is a characteristic of non-metals.

81. C is correct.

Hydrogen bonds are the strongest intermolecular forces (i.e., between molecules), followed by dipole–dipole, dipole–induced dipole, and van der Waals forces (i.e., London dispersion).

Hydrogens, bonded directly to N, F and O, participate in hydrogen bonds. The hydrogen is partial positive (i.e., delta plus or $\partial +$) due to the bond to electronegative atoms (i.e., F, O or N). The lone pair of electrons on the F, O or N interacts with the $\partial +$ hydrogen to form a hydrogen bond.

Polar compounds have a stronger dipole–dipole force between molecules because the partially positive end of one molecule bonds with the partial negative end of another molecule.

Nonpolar compounds have induced dipole-induced dipole intermolecular force, which is weak compared to dipole-dipole force. The strength of induced dipole-induced dipole interaction correlates with the number of electrons and protons of an atom.

82. B is correct.

The valence shell is the outermost shell (i.e., highest principal quantum number n) of an atom. Valence electrons are those electrons of the outermost electron shell that can participate in a chemical bond.

The number of valence electrons for an element can be determined by its group (i.e., vertical column) on the periodic table. Except for the transition metals (i.e., groups 3-12), the group number identifies how many valence electrons are associated with a particular element: elements of the same group have the same number of valence electrons.

Oxygen has 8 electrons.

The first shell ($1s$) holds 2 electrons, and the remaining 6 electrons are located within the orbitals of the next shell ($2s$, $2p$).

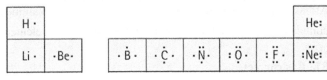

Sample Lewis dot structures for some elements

83. C is correct.

The nitrate ion is NO_3^- and has the following resonance forms:

Nitrate exists as a superposition of these three resonance forms

84. E is correct.

Ionic bonds form when the difference in electronegativity of atoms in a compound is greater than 1.7 Pauling units. Ionic bonds involve transferring an electron from the electropositive element (along the left-hand column/group) to the electronegative element (along the right-hand column/groups) on the periodic table.

85. C is correct.

Electronegativity is a chemical property that describes an atom's tendency to attract electrons to itself.

The most common use of electronegativity pertains to polarity along the *sigma* (single) bond.

The most electronegative atom is F, while the least electronegative atom is Fr. The trend for increasing electronegativity within the periodic table is up and toward the right (i.e., fluorine).

86. B is correct.

The octet rule states that atoms of main-group elements tend to combine so that each atom has eight electrons in its valence shell. This occurs because electron arrangements involving eight valence electrons are extremely stable, as is the case with noble gasses.

The valence shell is the outermost shell (i.e., highest principal quantum number n) of an atom. Valence electrons are those electrons of the outermost electron shell that can participate in a chemical bond.

The number of valence electrons for an element can be determined by its group (i.e., vertical column) on the periodic table. Except for the transition metals (i.e., groups 3-12), the group number identifies how many valence electrons are associated with a particular element: elements of the same group have the same number of valence electrons.

Atoms with a complete valence shell (i.e., containing the maximum number of electrons), such as noble gases, are the most non-reactive elements.

Atoms with only one electron in their valence shells (alkali metals) or those missing just one electron from a complete valence shell (halogens) are the most reactive elements.

87. B is correct.

To help determine the polarity of molecules, draw the Lewis structures of the molecules.

With SO_3, CO_2, CH_4, and CCl_4, the electrons in the central atom are bonded, and they are symmetrical. Therefore, they are nonpolar. SO_2 has a pair of nonbonding electrons on the central atom (S), which means the molecule is polar.

88. C is correct.

Intermolecular forces are between molecules.

Hydrogen bonds are the strongest intermolecular forces (i.e., between molecules), followed by dipole–dipole, dipole–induced dipole, and van der Waals forces (i.e., London dispersion).

Dipole–dipole interactions primarily occur between different molecules, whereas ionic bonds, covalent and coordinate covalent bonds hold atoms in a molecule (i.e., intramolecular forces).

Notes for active learning

Notes for active learning

Phases and Phase Equilibria – Detailed Explanations

==

Practice Set 1: Questions 1–20

==

1. A is correct.

Solids, liquids, and gases have a vapor pressure that increases from solid to gas.

Vapor pressure is the pressure exerted by a vapor in equilibrium with its condensed phases (i.e., solid or liquid) in a closed system at a given temperature.

Vapor pressure is a colligative property of a substance and depends only on the number of solutes present, not on their identity.

2. C is correct.

Charles' law (i.e., the law of volumes) explains how, at constant pressure, gases behave when the temperature changes:

$$V \alpha T$$

or

$$V / T = \text{constant}$$

or

$$(V_1 / T_1) = (V_2 / T_2)$$

Volume and temperature are proportional.

Doubling the temperature at constant pressure doubles the volume.

3. C is correct.

Colligative properties include: lowering of vapor pressure, the elevation of boiling point, depression of freezing point, and increased osmotic pressure.

The addition of solute to a pure solvent lowers the vapor pressure of the solvent; therefore, a higher temperature is required to bring the vapor pressure of the solution in an open container up to the atmospheric pressure. This increases the boiling point.

Because adding solute lowers the vapor pressure, the solution's freezing point decreases (e.g., automobile antifreeze).

4. A is correct.

At a pressure and temperature corresponding to the triple point (point D on the graph) of a substance, all three states (gas, liquid and solid) exist in equilibrium.

The critical point (point E on the graph) is the endpoint of the phase equilibrium curve where the liquid and its vapor become indistinguishable.

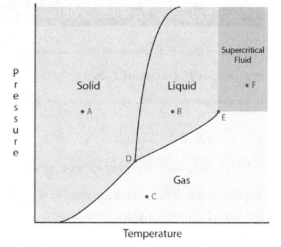

Phase diagram of pressure vs. temperature

5. A is correct.

In the van der Waals equation, *a* is the negative deviation due to attractive forces, and *b* is the positive deviation due to molecular volume.

6. B is correct.

R is the symbol for the ideal gas constant.

R is expressed in (L × atm) / mole × K and the value = 0.0821.

Convert to different units (torr and mL)

0.0821[(L × atm) / (mole × K)] × (760 torr/atm) × (1,000 mL/L)

R = 62,396 (torr × mL) / mole × K

7. B is correct.

Ideal gas law:

PV = nRT

where P is pressure, V is volume, n is the number of molecules, R is the ideal gas constant, and T is the temperature of the gas.

R and n are constant.

From this equation, if both pressure and temperature are halved, there would be no effect on the volume.

8. B is correct.

Kinetic molecular theory of gas molecules states that the average kinetic energy per molecule in a system is proportional to the temperature of the gas.

Since it is given that containers X and Y are at the same temperature and pressure, then molecules of both gases must possess the same amount of average kinetic energy.

9. E is correct.

Barometers and manometers are used to measure pressure.

Barometers are designed to measure atmospheric pressure, while a manometer can measure the pressure that is lower than atmospheric pressure.

A manometer has both ends of the tube open to the outside (while some may have one end closed), whereas a barometer is a type of closed-end manometer with one end of the glass tube closed and sealed with a vacuum.

The atmospheric pressure is 760 mmHg, so the barometer should be able to accommodate that.

10. D is correct.

Vapor pressure is the pressure exerted by a vapor in equilibrium with its condensed phases (i.e., solid or liquid) in a closed system at a given temperature.

Raoult's law states that the partial vapor pressure of each component of an ideal mixture of liquids is equal to the vapor pressure of the pure component multiplied by its mole fraction in the mixture.

In exothermic reactions, the vapor pressure deviates negatively from Raoult's law.

Depending on the ratios of the liquids in a solution, the vapor pressure could be lower than either or just lower than X because X is a higher boiling point, thus a lower vapor pressure.

The boiling point increases from adding Y to the mixture because the vapor pressure decreases.

11. B is correct.

$$2 \text{ Na } (s) + \text{Cl}_2 (g) \rightarrow 2 \text{ NaCl } (s)$$

In its elemental form, chlorine exists as a gas.
In its elemental form, Na exists as a solid.

12. C is correct.

The molecules of an ideal gas do not occupy a significant amount of space and exert no intermolecular forces. The molecules of a real gas do occupy space and do exert (weak attractive) intermolecular forces.

However, both an ideal gas and a real gas have pressure, which is created from molecular collisions with the walls of the container.

13. A is correct.

Boyle's law (i.e., pressure-volume law) states that pressure and volume are inversely proportional:

$$(P_1 V_1) = (P_2 V_2)$$

Solve for the final pressure:

$$P_2 = (P_1 V_1) / V_2$$

$$P_2 = [(0.950 \text{ atm}) \times (2.75 \text{ L})] / (0.450 \text{ L})$$

$$P_2 = 5.80 \text{ atm}$$

14. D is correct.

Avogadro's law is an experimental gas law relating the volume of gas to the amount of substance of gas present.

Avogadro's law states that equal volumes of gases at the same temperature and pressure have the same number of molecules.

For a given mass of an ideal gas, the volume and amount (i.e., moles) of the gas are directly proportional if the temperature and pressure are constant:

$$V \alpha\, n$$

where V = volume and n = number of moles of the gas.

Charles' law (i.e., the law of volumes) explains how, at constant pressure, gases behave when the temperature changes:

$$V \alpha\, T$$

or

$$V\,/\,T = constant$$

or

$$(V_1\,/\,T_1) = (V_2\,/\,T_2)$$

Volume and temperature are proportional. Therefore, an increase in one term increases the other.

Gay-Lussac's law (i.e., pressure-temperature law) states that pressure is proportional to temperature:

$$P \alpha\, T$$

or $\quad (P_1\,/\,T_1) = (P_2\,/\,T_2)$

or $\quad (P_1 T_2) = (P_2 T_1)$

or $\quad P\,/\,T = constant$

If the pressure of a gas increases, the temperature increases.

Boyle's law (i.e., pressure-volume law) states that pressure and volume are inversely proportional:

15. C is correct.

Ideal gas law:

$$PV = nRT$$

where P is pressure, V is volume, n is the number of molecules, R is the ideal gas constant, and T is the temperature of the gas.

R and n are constant.

If T is constant, the equation becomes PV = constant.

They are inversely proportional: if one of the values is reduced, the other increases.

16. A is correct.

Vaporization refers to the change of state from a liquid to a gas. There are two types of vaporization: boiling and evaporation, which are differentiated based on the temperature at which they occur.

Evaporation occurs at a temperature below the boiling point, while boiling occurs at or above the boiling point.

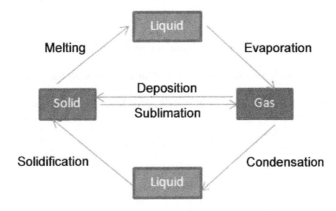

Interconversion of states of matter

17. C is correct.

Ideal gas law:

$$PV = nRT$$

from which simpler gas laws such as Boyle's, Charles', and Avogadro's laws are derived.

The value of n and R are constant.

The common format of the combined gas law:

$$(P_1V_1) / T_1 = (P_2V_2) / T_2$$

Try modifying the equations to recreate the formats of the equation provided by the problem.

$T_2 = T_1 \times P_1 / P_2 \times V_2 / V_1$ should be written as $T_2 = T_1 \times V_1/V_2 \times P_1/P_2$

18. C is correct.

Intermolecular forces act between neighboring molecules. Examples include hydrogen bonding, dipole-dipole, dipole-induced dipole, and van der Waals (i.e., London dispersion) forces.

A dipole-dipole attraction involves asymmetric, polar molecules (based on differences in electronegativity between atoms) that create a dipole moment (i.e., a net vector indicating the force).

Sulfur dioxide with indicated bond angle

19. C is correct.

Increasing the pressure of the gas above the liquid puts stress on the equilibrium of the system. Gas molecules start to collide with the liquid surface more often, which increases the rate of gas molecules entering the solution, thus increasing the solubility.

20. E is correct.

Avogadro's law states the correlation between volume and moles (n).

Avogadro's law is an experimental gas law relating the volume of a gas to the amount of gas present.

At the same pressure and temperature, equal volumes of gases have the same number of molecules.

$$V \propto n$$

or

$$V / n = k$$

where V is the volume of the gas, n is the number of moles of the gas, and k is a constant equal to RT/P (where R is the universal gas constant, T is the temperature in Kelvin and P is the pressure).

For comparing the same substance under two sets of conditions, the law is expressed as:

$$V_1 / n_1 = V_2 / n_2$$

===

Practice Set 2: Questions 21–40

===

21. E is correct.

Colligative properties of solutions depend on the ratio of the number of solute particles to the number of solvent molecules in a solution and not on the type of chemical species present.

Colligative properties include: lowering of vapor pressure, the elevation of boiling point, depression of freezing point, and increased osmotic pressure.

Boiling point (BP) elevation: $\Delta BP = iKm$

where i = the number of particles produced when the solute dissociates, K = boiling elevation constant and m = molality (moles/kg solvent).

In this problem, acid molality is not known.

22. A is correct.

An ideal gas has no intermolecular forces, indicating that its molecules have no attraction to each other. The molecules of a real gas, however, do have intermolecular forces, although these forces are extremely weak.

Therefore, the molecules of a real gas are slightly attracted to one another, although the attraction is nowhere near as strong as the attraction in liquids and solids.

23. E is correct.

Standard temperature and pressure (STP) has a temperature of 273.15 K (0 °C, 32 °F) and a pressure of 10^5 Pa (100 kPa, 750.06 mmHg, 1 bar, 14.504 psi, 0.98692 atm).

The mm of Hg was defined as the pressure generated by a column of mercury one millimeter high. The pressure of mercury depends on temperature and gravity.

This variation in mmHg and torr is a difference in units of about 0.000015%.

In general,

1 torr = 1 mm of Hg = 0.0013158 atm.

750.06 mmHg = 0.98692 atm.

24. D is correct.

At lower temperatures, the potential energy due to the intermolecular forces are more significant compared to the kinetic energy; this causes the pressure to be reduced because the gas molecules are attracted to each other.

---

25. D is correct.

Dalton's law (i.e., the law of partial pressures) states that *the pressure exerted by a mixture of gases is equal to the sum of the individual gas pressures*.

The pressure due to N_2 and CO_2:

(320 torr + 240 torr) = 560 torr

The partial pressure of O_2 is:

740 torr – 560 torr = 180 torr

180 torr / 740 torr = 24%

26. A is correct.

Ideal gas law:

$PV = nRT$

where P is pressure, V is volume, n is the number of molecules, R is the ideal gas constant, and T is the temperature of the gas.

If the volume is reduced by ½, the number of moles is reduced by ½.

Pressure is reduced to 90%, so the number of moles is reduced by 90%.

Therefore, the total reduction in moles is (½ × 90%) = 45%.

Mass is proportional to the number of moles for a given gas so that the mass reduction can be calculated directly:

New mass is 45% of 40 grams = (0.45 × 40 g) = 18 grams

27. D is correct.

Evaporation describes the phase change from liquid to gas.

The mass of the molecules and the attraction of the molecules with their neighbors (to form intermolecular attractions) determine their kinetic energy.

The increase in kinetic energy is required for individual molecules to move from the liquid to the gaseous phase.

28. B is correct.

The molecules of an ideal gas exert no attractive forces.

Therefore, a real gas behaves most nearly like an ideal gas when it is at high temperature and low pressure because under these conditions, the molecules are far apart from each other and exert little or no attractive forces on each other.

29. E is correct.

Hydroxyl (~OH) groups greatly increase the boiling point because they form hydrogen bonds with ~OH groups of neighboring molecules.

Hydrocarbons are nonpolar molecules, which means that the dominant intermolecular force is London dispersion. This force gets stronger as the number of atoms in each molecule increases. The stronger force increases the boiling point.

The branching of the hydrocarbon affects the boiling point. Straight molecules have slightly higher boiling points than branched molecules with the same number of atoms. The reason is that straight molecules can align parallel against each other, and atoms in the molecules are involved in the London dispersion forces.

Another factor is the presence of other heteroatoms (i.e., atoms other than carbon and hydrogen). For example, the electronegative oxygen atom between carbon groups or in an ether (C–O–C) slightly increases the boiling point.

30. A is correct.

Kinetic theory explains the macroscopic properties of gases (e.g., temperature, volume, and pressure) by their molecular composition and motion.

The gas pressure is due to the collisions on the walls of a container from molecules moving at different velocities.

Temperature = $\frac{1}{2}mv^2$

31. B is correct.

Methanol (CH_3OH) is alcohol that participates in hydrogen bonding.

Therefore, this gas experiences the strongest intermolecular forces.

32. D is correct.

Density = mass / volume

Gas molecules have a large amount of space between them; therefore, they can be pushed together, and thus gases are very compressible. Because there is such a large amount of space between each molecule in a gas, the extent to which the gas molecules can be pushed together is much greater than the extent to which liquid molecules can be pushed together. Therefore, gases have greater compressibility than liquids.

Gas molecules are further apart than liquid molecules, which is why gases have a smaller density.

33. C is correct.

Vapor pressure is the pressure exerted by a vapor in equilibrium with its condensed phases (i.e., solid or liquid) in a closed system at a given temperature.

Vapor pressure is inversely correlated with the strength of the intermolecular force.

The molecules are more likely to stick together in the liquid form with stronger intermolecular forces, and fewer participate in the liquid-vapor equilibrium.

The vapor pressure of a liquid decreases when a nonvolatile substance is dissolved into a liquid.

The decrease in the vapor pressure of a substance is proportional to the number of moles of the solute dissolved in a definite weight of the solvent. This is Raoult's law.

34. B is correct.

Dalton's law (i.e., the law of partial pressures) states that *the pressure exerted by a mixture of gases is equal to the sum of the individual gas pressures*. It is an empirical law that was observed by English chemist John Dalton and is related to the ideal gas laws.

35. C is correct.

Solids have a definite shape and volume. For example, a granite block does not change its shape or its volume regardless of the container in which it is placed.

Molecules in a solid are very tightly packed due to the strong intermolecular attractions, which prevent the molecules from moving around.

36. E is correct.

Ideal gas law: $PV = nRT$

where P is pressure, V is volume, n is the number of molecules, R is the ideal gas constant, and T is the temperature of the gas.

At STP (standard conditions for temperature and pressure), the pressure and temperature for the three flasks are the same. It is known that the volume is the same in each case – 2.0 L.

Therefore, since R is a constant, the number of molecules "n" must be the same for the ideal gas law to hold.

37. A is correct.

Hydrogens, bonded directly to F, O or N, participate in hydrogen bonds. The hydrogen is partially positive (i.e., delta plus: $\partial+$) due to the bond to these electronegative atoms. The lone pair of electrons on the F, O or N interacts with the partial positive ($\partial+$) hydrogen to form a hydrogen bond.

D: the hydrogen on the methyl carbon and that carbon is not attached to N, O, or F. Therefore, that particular hydrogen cannot form a hydrogen bond, even though there is available oxygen on the other methanol to form a hydrogen bond.

38. C is correct.

Boyle's law (Mariotte's law or the Boyle-Mariotte law) is an experimental gas law that describes how the volume of a gas increases as the pressure decreases (i.e., they are inversely proportional) if the temperature is constant.

Boyle's law (i.e., pressure-volume law) states that pressure and volume are inversely proportional:

$$P_1V_1 = P_2V_2$$

or $\quad P \times V = $ constant

If the volume of a gas increases, its pressure decreases proportionally.

Dalton's law (i.e., the law of partial pressures) states that *the pressure exerted by a mixture of gases is equal to the sum of the individual gas pressures.*

Charles' law (i.e., the law of volumes) explains how, at constant pressure, gases behave when the temperature changes:

$$V \alpha T$$

or $\quad V / T = $ constant

or $\quad (V_1 / T_1) = (V_2 / T_2)$

Volume and temperature are proportional: an increase in one results in an increase in the other.

Gay-Lussac's law (i.e., pressure-temperature law) states that pressure is proportional to temperature:

$$P \alpha T$$

or $\quad (P_1 / T_1) = (P_2 / T_2)$

or $\quad (P_1T_2) = (P_2T_1)$

or $\quad P / T = $ constant

If the pressure of a gas increases, the temperature increases.

Avogadro's law is an experimental gas law relating the volume of a gas to the amount of substance of gas present. It states that *equal volumes of all gases, at the same temperature and pressure, have the same number of molecules.*

39. B is correct.

As the automobile travels the highway, friction is generated between the road and its tires. The heat energy increases the temperature of the air inside the tires, causing the molecules to have more velocity. These fast-moving molecules collide with the walls of the tire at a higher rate, and thus the pressure is increased.

Gay-Lussac's law (i.e., pressure-temperature law) states that pressure is proportional to temperature:

$$P \alpha T$$

or $\quad (P_1 / T_1) = (P_2 / T_2)$

or $\quad (P_1T_2) = (P_2T_1)$

or $\quad P / T = $ constant

If the temperature of a gas is increased, the pressure increases.

40. C is correct.

The balanced chemical equation:

$$N_2 + 3 H_2 \rightarrow 2 NH_3$$

Use the balanced coefficients from the written equation, apply dimensional analysis to solve for the volume of H_2 needed to produce 12.5 L NH_3:

$V_{H2} = V_{NH3} \times (\text{mol } H_2 / \text{mol } NH_3)$

$V_{H2} = (12.5 \text{ L}) \times (3 \text{ mol} / 2 \text{ mol})$

$V_{H2} = 18.8 \text{ L}$

===

Practice Set 3: Questions 41–60

===

41. C is correct.

Dalton's law (i.e., the law of partial pressures) states that *the pressure exerted by a mixture of gases is equal to the sum of the individual gas pressures.*

Convert the masses of the gases into moles:

Moles of O_2:   16 g of O_2 ÷ 32 g/mole = 0.5 mole

Moles of N_2:   14 g of N_2 ÷ 28 g/mole = 0.5 mole

Mole of CO_2:   88 g of CO_2 ÷ 44 g/mole = 2 moles

Total moles: (0.5 mol + 0.5 mole + 2 mol) = 3 moles

The pressure of 1 atm (or 760 mmHg) has 38 mmHg (1 torr = 1 mmHg) contributed as H_2O vapor.

The partial pressure of CO_2:

    mole fraction × (total pressure of the gas mixture – H_2O vapor)

    (2 moles CO_2 / 3 moles total gas)] × (760 mmHg – 38 mmHg)

    partial pressure of CO_2 = 481 mmHg

42. E is correct.

Colligative properties are properties of solutions that depend on the ratio of the number of solute particles to the number of solvent molecules in a solution, not on the type of chemical species present.

Colligative properties include: lowering of vapor pressure, the elevation of boiling point, depression of freezing point, and increased osmotic pressure

Dissolving a solute into a solvent alters the solvent's freezing point, melting point, boiling point, and vapor pressure.

43. B is correct.

Boyle's law (i.e., pressure-volume law) states that pressure and volume are inversely proportional:

    $(P_1V_1) = (P_2V_2)$

or

    $P \times V = $ constant

If the volume of a gas increases, its pressure decreases proportionally.

44. E is correct.

Gases form homogeneous mixtures, regardless of the identities or relative proportions of the component gases. There is a relatively large distance between gas molecules (as opposed to solids or liquids where the molecules are much closer).

When pressure is applied to gas, its volume readily decreases, and thus gases are highly compressible.

There are no attractive forces between gas molecules, which is why molecules of gas can move about freely.

45. E is correct.

Solids have a definite shape and volume. For example, a granite block does not change its shape or volume regardless of the container in which it is placed.

Molecules in a solid are very tightly packed due to the strong intermolecular attractions, which prevent the molecules from moving around.

46. A is correct.

Standard temperature and pressure (STP) have a temperature of 273.15 K (0 °C, 32 °F) and a pressure of 10^5 Pa (100 kPa, 750.06 mmHg, 1 bar, 14.504 psi, 0.98692 atm).

The mm of Hg was defined as the pressure generated by a column of mercury one millimeter high. The pressure of mercury depends on temperature and gravity.

This variation in mmHg and torr is a difference in units of about 0.000015%.

In general,

        1 torr = 1 mm of Hg = 0.0013158 atm.

        750.06 mmHg = 0.98692 atm.

47. C is correct.

Intermolecular forces act between neighboring molecules. Examples include hydrogen bonding, dipole-dipole, dipole-induced dipole, and van der Waals (i.e., London dispersion) forces.

Stronger force results in a higher boiling point.

CH_3COOH is a carboxylic acid that can form two hydrogen bonds. Therefore, it has the highest boiling point.

Ethanoic acid with the two hydrogen bonds indicated on the structure

48. C is correct.

Molecules in solids have the most attraction to their neighbors, followed by liquids (significant motion between the individual molecules) and then gas.

A molecule in an ideal gas has no attraction to other gas molecules. For a gas experiencing low pressure, the particles are far enough apart for no attractive forces between the individual gas molecules.

49. C is correct.

Ideal gas law:

$$PV = nRT$$

where P is pressure, V is volume, n is the number of molecules, R is the ideal gas constant, and T is the temperature of the gas.

Set the initial and final P, V and T conditions equal:

$$(P_1V_1 / T_1) = (P_2V_2 / T_2)$$

Solve for the final volume of N_2:

$$(P_2V_2 / T_2) = (P_1V_1 / T_1)$$

$$V_2 = (T_2 P_1V_1) / (P_2T_1)$$

$$V_2 = [(295 \text{ K}) \times (750.06 \text{ mmHg}) \times (0.190 \text{ L } N_2)] / [(660 \text{ mmHg}) \times (298.15 \text{ K})]$$

$$V_2 = 0.214 \text{ L}$$

50. B is correct.

Volatility is the tendency of a substance to vaporize (phase change from liquid to vapor).

Volatility is directly related to a substance's vapor pressure. At a given temperature, a substance with higher vapor pressure vaporizes more readily than a substance with lower vapor pressure.

Molecules with weak intermolecular attraction can increase their kinetic energy by transferring less heat due to a smaller molecular mass. The increase in kinetic energy is required for individual molecules to move from the liquid to the gaseous phase.

51. B is correct.

Barometer and manometer are used to measure pressure.

Barometers are designed to measure atmospheric pressure, while a manometer can measure the pressure that is lower than atmospheric pressure.

A manometer has both ends of the tube open to the outside (while some may have one end closed), whereas a barometer is a type of closed-end manometer with one end of the glass tube closed and sealed with a vacuum.

The difference in mercury height on both necks indicates the capacity of a manometer.

$$820 \text{ mm} - 160 \text{ mm} = 660 \text{ mm}$$

Historically, the pressure unit of torr is set to equal 1 mmHg or the rise/dip of 1 mm of mercury in a manometer.

Because the manometer uses mercury, the height difference (660 mm) equals its measuring capacity in torr (660 torrs).

52. C is correct.

The conditions of the ideal gases are the same, so the number of moles (i.e., molecules) is equal.

At STP, the temperature is the same, so the kinetic energy of the molecules is the same.

However, the molar mass of oxygen and nitrogen are different. Therefore, the density is different.

53. A is correct.

Vapor pressure is the pressure exerted by a vapor in equilibrium with its condensed phases (i.e., solid or liquid) in a closed system at a given temperature.

The compound exists as a liquid if the external pressure > compound's vapor pressure.

A substance boils when vapor pressure = external pressure.

54. D is correct.

Gas molecules have a large amount of space between them; therefore, they can be pushed together, and gases are thus very compressible.

Molecules in solids and liquids are already close together; therefore, they cannot get significantly closer and are nearly incompressible.

55. C is correct.

Charles' law (i.e., the law of volumes) explains how, at constant pressure, gases behave when the temperature changes:

$V \alpha T$

or

$V / T = \text{constant}$

or

$(V_1 / T_1) = (V_2 / T_2)$

Volume and temperature are proportional. Therefore, an increase in one results in an increase in the other.

Gay-Lussac's law (i.e., pressure-temperature law) states that pressure is proportional to temperature:

$P \alpha T$

or

$(P_1 / T_1) = (P_2 / T_2)$

or

$(P_1 T_2) = (P_2 T_1)$

or

$P / T = \text{constant}$

If the pressure of a gas increases, the temperature increases.

Dalton's law (i.e., the law of partial pressures) states that *the pressure exerted by a mixture of gases is equal to the sum of the individual gas pressures*

Boyle's law (i.e., pressure-volume law) states that pressure and volume are inversely proportional:

$$(P_1V_1) = (P_2V_2)$$

or

$$P \times V = constant$$

If the volume of a gas increases, its pressure decreases proportionally.

Avogadro's law is an experimental gas law relating the volume of a gas to the amount of substance of gas present.

56. B is correct.

Sublimation is the direct change of state from a solid to a gas, skipping the intermediate liquid phase.

An example of a compound that undergoes sublimation is solid carbon dioxide (i.e., dry ice). CO_2 changes phases from solid to gas (i.e., bypasses the liquid phase) and is often used as a cooling agent.

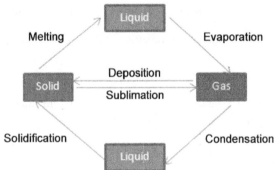

Interconversion of states of matter

57. B is correct.

Graham's law of effusion states that the rate of effusion (i.e., escaping through a small hole) of a gas is inversely proportional to the square root of the molar mass of its particles.

$$Rate\ 1\ /\ Rate\ 2 = \sqrt{(molar\ mass\ gas\ 2\ /\ molar\ mass\ gas\ 1)}$$

The diffusion rate is the inverse root of the molecular weights of the gases.

Therefore, the rate of effusion is:

$$O_2\ /\ H_2 = \sqrt{(2\ /\ 32)}$$

rate of diffusion = 1:4

58. B is correct.

To calculate the number of molecules, calculate the moles of gas using the ideal gas law:

$$PV = nRT$$

$$n = PV\ /\ RT$$

Because the gas constant R is in $L \cdot atm\ K^{-1}\ mol^{-1}$, pressure must be converted into atm:

$$320\ mmHg \times (1\ /\ 760\ atm/mmHg) = 320\ mmHg\ /\ 760\ atm$$

(Leave it in this fraction form because the answer choices are in this format.)

Convert the temperature to Kelvin:

$$10 \,°C + 273 = 283 \text{ K}$$

Substitute those values into the ideal gas equation:

$$n = PV / RT$$

$$n = (320 \text{ mmHg} / 760 \text{ atm}) \times 6 \text{ L} / (0.0821 \text{ L·atm K}^{-1}\text{mol}^{-1} \times 283 \text{ K})$$

This expression represents the number of gas moles present in the container.

Calculate the number of molecules:

number of molecules = moles × Avogadro's number

number of molecules = $(320 \text{ mmHg} / 760 \text{ atm}) \times 6 / (0.0821 \times 283) \times 6.02 \times 10^{23}$

number of molecules = $(320 / 760)·(6)·(6 \times 10^{23}) / (0.0821)·(283)$

59. A is correct.

At STP (standard conditions for temperature and pressure), the pressure and temperature are the same regardless of the gas.

The ideal gas law:

$$PV = nRT$$

where P is pressure, V is volume, n is the number of molecules, R is the ideal gas constant, and T is the temperature of the gas.

Therefore, one molecule of each gas occupies the same volume at STP. Since CO_2 molecules have the largest mass, CO_2 gas has a greater mass in the same volume, and thus it has the greatest density.

60. C is correct.

Hydrogens, bonded directly to F, O or N, participate in hydrogen bonds.

The hydrogen is partially positive (i.e., delta plus: $\partial+$) due to the bond to these electronegative atoms. The lone pair of electrons on the F, O or N interacts with the partial positive ($\partial+$) hydrogen to form a hydrogen bond.

==

Practice Set 4: Questions 61–80

==

61. D is correct.

Calculate the moles of each gas:

moles = mass / molar mass

moles H_2 = 9.50 g / (2 × 1.01 g/mole)

moles H_2 = 4.70 moles

moles Ne = 14.0 g / (20.18 g/mole)

moles Ne = 0.694 moles

Calculate the mole fraction of H_2:

Mole fraction of H_2 = moles of H_2 / total moles in mixture

Mole fraction of H_2 = 4.70 moles / (4.70 moles + 0.694 moles)

Mole fraction of H_2 = 4.70 moles / (5.394 moles)

Mole fraction of H_2 = 0.87 moles

62. E is correct.

Vapor pressure is the pressure exerted by a vapor in equilibrium with its condensed phases (i.e., solid or liquid) in a closed system at a given temperature.

Boiling occurs when the vapor pressure of a liquid equals atmospheric pressure.

Vapor pressure increases as the temperature increases.

Liquid A boils at a lower temperature than B because the vapor pressure of Liquid A is closer to the atmospheric pressure.

63. B is correct.

Colligative properties of solutions depend on the ratio of the number of solute particles to the number of solvent molecules in a solution and not on the type of chemical species present.

Colligative properties include: lowering of vapor pressure, the elevation of boiling point, depression of freezing point, and increased osmotic pressure.

Freezing point (FP) depression:

$\Delta FP = -iKm$

where i = the number of particles produced when the solute dissociates, K = freezing point depression constant and m = molality (moles/kg solvent).

$$-iKm = -2 \text{ K}$$

$$-(1)\cdot(40)\cdot(x) = -2$$

$$x = -2 / -1(40)$$

$$x = 0.05 \text{ molal}$$

Assume that compound x does not dissociate.

$$0.05 \text{ mole compound } (x) / \text{kg camphor} = 25 \text{ g} / \text{kg camphor}$$

Therefore, if 0.05 mole = 25 g

$$1 \text{ mole} = 500 \text{ g}$$

64. E is correct.

Boiling occurs when the vapor pressure of a liquid equals atmospheric pressure.

Vapor pressure is the pressure exerted by a vapor in equilibrium with its condensed phases (i.e., solid or liquid) in a closed system at a given temperature.

Atmospheric pressure is the pressure exerted by the weight of air in the atmosphere.

Vapor pressure is inversely correlated with the strength of the intermolecular force.

In liquid form, the molecules are more likely to stick together due to stronger intermolecular forces. Few of them participate in the liquid-vapor equilibrium; therefore, the molecule would boil at a higher temperature.

65. D is correct.

Van der Waals equation describes factors that must be accounted for when the ideal gas law calculates values for nonideal gases.

The terms that affect the pressure and volume of the ideal gas law are intermolecular forces and the volume of nonideal gas molecules.

66. E is correct.

Ideal gas law:

$$PV = nRT$$

where P is pressure, V is volume, n is the number of molecules, R is the ideal gas constant, and T is the temperature of the gas.

Units of R can be calculated by rearranging the expression:

$$R = PV/nT$$

$$R = \text{atm}\cdot\text{L/mol}\cdot\text{K}$$

67. A is correct.

Ideal gas law:

$$PV = nRT$$

where P is pressure, V is volume, n is the number of molecules, R is the ideal gas constant, and T is the temperature of the gas.

Set the initial and final P/V/T conditions equal:

$$(P_1V_1 / T_1) = (P_2V_2 / T_2)$$

STP condition is the temperature of 0 °C (273 K) and pressure of 1 atm.

Solve for the final temperature:

$$(P_2V_2 / T_2) = (P_1V_1 / T_1)$$

$$T_2 = (P_2V_2T_1) / (P_1V_1)$$

$$T_2 = [(0.80 \text{ atm}) \times (0.155 \text{ L}) \times (273 \text{ K})] / [(1.00 \text{ atm}) \times (0.120 \text{ L})]$$

$$T_2 = 282.1 \text{ K}$$

Then convert temperature units to degrees Celsius:

$$T_2 = (282.1 \text{ K} - 273 \text{ K}) = 9.1 \text{ °C}$$

68. C is correct.

Atmospheric pressure is the pressure exerted by the weight of air in the atmosphere. Boiling occurs when the vapor pressure of the liquid is higher than the atmospheric pressure.

At standard atmospheric pressure and 22 °C, water vapor pressure is less than the atmospheric pressure, and it does not boil. However, when a vacuum pump is used, the atmospheric pressure is reduced until it has a lower vapor pressure than water, allowing water to boil at a much lower temperature.

69. E is correct.

The kinetic theory of gases describes a gas as a large number of small particles in constant rapid motion. These particles collide with each other and with the walls of the container.

Their average kinetic energy depends only on the absolute temperature of the system.

$$\text{temperature} = \tfrac{1}{2}mv^2$$

At high temperatures, the particles are moving greater velocity, and at a temperature of absolute zero (i.e., 0 K), there is no movement of gas particles.

Therefore, as temperature decreases, kinetic energy decreases, and so does the velocity of the gas molecules.

70. A is correct.

Dalton's law (i.e., the law of partial pressures) states that *the pressure exerted by a mixture of gases is equal to the sum of the individual gas pressures.*

The partial pressure of molecules in a mixture is proportional to their molar ratios.

Use the coefficients of the reaction to determine the molar ratio.

Based on that information, calculate the partial pressure of O_2:

> (coefficient O_2) / (sum of coefficients in the mixture) × total pressure

> $[1 / (2 + 1)]$ × 1,250 torr

> 417 torr = partial pressure of O_2

71. D is correct.

> $2 \text{ Na } (s) + \text{Cl}_2 (g) \rightarrow 2 \text{ NaCl } (s)$

In its elemental form, Na exists as a solid.

In its elemental form, chlorine exists as a gas.

72. B is correct.

Charles' law (i.e., the law of volumes) explains how, at constant pressure, gases behave when the temperature changes:

> $V \propto T$

or

> $V / T = \text{constant}$

or

> $(V_1 / T_1) = (V_2 / T_2)$

Volume and temperature are proportional.

Therefore, an increase in one term increases the other.

73. E is correct.

Gay-Lussac's law (i.e., pressure-temperature law) states that pressure is proportional to temperature:

> $P \propto T$

or $(P_1 / T_1) = (P_2 / T_2)$

or $(P_1 T_2) = (P_2 T_1)$

or $P / T = \text{constant}$

If the pressure of a gas increases, the temperature is increased proportionally.

Quadrupling the temperature increases the pressure fourfold.

Boyle's law (i.e., pressure-volume law) states that pressure and volume are inversely proportional:

$$(P_1V_1) = (P_2V_2)$$

or $P \times V = constant$

Reducing the volume by half increases pressure twofold.

Therefore, the increase in pressure would be by a factor of $4 \times 2 = 8$.

74. C is correct.

Scientists found that the relationships between pressure, temperature, and volume of a sample of gas hold for gases, and the gas laws were developed.

Boyle's law (i.e., pressure-volume law) states that pressure and volume are inversely proportional:

$$(P_1V_1) = (P_2V_2)$$

or

$$P \times V = constant$$

If the volume of a gas increases, its pressure decreases proportionally.

Charles' law (i.e., the law of volumes) explains how, at constant pressure, gases behave when the temperature changes:

$$(V_1 / T_1) = (V_2 / T_2)$$

Gay-Lussac's law (i.e., pressure-temperature law) states that pressure is proportional to temperature:

$$P \: \alpha \: T$$

or

$$(P_1 / T_1) = (P_2 / T_2) \text{ or } (P_1T_2) = (P_2T_1)$$

or

$$P / T = constant$$

If the pressure of a gas increases, the temperature increases.

75. A is correct.

Sublimation is the direct change of state from a solid to a gas, skipping the intermediate liquid phase.

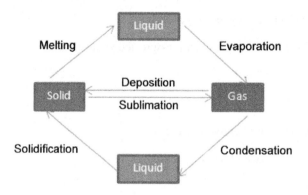

Interconversion of states of matter

An example of a compound that undergoes sublimation is solid carbon dioxide (i.e., dry ice). CO_2 changes phases from solid to gas (i.e., bypasses the liquid phase) and is often used as a cooling agent.

76. B is correct.

Graham's law of effusion states that the rate of effusion (i.e., escaping through a small hole) of a gas is inversely proportional to the square root of the molar mass of its particles.

Rate 1 / Rate 2 = √(molar mass gas 2 / molar mass gas 1)

Set the rate of effusion of krypton over the rate of effusion of methane:

$Rate_{Kr}$ / $Rate_{CH4}$ = √ [(M_{Kr}) / (M_{CH4})]

Solve for the ratio of effusion rates:

$Rate_{Kr}$ / $Rate_{CH4}$ = √ [(83.798 g/mol) / (16.04 g/mol)]

$Rate_{Kr}$ / $Rate_{CH4}$ = 2.29

Larger gas molecules must effuse at a slower rate; the effusion rate of Kr gas molecules must be slower than methane molecules:

$Rate_{Kr}$ = [$Rate_{CH4}$ / (2.29)]

$Rate_{Kr}$ = [(631 m/s) / (2.29)]

$Rate_{Kr}$ = 276 m/s

77. C is correct.

Observe information about gas C. The temperature, pressure, and volume of the gas are indicated.

Substitute these values into the ideal gas law equation to determine the number of moles:

PV = nRT

n = PV / RT

Convert units to the units indicated in the gas constant:

volume = 668.5 mL × (1 L / 1,000 mL)

volume = 0. 669 L

pressure = 745.5 torr × (1 atm / 760 torr)

pressure = 0.981 atm

temperature = 32.0 °C + 273.15 K

temperature = 305.2 K

n = PV / RT

n = (0.981 atm × 0.669 L) / (0.0821 L·atm K^{-1} mol^{-1} × 305.2 K)

n = 0.026 mole

The *law of mass conservation* states that the mass of products = the mass of reactants.

Based on this law, the mass of gas C can be determined:

mass product = mass reactant

mass A = mass B + mass C

5.2 g = 3.8 g + mass C

mass C = 1.4 g

Calculate the molar mass of C:

molar mass = 1.4 g / 0.026 mole

molar mass = 53.9 g / mole

78. E is correct.

Liquids take the shape of their container (i.e., they have an indefinite shape). This can be visualized by considering a liter of water poured into a cylindrical bucket or a square box. In both cases, it takes up the shape of the container.

Liquids have a definite volume. The volume of water in the given example is 1 L, regardless of its container.

79. D is correct.

Intermolecular forces act between neighboring molecules. Examples include hydrogen bonding, dipole-dipole, dipole-induced dipole, and van der Waals (i.e., London dispersion) forces.

Hydrogens, bonded directly to F, O or N, participate in hydrogen bonds. The hydrogen is partially positive (i.e., delta plus: $\partial+$) due to the bond to these electronegative atoms. The lone pair of electrons on the F, O or N interacts with the partial positive ($\partial+$) hydrogen to form a hydrogen bond.

80. D is correct.

When the balloon is placed in a freezer, the temperature of the helium in the balloon decreases because the surroundings are colder than the balloon, and heat is transferred from the balloon to the surrounding air until a thermal equilibrium is reached.

When the temperature of a gas decreases, the molecules move more slowly and become closer together, causing the volume of the balloon to decrease.

From the ideal gas law,

$$PV = nRT$$

If temperature decreases, the volume must decrease (assuming that pressure remains constant)

===

Practice Set 5: Questions 81–100

===

81. A is correct.

Boyle's law (i.e., pressure-volume law) states that pressure and volume are inversely proportional:

$$(P_1V_1) = (P_2V_2)$$

or

$$P \times V = \text{constant}$$

If the pressure of a gas increases, its volume decreases proportionally.

Doubling the pressure reduces the volume by half.

82. C is correct.

Vapor pressure is the pressure exerted by a vapor in equilibrium with its condensed phases (i.e., solid or liquid) in a closed system at a given temperature.

The pressure of a gas is the force that it exerts on the walls of its container. This is essentially the frequency and energy with which the gas molecules collide with the container's walls.

Partial pressure refers to the individual pressure of a gas that is in a mixture of gases.

Vapor pressure is the pressure of a vapor in thermodynamic equilibrium with its solid or liquid phases.

83. B is correct.

Colligative properties of solutions depend on the ratio of the number of solute particles to the number of solvent molecules in a solution and not on the type of chemical species present.

Colligative properties include: lowering of vapor pressure, the elevation of boiling point, depression of freezing point, and increased osmotic pressure.

Freezing point (FP) depression:

$$\Delta FP = -iKm$$

where i = the number of particles produced when the solute dissociates, K = freezing point depression constant, and m = molality (moles/kg solvent).

Given in the question stem:

$$\Delta FP = -10 \text{ K}$$

$$-iKm = -10 \text{ K}$$

Toluene is the solvent and does not dissociate; the K value is not used.

FP depression from the addition of benzene:

$i = 10 \text{ K / Km}$

$i = 1 - (1) \cdot (5.0) \cdot (x)$

K value of solvent:

$x = 2$ molal

2 moles toluene/kg benzene $= x$ moles / 0.1 kg benzene

x moles / 0.1 kg benzene $= 2$ moles toluene/kg benzene

x moles $= 2$ moles toluene/kg benzene $\times$ 0.1 kg benzene

toluene $= 0.2$ mole

84. E is correct.

Considering the ideal gas law, there is no difference between He and Ne since pressure, volume, temperature, and the number of moles are known quantities.

The identity of the gas would only be relevant to convert from moles to mass or vice-versa.

85. D is correct.

Gas molecules transfer kinetic energy between each other.

86. D is correct.

Assume a is the equivalent temperature (i.e., has the same value in °C and °F).

Set up a formula with the equivalent temperature on one side and the conversion factor to the other unit on the other side.

For example, the left side is °C, while the right side is the conversion to °F:

$a = [(9/5) \times a] + 32$

$a - (9/5)a = 32$

$(-4/5)a = 32$

$-4a = 160$

$a = -40$

At –40, the temperature is the same in °C and °F.

87. D is correct.

At low elevations (i.e., sea level), the boiling point of water is 100 °C.

Atmospheric pressure is the pressure exerted by the weight of air in the atmosphere.

The boiling point decreases with increasing altitude due to the reduced atmospheric pressure above the water at high altitudes.

88. A is correct.

Hydrocarbons are nonpolar molecules, which means that the dominant intermolecular force is London dispersion. This force increases as the number of atoms in each molecule increases.

Stronger force results in a higher boiling point.

CH_4 has the least atoms and therefore has the lowest boiling point.

89. C is correct.

The melting point of a substance depends on the pressure that it is at.

The triple point (point D on the graph) is the pressure and temperature at which a substance exists as a solid, liquid, and gas.

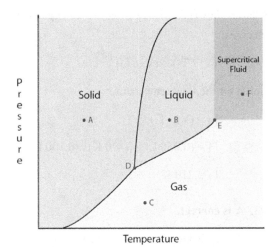

The critical point (point E on the graph) is the endpoint of the phase equilibrium curve where the liquid and its vapor become indistinguishable.

Generally, melting points are given at standard pressure. If the substance is denser in the solid than in the liquid state (which is true of most substances), the melting point increases when pressure is increased.

Phase diagram of pressure vs. temperature

This is observed with the phase diagram for CO_2. However, with certain substances, including water, the solid is less dense than the liquid state (e.g., ice is less dense than water), so the reverse is true, and the melting point decreases when pressure increases.

90. E is correct.

Dalton's law (i.e., the law of partial pressures) states that *the pressure exerted by a mixture of gases is equal to the sum of the individual gas pressures*.

The partial pressure of molecules in a mixture is proportional to their molar ratios.

Moles of CH_4: 32 g × 16 g/mol = 2 moles of CH_4

Moles of NH_3: 12.75 g × 17 g/mole = 0.75 mole of NH_3

Mole fraction of NH_3 gas:

   mole fraction of NH_3 = (moles of NH_3) / (total moles in mixture)

   mole fraction of NH_3 = (0.75 mole) / (2.75 moles)

   mole fraction of NH_3 = 0.273

Partial pressure of NH_3 gas:

   mole fraction of NH_3 × total pressure

   (0.273)·(2.4 atm) = 0.66 atm

91. B is correct.

Liquids take the shape of the container they are in (i.e., they have an indefinite shape). This can be visualized by considering a liter of water poured into a cylindrical bucket or a square box. In both cases, it takes up the shape of the container.

However, liquids have a definite volume. The volume of water in the example is 1 L, regardless of its container.

92. E is correct.

Charles' law (i.e., the law of volumes) explains how, at constant pressure, gases behave when the temperature changes:

$$(V_1 / T_1) = (V_2 / T_2)$$

Solve for the final temperature:

$$T_2 = (V_1 / T_1) / V_2$$

$$T_2 = (0.050 \text{ L}) \times [(420 \text{ K}) / (0.100 \text{ L})]$$

$$T_2 = 210 \text{ K}$$

93. A is correct.

Gay-Lussac's law (i.e., pressure-temperature law) states that pressure is proportional to temperature:

$$P \ \alpha \ T$$

or　　$$(P_1 / T_1) = (P_2 / T_2)$$

or　　$$(P_1 T_2) = (P_2 T_1)$$

or　　$$P / T = \text{constant}$$

If the pressure of a gas increases, the temperature increases.

Dalton's law (i.e., the law of partial pressures) states that *the pressure exerted by a mixture of gases is equal to the sum of the individual gas pressures*.

Charles' law (i.e., the law of volumes) explains how, at constant pressure, gases behave when the temperature changes:

$$V \ \alpha \ T \text{ or } V / T = \text{constant}$$

Volume and temperature are proportional.

Therefore, an increase in one term increases the other.

Boyle's law (i.e., pressure-volume law) states that pressure and volume are inversely proportional:

$$(P_1 V_1) = (P_2 V_2) \text{ or } P \times V = \text{constant}$$

If the volume of a gas increases, its pressure decreases proportionally.

94. E is correct.

Gay-Lussac's law (i.e., pressure-temperature law) states that pressure is proportional to temperature:

$P \alpha T$

or

$(P_1 / T_1) = (P_2 / T_2)$

or

$(P_1 T_2) = (P_2 T_1)$

or

$P / T = constant$

If the pressure of a gas increases, the temperature increases.

The kinetic theory of gases describes a gas as a large number of small particles in constant rapid motion. These particles collide with each other and with the walls of the container.

Their average kinetic energy depends only on the absolute temperature of the system.

$$temperature = \tfrac{1}{2}mv^2$$

At high temperatures, the particles are moving with greater velocity.

95. C is correct.

Vaporization refers to the change of state from a liquid to a gas. There are two types of vaporization: boiling and evaporation, which are differentiated based on the temperature at which they occur.

Evaporation occurs at a temperature below the boiling point, while boiling occurs at or above the boiling point.

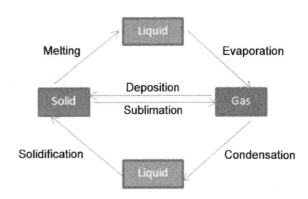

Interconversion of states of matter

96. C is correct.

Intermolecular forces act between neighboring molecules. Examples include hydrogen bonding, dipole-dipole, dipole-induced dipole, and van der Waals (i.e., London dispersion) forces.

A dipole-dipole attraction involves asymmetric, polar molecules (based on differences in electronegativity between atoms) that create a dipole moment (i.e., a net vector indicating the force).

Dimethyl ether with an indicated bond angle

97. E is correct.

Ideal gas law:

PV = nRT

where P is pressure, V is volume, n is the number of molecules, R is the ideal gas constant, and T is the temperature of the gas.

R and n are constant.

Therefore, decreasing the volume, increasing temperature, or increasing the number of molecules increases the pressure of a gas.

98. E is correct.

Avogadro's law is an experimental gas law relating the volume of a gas to the amount of substance of gas present. It states that "equal volumes of all gases, at the same temperature and pressure, have the same number of molecules."

For a given mass of an ideal gas, the gas's volume and amount (moles) are directly proportional if the temperature and pressure are constant.

V α n

or

V / n = k

where V = volume of the gas, n is the amount of substance of the gas (measured in moles).

k is a constant equal to RT / P,

where R is the universal gas constant, T is the Kelvin temperature, and P is the pressure.

As temperature and pressure are constant, RT / P is constant and represented as k (derived from the ideal gas law).

At the same temperature and pressure, equal volumes of gases have the same number of molecules.

Comparing the same substance under two different sets of conditions:

V_1 / n_1 = V_2 / n_2

As the number of moles of gas increases, the volume of the gas increases in proportion.

Similarly, if the number of moles of gas is decreased, then the volume decreases.

Thus, the number of molecules or atoms in a specific volume of an ideal gas is independent of their size or the molar mass of the gas.

99. B is correct.

The ideal gas law: PV = nRT

Using proportions, if the temperature is increased by a factor of 1.5 and the pressure is increased by a factor of 1.5 (constant number of moles), volume remains the same.

100. C is correct.

Since both He and Ne are at the same temperature, pressure, and volume, they have equal moles of gas.

Even though there are equal moles of each gas of He and Ne, use the molar masses given in the periodic table and notice that the samples contain different masses.

Notes for active learning

Stoichiometry – Detailed Explanations

==

Practice Set 1: Questions 1–20

==

1. C is correct.

One mole of an ideal gas occupies 22.71 L at STP.

$$3 \text{ mole} \times 22.71 \text{ L} = 68.13 \text{ L}$$

2. C is correct.

Use the mnemonic OIL RIG: <u>O</u>xidation <u>I</u>s <u>L</u>oss, <u>R</u>eduction <u>I</u>s <u>G</u>ain (of electrons).

Oxidation is the loss of electrons, while reduction is the gain of electrons.

An oxidizing agent undergoes reduction, while a reducing agent undergoes oxidation.

Check the oxidation numbers of each atom.

The oxidation number for S is +6 as a reactant and +4 as a product, which means that it is reduced.

If a substance is reduced in a redox reaction, it is the oxidizing agent because it causes the other reagent (HI) to be oxidized.

3. C is correct.

Balanced equation (combustion): $2 C_6H_{14} + 19 O_2 \rightarrow 12 CO_2 + 14 H_2O$

4. E is correct.

An oxidation number is a charge on an atom that is not present in the elemental state when the element is neutral. Electronegativity refers to the attraction an atom has for additional electrons.

5. C is correct.

CH_3COOH has 2 carbons, 4 hydrogens, and 2 oxygens: $C_2O_2H_4$

Reduce to the smallest coefficients by dividing by two: COH_2

6. B is correct.

Oxidation numbers of Cl in each compound:

$NaClO_2$ +3

$Al(ClO_4)_3$ +7

$Ca(ClO_3)_2$ +5

$LiClO_3$ +5

---

7. A is correct.

Formula mass is a synonym for molecular mass/molecular weight (MW).

Molecular mass = (atomic mass of C) + (2 × atomic mass of O)

Molecular mass = (12.01 g/mole) + (2 × 16.00 g/mole)

Molecular mass = 44.01 g/mole

Note: 1 amu = 1 g/mole

8. C is correct.

Cadmium is located in group II, so its oxidation number is +2.

Sulfur tends to form a sulfide ion (oxidation number –2).

The net oxidation number is zero, and therefore the compound is CdS.

9. B is correct.

Balanced reaction (single replacement):

$$Cl_2 \, (g) + 2 \, NaI \, (aq) \rightarrow I_2 \, (s) + 2 \, NaCl \, (aq)$$

10. C is correct.

Use the mnemonic OIL RIG: Oxidation Is Loss, Reduction Is Gain (of electrons).

Oxidation is the loss of electrons, while reduction is the gain of electrons.

An oxidizing agent undergoes reduction, while a reducing agent undergoes oxidation.

When there are 2 reactants in a redox reaction, one will be oxidized and the other reduced. The oxidized species is the reducing agent and vice versa.

Determine the oxidation number of all atoms:

I changes its oxidation state from –1 as a reactant to 0 as a product.

I is oxidized; therefore, the compound that contains I (NaI) is the reducing agent. Species in their elemental state have an oxidation number of 0.

Since the reaction is spontaneous, NaI is the *strongest* reducing agent because it can reduce another species spontaneously without needing any external energy for the reaction to proceed.

11. C is correct.

To determine the remainder of a reactant, calculate how many moles of that reactant are required to produce the specified amount of product.

Use the coefficients to calculate the moles of the reactant:

$$\text{moles } N_2 = (1 \, / \, 2) \times 18 \text{ moles} = 9 \text{ moles } N_2$$

Therefore, the remainder is: (14.5 moles N_2) – (9 moles N_2) = 5.5 moles N_2 remaining

12. D is correct.

$CaCO_3 \rightarrow CaO + CO_2$ is a decomposition reaction because one complex compound is being broken down into two or more parts. However, the redox part is incorrect because the reactants and products are compounds (no single elements), and there is no change in the oxidation state (i.e., no gain or loss of electrons).

A: $AgNO_3 + NaCl \rightarrow AgCl + NaNO_3$ is correctly classified as a double-replacement reaction (sometimes referred to as double-displacement) because parts of two ionic compounds are exchanged to make two new compounds. It is a non-redox reaction because the reactants and products are compounds (no single elements), and there is no change in oxidation state (i.e., no gain or loss of electrons).

B: $Cl_2 + F_2 \rightarrow 2\ ClF$ is correctly classified as a synthesis reaction because two species are combining to form a more complex chemical compound as the product. It is a redox reaction because the F atoms in F_2 are reduced (i.e., gain electrons), and the Cl atoms in Cl_2 are oxidized (i.e., lose electrons), forming a covalent compound.

C: $H_2O + SO_2 \rightarrow H_2SO_3$ is correctly classified as a synthesis reaction because two species are combining to form a more complex chemical compound as the product. It is a non-redox reaction because the reactants and products are compounds (no single elements), and there is no change in oxidation state (i.e., no gain or loss of electrons).

13. C is correct.

Hydrogen is being oxidized (or ignited) to produce water.

The balanced equation that describes the ignition of hydrogen gas in the air to produce water is:

$$\tfrac{1}{2}\ O_2 + H_2 \rightarrow H_2O$$

This equation suggests that for every mole of water produced in the reaction, 1 mole of hydrogen and a ½ mole of oxygen gas is needed.

The maximum amount of water produced in the reaction is determined by the amount of limiting reactant available. This requires identifying which reactant is the limiting reactant and can be done by comparing the number of moles of oxygen and hydrogen.

From the question, there are 10 grams of oxygen gas and 1 gram of hydrogen gas. This is equivalent to 0.3125 moles of oxygen and 0.5 moles of hydrogen.

Because the reaction requires twice as much hydrogen as oxygen gas, the limiting reactant is hydrogen gas (0.3215 moles of O_2 requires 0.625 moles of hydrogen, but only 0.5 moles of H_2 are available).

Since one equivalent of water is produced for every equivalent of hydrogen that is burned, the amount of water produced is:

$$(18 \text{ grams/mol } H_2O) \times (0.5 \text{ mol } H_2O) = 9 \text{ grams of } H_2O$$

14. D is correct.

Calculate the moles of Cl^- ion:

Moles of Cl^- ion = number of Cl^- atoms / Avogadro's number

Moles of Cl^- ion = 6.8×10^{22} atoms / 6.02×10^{23} atoms/mole

Moles of Cl^- ion = 0.113 moles

In a solution, $BaCl_2$ dissociates:

$BaCl_2 \rightarrow Ba^{2+} + 2\ Cl^-$

The moles of from Cl^- previous calculation can be used to calculate the moles of Ba^{2+}:

Moles of Ba^{2+} = (coefficient of Ba^{2+} / coefficient Cl^-) × moles of Cl^-

Moles of Ba^{2+} = (½) × 0.113 moles

Moles of Ba^{2+} = 0.0565 moles

Calculate the mass of Ba^{2+}:

Mass of Ba^{2+} = moles of Ba^{2+} × atomic mass of Ba

Mass of Ba^{2+} = 0.0565 moles × 137.33 g/mole

Mass of Ba^{2+} = 7.76 g

15. D is correct.

Na	Br	O_3
+1	x	(3×-2)
+1	x	-6

The sum of charges in a neutral molecule is zero:

1 + oxidation number of Br:

$1 + x + (-6) = 0$

$-5 + x = 0$

$x = +5$

oxidation number of Br = +5

16. E is correct.

Balanced equation:

$2\ Al + Fe_2O_3 \rightarrow 2\ Fe + Al_2O_3$

The sum of the coefficients of the products = 3.

17. B is correct.

$2 H_2O_2 (s) \rightarrow 2 H_2O (l) + O_2 (g)$ is incorrectly classified. The decomposition part of the classification is correct because one complex compound (H_2O_2) is being broken down into two or more parts (H_2O and O_2).

However, it is a redox reaction because oxygen is being lost from H_2O_2, one of the three indicators of a redox reaction (i.e., electron loss/gain, hydrogen loss/gain, oxygen loss/gain). Additionally, one of the products is a single element (O_2), which is an indication that it is a redox reaction.

$AgNO_3 (aq) + KOH (aq) \rightarrow KNO_3 (aq) + AgOH (s)$ is correctly classified as a non-redox reaction because the reactants and products are compounds (no single elements), and there is no change in oxidation state (i.e., no gain or loss of electrons). It is also a precipitation reaction because the chemical reaction occurs in an aqueous solution, and one of the products formed (AgOH) is insoluble, which makes it a precipitate.

$Pb(NO_3)_2 (aq) + 2 Na (s) \rightarrow Pb (s) + 2 NaNO_3 (aq)$ is correctly classified as a redox reaction because the nitrate (NO_3^-), which dissociates, is oxidized (loses electrons), the Na^+ is reduced (gains electrons). The two combine to form the ionic compound $NaNO_3$. It is a single-replacement reaction because one element is substituted for another element in a compound, making a new compound ($2NaNO_3$) and an element (Pb).

$HNO_3 (aq) + LiOH (aq) \rightarrow LiNO_3 (aq) + H_2O (l)$ is correctly classified as a non-redox reaction because the reactants and products are compounds (no single elements), and there is no change in oxidation state (i.e., no gain or loss of electrons). It is a double-replacement reaction because parts of two ionic compounds are exchanged to make two new compounds.

18. C is correct.

At STP, 1 mole of gas has a volume of 22.4 L.

Calculate moles of O_2:

moles of O_2 = (15.0 L) / (22.4 L/mole)

moles of O_2 = 0.67 mole

Calculate the moles of O_2, using the fact that 1 mole of O_2 has 6.02×10^{23} O_2 molecules:

molecules of O_2 = 0.67 mole × (6.02×10^{23} molecules/mole)

molecules of O_2 = 4.03×10^{23} molecules

19. E is correct.

20. D is correct.

Use the mnemonic OIL RIG: Oxidation Is Loss, Reduction Is Gain (of electrons).

Oxidation is the loss of electrons, while reduction is the gain of electrons.

An oxidizing agent undergoes reduction, while a reducing agent undergoes oxidation.

Co^{2+} is the starting reactant because it has to lose electrons and produce an ion with a higher oxidation number.

===

Practice Set 2: Questions 21–40

===

21. D is correct.

Double replacement reaction:

HCl has a molar mass of 36.5 g/mol:

  365 grams HCl = 10 moles HCl

  10 moles are 75% of the yield.

  10 moles PCl_3 / 0.75 = 13.5 moles of HCl.

Apply the mole ratios:

  Coefficients: 3 HCl = PCl_3

  13.5 / 3 = moles of PCl_3

  13.5 moles of HCl is produced by 4.5 moles of PCl_3

22. A is correct.

All elements in their free state have an oxidation number of zero.

23. B is correct.

Balanced chemical equation:

  $C_3H_8 + 5\ O_2 \rightarrow 3\ CO_2 + 4\ H_2O$

By using the coefficients from the balanced equation, apply dimensional analysis to solve for the volume of H_2O produced from the 2.6 L C_3H_8 reacted:

  $V_{H2O} = V_{C3H8} \times$ (mol H_2O / mol C_3H_8)

  $V_{H2O} = (2.6\ L) \times (4\ mol\ /\ 1\ mol)$

  $V_{H2O} = 10.4\ L$

24. C is correct.

Use the mnemonic OIL RIG: Oxidation Is Loss, Reduction Is Gain (of electrons).

Oxidation is the loss of electrons, while reduction is the gain of electrons.

An oxidizing agent undergoes reduction, while a reducing agent undergoes oxidation.

The oxidation number of Hg decreases (i.e., reduction) from +2 as a reactant ($HgCl_2$) to +1 as a product (Hg_2Cl_2). This means that Hg gained one electron.

25. C is correct.

Balanced equation (synthesis):

$$4\ P\ (s) + 5\ O_2\ (g) \rightarrow 2\ P_2O_5\ (s)$$

26. E is correct.

27. A is correct.

Formula mass is a synonym for molecular mass/molecular weight (MW).

Start by calculating the mass of the formula unit (CH_2O)

$$CH_2O = (12.01\ g/mol + 2.02\ g/mol + 16.00\ g/mol) = 30.03\ g/mol$$

Divide the molar mass with the formula unit mass:

$$180\ g/mol\ /\ 30.03\ g/mol = 5.994$$

Round to the closest whole number: 6

Multiply the formula unit by 6:

$$(CH_2O)_6 = C_6H_{12}O_6$$

28. E is correct.

Calculate the moles of LiI present. The molecular weight of LiI is 133.85 g/mol.

Moles of LiI:

$$6.45\ g\ /\ 133.85\ g/mole = 0.0482\ moles$$

Each mole of LiI contains 6.02×10^{23} molecules.

Number of molecules:

$$0.0482 \times 6.02 \times 10^{23} = 2.90 \times 10^{22}\ molecules$$

1 molecule is equal to 1 formula unit.

29. E is correct.

A combustion engine creates various nitrogen oxide compounds, often referred to as NOx gases because each gas would have a different value of x in its formula.

30. C is correct.

Balanced equation (synthesis):

$$4\ P\ (s) + 3\ O_2\ (g) \rightarrow 2\ P_2O_3\ (s)$$

31. A is correct.

Use the mnemonic OIL RIG: $\underline{O}$xidation $\underline{I}$s $\underline{L}$oss, $\underline{R}$eduction $\underline{I}$s $\underline{G}$ain (of electrons).

Oxidation is the loss of electrons, while reduction is the gain of electrons.

An oxidizing agent undergoes reduction, while a reducing agent undergoes oxidation.

Because the forward reaction is spontaneous, the reverse reaction is not spontaneous.

Sn cannot reduce Mg^{2+}; therefore, Sn is the weakest reducing agent.

32. C is correct.

Balancing Redox Equations

From the balanced equation, the coefficient for the proton can be determined. Balancing a redox equation requires balancing the atoms in the equation, but the charges must be balanced as well.

The equations must be separated into two different half-reactions. One of the half-reactions addresses the oxidizing component, and the other addresses the reducing component.

Magnesium is being oxidized; therefore, the unbalanced oxidation half-reaction will be:

$$Mg\ (s) \rightarrow Mg^{2+}\ (aq)$$

Furthermore, the nitrogen is being reduced; therefore, the unbalanced reduction half-reaction will be:

$$NO_3^-\ (aq) \rightarrow NO_2\ (aq)$$

At this stage, each half-reaction needs to be balanced for each atom, and the net electric charge on each side of the equations must be balanced.

Order of operations for balancing half-reactions:

1) Balance atoms except for oxygen and hydrogen.

2) Balance the oxygen atom count by adding water.

3) Balance the hydrogen atom count by adding protons.

   a) If in basic solution, add equal amounts of hydroxide to each side to cancel the protons.

4) Balance the electric charge by adding electrons.

5) If necessary, multiply the coefficients of one half-reaction equation by a factor that cancels the electron count when both equations are combined.

6) Cancel any ions or molecules that appear on both sides of the overall equation.

After determining the balanced overall redox reaction, stoichiometry indicates the moles of protons that are involved.

For magnesium:

$$Mg\ (s) \rightarrow Mg^{2+}\ (aq)$$

The magnesium is already balanced with a coefficient of 1. There are no hydrogen or oxygen atoms present in the equation. The magnesium cation has a +2 charge, so to balance the charge, 2 moles of electrons should be added to the right side.

The balanced half-reaction for oxidation:

$$Mg\,(s) \rightarrow Mg^{2+}\,(aq) + 2\,e^-$$

For nitrogen:

$$NO_3^-\,(aq) \rightarrow NO_2\,(aq)$$

The equation is already balanced for nitrogen because one nitrogen atom appears on both sides of the reaction. The nitrate reactant has three oxygen atoms, while the nitrite product has two oxygen atoms.

To balance the oxygen, one mole of water should be added to the right side:

$$NO_3^-\,(aq) \rightarrow NO_2\,(aq) + H_2O$$

Adding water to the right side of the equation introduces hydrogen atoms to that side. Therefore, the hydrogen atom count needs to be balanced. Water possesses two hydrogen atoms; therefore, two protons need to be added to the left side of the reaction:

$$NO_3^-\,(aq) + 2\,H^+ \rightarrow NO_2\,(aq) + H_2O$$

The reaction is occurring in acidic conditions. If the reaction were basic, then OH^- would need to be added to both sides to cancel the protons.

Next, the net charge must be balanced. The left side has a net charge of +1 (+2 from the protons and –1 from the electron), while the right side is neutral. Therefore, one electron should be added to the left side:

$$NO_3^-\,(aq) + 2\,H^+ + e^- \rightarrow NO_2\,(aq) + H_2O$$

When half-reactions are recombined, the electrons in the overall reaction must cancel.

The reduction half-reaction contributes one electron to the left side of the overall equation, while the oxidation half-reaction contributes two electrons to the product side.

Therefore, the coefficients of the reduction half-reaction should be doubled:

$$2 \times [NO_3^-\,(aq) + 2\,H^+ + e^- \rightarrow NO_2\,(aq) + H_2O] = 2\,NO_3^-\,(aq) + \mathbf{4\,H^+} + 2\,e^- \rightarrow 2\,NO_2\,(aq) + 2\,H_2O$$

The answer is 4 at this step.

Combining both half-reactions gives:

$$Mg\,(s) + 2\,NO_3^-\,(aq) + \mathbf{4\,H^+} + 2\,e^- \rightarrow Mg^{2+}\,(aq) + 2\,e^- + 2\,NO_2\,(aq) + 2\,H_2O$$

Cancel the electrons that appear on both sides of the reaction in the following balanced net equation:

$$Mg\,(s) + 2\,NO_3^-\,(aq) + \mathbf{4\,H^+} \rightarrow Mg^{2+}\,(aq) + 2\,NO_2\,(aq) + 2\,H_2O$$

This equation is now fully balanced for mass, oxygen, hydrogen and electric charge.

33. E is correct.

To obtain moles, divide sample mass by the molecular mass.

Because the options have the same mass, a comparison of molecular mass provides the answer without performing the actual calculation.

The molecule with the smallest molecular mass has the greatest number of moles.

34. A is correct.

The law of constant composition states that *all samples of a given chemical compound have the same chemical composition by mass.* This is true for compounds, including ethyl alcohol.

This law is the *law of definite proportions* or *Proust's Law* because this observation was first made by the French chemist Joseph Proust.

35. A is correct.

The coefficients in balanced reactions refer to moles (or molecules) but not grams.

1 mole of N_2 gas reacts with 3 moles of H_2 gas to produce 2 moles of NH_3 gas.

36. C is correct.

Calculate the number of moles of Al_2O_3:

$$2 \text{ Al} / 1 \text{ Al}_2O_3 = 0.2 \text{ moles Al} / x \text{ moles Al}_2O_3$$

$$x \text{ moles Al}_2O_3 = 0.2 \text{ moles Al} \times 1 \text{ Al}_2O_3 / 2 \text{ Al}$$

$$x = 0.1 \text{ moles Al}_2O_3$$

Therefore:

$$(0.1 \text{ mole Al}_2O_3) \cdot (102 \text{ g/mole Al}_2O_3) = 10.2 \text{ g Al}_2O_3$$

37. D is correct.

Because Na is a metal, Na is more electropositive than H.

Therefore, the positive charge is on Na (i.e., +1), which means that the charge on H is –1.

38. A is correct.

$PbO + C \rightarrow Pb + CO$ is a single replacement reaction because only one element is being transferred from one reactant to another.

However, it is a redox reaction because the lead (Pb^{2+}), which dissociates, is reduced (gains electrons), the oxygen (O^{2-}) is oxidized (loses electrons). Additionally, one of the products is a single element (Pb), which is an indication that it is a redox reaction.

C: the reaction is double-replacement and not combustion because the combustion reactions must have hydrocarbon and O_2 as reactants. Double-replacement reactions often result in the formation of a solid, water, and gas.

39. B is correct.

The sum of the oxidation numbers in any neutral molecule must be zero.

The oxidation number of each H is +1, and the oxidation number of each O is –2.

If x denotes the oxidation number of S in H_2SO_4, then:

$$2(+1) + x + 4(-2) = 0$$

$$+2 + x + -8 = 0$$

$$x = +6$$

In H_2SO_4, the oxidation sate of sulfur is +6.

40. A is correct.

To calculate the number of molecules, start by calculating the moles of gas using the ideal gas equation:

$$PV = nRT$$

Rearrange to isolate the number of moles:

$$n = PV / RT$$

Because the gas constant R is in $L \cdot atm \ K^{-1} \ mol^{-1}$, the pressure has to be converted into atm:

$$780 \text{ torr} \times (1 \text{ atm} / 760 \text{ torr}) = 1.026 \text{ atm}$$

Convert the volume to liters:

$$500 \text{ mL} \times 0.001 \text{ L/mL} = 0.5 \text{ L}$$

Substitute into the ideal gas equation:

$$n = PV / RT$$

$$n = 1.026 \text{ atm} \times 0.5 \text{ L} / (0.08206 \ L \cdot atm \ K^{-1} \ mol^{-1} \times 320 \text{ K})$$

$$n = 0.513 \ L \cdot atm / (26.259 \ L \cdot atm \ mol^{-1})$$

$$n = 0.0195 \text{ mol}$$

===

Practice Set 3: Questions 41–60

===

41. B is correct.

Use the mnemonic OIL RIG: <u>O</u>xidation <u>I</u>s <u>L</u>oss, <u>R</u>eduction <u>I</u>s <u>G</u>ain (of electrons).

The oxidizing and reducing agents are reactants, not products, in a redox reaction.

The oxidizing agent is the species that gets reduced (i.e., gains electrons) and causes oxidation of the other species.

The reducing agent is the species oxidized (i.e., loses electrons) and causes a reduction of the other species.

42. E is correct.

Balancing a redox equation in acidic solution by the half-reaction method involves each of the steps described.

43. D is correct.

Formula mass is a synonym for molecular mass/molecular weight (MW).

MW of $C_6H_{12}O_6$ = (6 × atomic mass C) + (12 × atomic mass H) + (6 × atomic mass O)

MW of $C_6H_{12}O_6$ = (6 × 12.01 g/mole) + (12 × 1.01 g/mole) + (6 × 16.00 g/mole)

MW of $C_6H_{12}O_6$ = (72.06 g/mole) + (12.12 g/mole) + (96.00 g/mole)

MW of $C_6H_{12}O_6$ = 180.18 g/mole

44. E is correct.

The molecular mass of oxygen is 16 g/mole.

The atomic mass unit (amu) or dalton (Da) is the standard unit for indicating mass on an atomic or molecular scale (atomic mass). One amu is approximately the mass of one nucleon (either a single proton or neutron) and is numerically equivalent to 1 g/mol.

45. D is correct.

Balanced equation (combustion):

$$2\ C_2H_6 + 7\ O_2 \rightarrow 4\ CO_2 + 6\ H_2O$$

46. E is correct.

Use the mnemonic OIL RIG: <u>O</u>xidation <u>I</u>s <u>L</u>oss, <u>R</u>eduction <u>I</u>s <u>G</u>ain (of electrons).

Oxidation is the loss of electrons, while reduction is the gain of electrons.

An oxidizing agent undergoes reduction, while a reducing agent undergoes oxidation.

Evaluate the oxidation number in each atom.

S changes from +6 as a reactant to –2 as a product, reducing it because the oxidation number decreases.

47. C is correct.

Balanced equation (double replacement):

$$C_2H_5OH\ (g) + 3\ O_2\ (g) \rightarrow 2\ CO_2\ (g) + 3\ H_2O\ (g)$$

48. E is correct.

Note: 1 amu (atomic mass unit) = 1 g/mole.

Moles of C = mass of C / atomic mass of C

Moles of C = (4.50 g) / (12.01 g/mole)

Moles of C = 0.375 mole

49. C is correct.

According to the law of mass conservation, there should be equal amounts of carbon in the product (CO_2) and reactant (the hydrocarbon sample).

Start by calculating the amount of carbon in the product (CO_2).

First, calculate the mass % of carbon in CO_2:

Molecular mass of CO_2 = atomic mass of carbon + (2 × atomic mass of oxygen)

Molecular mass of CO_2 = 12.01 g/mole + (2 × 16.00 g/mole)

Molecular mass of CO_2 = 44.01 g/mole

Mass % of carbon in CO_2 = (mass of carbon / molecular mass of CO_2) × 100%

Mass % of carbon in CO_2 = (12.01 g/mole / 44.01 g/mole) × 100%

Mass % of carbon in CO_2 = 27.3%

Calculate the mass of carbon in the CO_2:

Mass of carbon = mass of CO_2 × mass % of carbon in CO_2

Mass of carbon = 8.98 g × 27.3%

Mass of carbon = 2.45 g

The mass of carbon in the starting reactant (the hydrocarbon sample) should also be 2.45 g.

Mass % carbon in hydrocarbon = (mass carbon in sample / total mass sample) × 100%

Mass % of carbon in hydrocarbon = (2.45 g / 6.84 g) × 100%

Mass % of carbon in hydrocarbon = 35.8%

50. E is correct.

The number of each atom on the reactants' side must be the same and equal in number to the number of atoms on the product side of the reaction.

51. D is correct.

Using ½ reactions, the balanced reaction is:

$$6 \, (Fe^{2+} \rightarrow Fe^{3} + 1 \, e^{-})$$

$$\underline{Cr_2O_7^{2-} + 14 \, H^+ + 6 \, e^- \rightarrow 2 \, Cr^{3+} + 7 \, H_2O}$$

$$Cr_2O_7^{2-} + 14 \, H^+ + 6 \, Fe^{2+} \rightarrow 2 \, Cr^{3+} + 6 \, Fe^{3+} + 7 \, H_2O$$

The sum of coefficients in the balanced reaction is 36.

52. E is correct.

Balanced reaction:

$$4 \, RuS \, (s) + 9 \, O_2 + 4 \, H_2O \rightarrow 2 \, Ru_2O_3 \, (s) + 4 \, H_2SO_4$$

Calculate the moles of H_2SO_4:

49 g × 1 mol / 98 g = 0.5 mole H_2SO_4

If 0.5 mole H_2SO_4 is produced, set a ratio for O_2 consumed.

$9 \, O_2 / 4 \, H_2SO_4 = x$ moles $O_2 / 0.5$ mole H_2SO_4

x moles $O_2 = 9 \, O_2 / 4 \, H_2SO_4 × 0.5$ mole H_2SO_4

$x = 1.125$ moles $O_2 × 22.4$ liters / 1 mole

$x = 25.2$ liters of O_2

53. C is correct.

Identify the limiting reactant:

Al is the limiting reactant because 2 atoms of Al combine with 1 molecule of ferric oxide to produce the products. If the reaction starts with equal numbers of moles of each reactant, Al is depleted.

Use moles of Al and coefficients to calculate moles of other products/reactants.

Coefficient of Al = coefficient of Fe

Moles of iron produced = 0.20 moles.

54. B is correct.

$$S_2 \qquad O_8$$

$$2x \qquad (8 \times -2)$$

The sum of charges on the molecule is –2:

$$2x + (8 \times -2) = -2$$

$$2x + (-16) = -2$$

$$2x = +14$$

oxidation number of S = +7

55. D is correct.

Na is in group IA, so its oxidation number is +1.

Oxygen is usually –2, with some exceptions, such as peroxide (H_2O_2), where it is –1.

Cr is a transition metal and could have more than one possible oxidation number.

To determine Cr's oxidation number, use the known oxidation numbers:

$$2(+1) + Cr + 4(-2) = 0$$

$$2 + Cr - 8 = 0$$

$$Cr = 6$$

56. B is correct.

The net charge of $H_2SO_4 = 0$

Hydrogen has a common oxidation number of +1, whereas each oxygen atom has an oxidation number of –2.

$$H_2 \qquad\qquad S \qquad\qquad O_4$$

$$(2 \times 1) \qquad\qquad x \qquad\qquad (4 \times -2)$$

$$2 + x + -8 = 0$$

$$x = +6$$

In H_2SO_4, the oxidation of sulfur is +6.

57. C is correct.

Use the mnemonic OIL RIG: Oxidation Is Loss, Reduction Is Gain (of electrons).

Oxidation is the loss of electrons, while reduction is the gain of electrons.

An oxidizing agent undergoes reduction, while a reducing agent undergoes oxidation.

The oxidation number for Sn is +2 as a reactant and +4 as a product; an increase in oxidation number means that Sn^{2+} is oxidized.

58. D is correct.

Na_2Cl_2 and 2 NaCl are not the same.

Na_2Cl_2 is a molecule of 2 Na and 2 Cl.

NaCl is a compound that consists of 1 Na and 1 Cl.

59. E is correct.

I: represents the volume of 1 mol of an ideal gas at STP

II: mass of 1 mol of PH_3

III: number of molecules in 1 mol of PH_3

60. D is correct.

Use the mnemonic OIL RIG: <u>O</u>xidation <u>I</u>s <u>L</u>oss, <u>R</u>eduction <u>I</u>s <u>G</u>ain (of electrons).

Oxidation is the loss of electrons, while reduction is the gain of electrons.

An oxidizing agent undergoes reduction, while a reducing agent undergoes oxidation.

Co^{2+} is the starting reactant because it has to lose electrons and produce an ion with a higher oxidation number.

===

Practice Set 4: Questions 61–80

===

61. E is correct.

Use the mnemonic OIL RIG: $\underline{O}$xidation $\underline{I}$s $\underline{L}$oss, $\underline{R}$eduction $\underline{I}$s $\underline{G}$ain (of electrons).

The oxidizing and reducing agents are reactants, not products, in a redox reaction.

The oxidizing agent is the species that is reduced (i.e., gains electrons).

The reducing agent is the species that is oxidized (i.e., loses electrons).

62. D is correct.

Calculations:

% mass Cl = (molecular mass Cl) / (molecular mass of compound) × 100%

% mass Cl = 4(35.5 g/mol) / [12 g/mol + 4(35.5 g/mol)] × 100%

% mass Cl = (142 g/mol) / (154 g/mol) × 100%

% mass Cl = 0.922 × 100% = 92%

63. E is correct.

I: ionic charge of reactants equals the ionic charge of products, which means that no electrons can "disappear" from the reaction; they can only be transferred from one atom to another.

Therefore, the total charge will be the same.

II: atoms of each reactant must equal atoms of product, which is true for chemical reactions because atoms in chemical reactions cannot be created or destroyed.

III: another atom must lose any electrons gained by one atom.

64. B is correct.

Determine the molecular weight of each compound.

Note, unlike % by mass, there is no division.

$Cl_2C_2H_4$ = 2(35.5 g/mol) + 2(12 g/mol) + 4(1 g/mol)

$Cl_2C_2H_4$ = 71 g/mol + 24 g/mol + 4 g/mol

$Cl_2C_2H_4$ = 99 g/mol

65. E is correct.

Formula mass is a synonym for molecular mass/molecular weight (MW).

Start by calculating the mass of the formula unit (CH):

CH = (12.01 g/mol + 1.01 g/mol) = 13.02 g/mol

Divide the molar mass by the formula unit mass:

78 g/mol / 13.02 g/mol = 5.99

Round to the closest whole number: 6

Multiply the formula unit by 6:

$(CH)_6 = C_6H_6$

66. D is correct.

There is 1 oxygen atom from GaO plus 6 oxygen atoms from $(NO_3)_2$, so the total is 7.

67. B is correct.

Ethanol (C_2H_5OH) undergoes combination with oxygen to produce carbon dioxide and water.

68. C is correct.

Balanced reaction (combustion): $2\ C_3H_7OH + 9\ O_2 \rightarrow 6\ CO_2 + 8\ H_2O$

69. C is correct.

Balanced equation (double replacement):

$Co_2O_3\ (s) + 3\ CO\ (g) \rightarrow 2\ Co\ (s) + 3\ CO_2\ (g)$

70. B is correct.

The molar volume (V_m) is the volume occupied by one mole of a substance (i.e., element or compound) at a given temperature and pressure.

71. A is correct.

Use the mnemonic OIL RIG: Oxidation Is Loss, Reduction Is Gain (of electrons).

Oxidation is the loss of electrons, while reduction is the gain of electrons.

An oxidizing agent undergoes reduction, while a reducing agent undergoes oxidation.

Because the forward reaction is spontaneous, the reverse reaction is not spontaneous.

Mg^{2+} cannot oxidize Sn; therefore, Mg^{2+} is the weakest oxidizing agent.

72. D is correct.

H_2 is the limiting reactant.

24 moles of H_2 should produce: $24 \times (2 / 3) = 16$ moles of NH_3

If only 13.5 moles are produced, the yield:

13.5 moles / 16 moles $\times$ 100% = 84%

73. B is correct.

Balanced reaction:

$H_2 + \frac{1}{2} O_2 \rightarrow H_2O$

Multiply the equations by 2 to remove the fraction:

$2 H_2 + O_2 \rightarrow 2 H_2O$

Find the limiting reactant by calculating the moles of each reactant:

Moles of hydrogen = mass of hydrogen / (2 $\times$ atomic mass of hydrogen)

Moles of hydrogen = 25 g / (2 $\times$ 1.01 g/mole)

Moles of hydrogen = 12.38 moles

Moles of oxygen = mass of oxygen / (2 $\times$ atomic mass of oxygen)

Moles of oxygen = 225 g / (2 $\times$ 16.00 g/mole)

Moles of oxygen = 7.03 moles

Divide the number of moles of each reactant by its coefficient.

The reactant with a smaller number of moles is the limiting reactant.

Hydrogen = 12.38 moles / 2

Hydrogen = 6.19 moles

Oxygen = 7.03 moles / 1

Oxygen = 7.03 moles

Because hydrogen has a smaller number of moles after the division, hydrogen is the limiting reactant, and hydrogen will be depleted in the reaction.

$2 H_2 + O_2 \rightarrow 2 H_2O$

Hydrogen and water have coefficients of 2; therefore, these have the same number of moles.

There are 12.38 moles of hydrogen, which means 12.38 moles of water are produced by this reaction.

Molecular mass of H_2O = (2 $\times$ molecular mass of hydrogen) + atomic mass of oxygen

Molecular mass of H_2O = (2 $\times$ 1.01 g/mole) + 16.00 g/mole

Molecular mass of H_2O = 18.02 g/mole

Mass of H_2O = moles of H_2O × molecular mass of H_2O

Mass of H_2O = 12.38 moles × 18.02 g/mole

Mass of H_2O = 223 g

74. C is correct.

Li	Cl	O_2
+1	x	(2×-2)

The sum of charges in a neutral molecule is zero:

$1 + x + (2 \times -2) = 0$

$1 + x + (-4) = 0$

$x = -1 + 4$

oxidation number of Cl = +3

75. D is correct.

Using half-reactions, the balanced reaction is:

To balance hydrogen	To balance oxygen
↓	↓

$$4[5\ e^- + 8\ H^+ + MnO_4^- \rightarrow Mn^{2+} + 4\ (H_2O)]$$

$$5[(H_2O) + C_3H_7OH \rightarrow C_2H_5CO_2H + 4\ H^+ + 4\ e^-]$$

$$\overline{12\ H^+ + 4\ MnO_4^- + 5\ C_3H_7OH \rightarrow 11\ H_2O + 4\ Mn^{2+} + 5\ C_2H_5CO_2H}$$

Multiply by a common multiple of 4 and 5:

The sum of the product's coefficients: $11 + 4 + 5 = 20$.

76. A is correct.

PbO (s) + C (s) → Pb (s) + CO (g) is a single-replacement reaction because only one element (i.e., oxygen) is being transferred from one reactant to another.

77. E is correct.

The reactants in a chemical reaction are on the left side of the reaction arrow; the products are on the right side.

In this reaction, the reactants are $C_6H_{12}O_6$, H_2O and O_2.

There is a distinction between the terms *reactant* and *reagent*.

A reactant is a substance consumed in the course of a chemical reaction.

A reagent is a substance (e.g., solvent) added to a system to cause a chemical reaction.

78. C is correct.

Determine the number of moles of He:

$$4 \text{ g} \div 4.0 \text{ g/mol} = 1 \text{ mole}$$

Each mole of He has 2 electrons since He has atomic number 2 (# electrons = # protons for neutral atoms).

Therefore, 1 mole of He × 2 electrons / mole = 2 mole of electrons

79. D is correct.

In this molecule, Br has the oxidation number of –1 because it is a halogen, and the gaining of one electron results in a complete octet for bromine.

The sum of charges in a neutral molecule = 0

$$0 = (\text{oxidation state of Fe}) + (3 \times \text{oxidation state of Br})$$

$$0 = (\text{oxidation state of Fe}) + (3 \times -1)$$

$$0 = \text{oxidation state of Fe} - 3$$

$$\text{oxidation state of Fe} = +3$$

80. B is correct.

A mole is a unit of measurement used to express amounts of a chemical substance.

The number of molecules in a mole is 6.02×10^{23}, which is Avogadro's number.

However, Avogadro's number relates to the number of molecules, not the amount of substance.

Molar mass refers to the mass per mole of a substance.

Formula mass is a term that is sometimes used to mean molecular mass or molecular weight, and it refers to the mass of a certain molecule.

===

Practice Set 5: Questions 81–100

===

81. B is correct.

Use the mnemonic OIL RIG: <u>O</u>xidation <u>I</u>s <u>L</u>oss, <u>R</u>eduction <u>I</u>s <u>G</u>ain (of electrons).

To determine which species undergoes oxidation, determine the oxidation number of species in the reaction.

Reactants: CuBr

Br ion is –1, which means that Cu is +1.

Products: Cu = 0 (it's an element)

　　　　　Br_2 = 0 (it's an element)

Br is the species that undergoes oxidation. It loses electrons and has its oxidation number increase.

82. C is correct.

Use the mnemonic OIL RIG: <u>O</u>xidation <u>I</u>s <u>L</u>oss, <u>R</u>eduction <u>I</u>s <u>G</u>ain (of electrons).

Oxidation is the loss of electrons, while reduction is the gain of electrons.

An oxidizing agent undergoes reduction, while a reducing agent undergoes oxidation.

Each formula unit of $CuBr_2$ is converted into Cu.

The Br ion does not undergo any changes (i.e., neither oxidation nor reduction); it just moves to Zn to form another compound.

In $CuBr_2$, the oxidation number of Cu is +2.

Therefore, when it is reduced, it gains 2 electrons to form Cu.

83. B is correct.

Calculate the molecular mass of CO_2:

　　　　Molecular mass of CO_2 = atomic mass of C + (2 × atomic mass of O)

　　　　Molecular mass of CO_2 = 12.01 g/mole + (2 × 16.00 g/mole)

　　　　Molecular mass of CO_2 = 44.01 g/mole

Calculate the moles of CO_2:

　　　　Moles of CO_2 = mass of CO_2 / molecular mass of CO_2

　　　　Moles of CO_2 = 168 g / (44.01 g/mole)

　　　　Moles of CO_2 = 3.82 moles

One mole of atoms has 6.02×10^{23} atoms (Avogadro's number, or N_A).

Number of CO_2 atoms = moles of CO_2 × Avogadro's number

Number of CO_2 atoms = 3.82 moles × (6.02×10^{23} atoms/mole)

Number of CO_2 atoms = 2.30×10^{24} atoms

84. E is correct.

For calculations, assume 100 g of the compound:

64 g of Ag, 8 g of N and 28 g of O.

Find the number of moles of each of these elements:

Ag: (64 g) / (108 g/mol) = 0.6 mol

N: (8 g) / (14 g/ mol) = 0.6 mol

O: (28 g) / (16 g/mol) = 1.8 mol

Reduce to the smallest coefficients by dividing by 0.6: $AgNO_3$

85. D is correct.

To solve for the density of a gas, use a modification of the ideal gas law.

Ideal gas law:

PV = nRT

where n is the number of moles, which is equal to mass/molecular weight (MW). Substitute that expression into the ideal gas equation:

PV = (mass / MW) × RT

density (ρ) = mass / volume

Rearrange the equation to get mass / volume on one side:

mass / volume = (P × MW) / RT

density (ρ) = (P × MW) / RT

Now rearrange the equation to solve for molecular weight (MW):

MW = ρRT / P

Here, the gas is in STP condition, which has T of 0 °C = 273.15 K and P of 1 atm.

Calculate MW:

MW = 1.34 g/L × 0.0821 L atm K^{-1} mol^{-1} × 273.15 K / 1 atm

MW = 30.1 g

86. C is correct.

Balanced equation (double replacement):

$$C_3H_8 \, (g) + 5 \, O_2 \, (g) \rightarrow 3 \, CO_2 \, (g) + 4 \, H_2O \, (g)$$

87. D is correct.

To find what the reaction yields, look at the right side of the equation.

There are 6 CO_2 atoms: $6 \times 2 = 12$

There are 12 H_2O atoms: $12 \times 1 = 12$

$12 + 12 = 24$ atoms of oxygen.

88. C is correct.

Balanced equation (single replacement):

$$2 \, Al_2O_3 \, (s) + 6 \, Cl_2 \, (g) \rightarrow 4 \, AlCl_3 \, (aq) + 3 \, O_2 \, (g)$$

89. C is correct.

The given half-reaction equation for the reduction of chromium is:

$$Cr_2O_7{}^{2-} \, (aq) \rightarrow Cr^{3+} \, (aq)$$

where the oxidation state of oxygen in $Cr_2O_7{}^{2-}$ is –2, and the oxidation state of chromium is +6:

2(charge of metal) + 7(charge of oxygen) = net charge of the molecule or ion

$2(Cr^{n+}) + 7(-2) = -2$

$2(Cr^{n+}) - 14 = -2$

$2(Cr^{n+}) = 12$

$Cr^{n+} = +6$

To balance this reduction half-reaction:

1) Balance the metal count.

2) Balance oxygen count.

3) Balance hydrogen count.

4) Add hydroxide if the reaction is done in base.

5) Balance the charge.

6) Optional step: multiply the coefficients by a factor to cancel electrons, if the reaction is to be combined with the oxidation half-reaction

Applying these operations to the half-reaction:

$$Cr_2O_7^{2-} \rightarrow Cr^{3+}$$

1) Balance the metal count on both sides of the equation by adding the appropriate coefficients:

$$Cr_2O_7^{2-} \rightarrow 2\ Cr^{3+}$$

2) Balance the oxygen count by adding water:

$$Cr_2O_7^{2-} \rightarrow 2\ Cr^{3+} + 7\ H_2O$$

3) Balance the hydrogen count by adding protons:

$$14\ H^+ + Cr_2O_7^{2-} \rightarrow 2\ Cr^{3+} + 7\ H_2O$$

Skip step 4.

5) Balance the charge by adding electrons (there should be a net charge of +6 on both sides):

$$14\ H^+ + Cr_2O_7^{2-} + 6\ e^- \rightarrow 2\ Cr^{3+} + 7\ H_2O$$

90. B is correct.

Na	H	C	O_3
+1	+1	x	(3×-2)

The sum of charges in a neutral molecule = 0

$$1 + 1 + x + (3 \times -2) = 0$$
$$1 + 1 + x + (-6) = 0$$
$$x = -1 - 1 + 6$$
$$x = +4$$

oxidation number of C = +4

91. D is correct.

Avogadro's law states that the relationship between the masses of the same volume of the same gases (at the same temperature and pressure) corresponds to the relationship between their respective molecular weights. Hence, the relative molecular mass of a gas can be calculated from the mass of a sample of known volume.

The flaw in Dalton's theory was corrected in 1811 by Avogadro, who proposed that equal volumes of any two gases, at equal temperature and pressure, contain equal numbers of molecules (i.e., the mass of a gas's particles does not affect the volume that it occupies).

Avogadro's law allowed him to deduce the diatomic nature of numerous gases by studying the volumes at which they reacted. For example, two liters of hydrogen react with just one liter of oxygen to produce two liters of water vapor (at constant pressure and temperature). This means that a single oxygen molecule (O_2) splits to form two particles of water.

Thus, Avogadro was able to offer more accurate estimates of the atomic mass of oxygen and various other elements and distinguished between molecules and atoms.

92. B is correct.

Balanced reaction: $BaCl_2$ (*aq*) + K_2SO_4 (*aq*) → $BaSO_4$ and 2 KCl

A double replacement reaction indicates an exchange of cations and anions between the reactants.

Separate the reactants into ions and then exchange the cation and anion pairings.

93. B is correct.

% mass Cl = (mass of chlorine in molecule / molecular mass) × 100%

% mass Cl = (2 × 35.45 g/mol) / 159.09 g/mol × 100%

% mass Cl = 0.446 × 100% = 44.6%

94. A is correct.

Use the mnemonic OIL RIG: Oxidation Is Loss, Reduction Is Gain (of electrons).

Oxidation is the loss of electrons, while reduction is the gain of electrons.

An oxidizing agent undergoes reduction, while a reducing agent undergoes oxidation.

Because the forward reaction is spontaneous, the reverse reaction is not spontaneous.

I_2 cannot oxidize $FeCl_2$; therefore, I_2 is the weakest oxidizing agent.

NaCl is not the answer because Na and Cl ions do not participate in the redox reaction; each has an identical oxidation number on both sides of the reaction.

95. C is correct.

Balanced reaction:

$$4\ RuS\ (s) + 9\ O_2 + 4\ H_2O \rightarrow 2\ Ru_2O_3\ (s) + 4\ H_2SO_4$$

Find moles of RuS:

22.4 liters O_2 at STP = 1 mole O_2

67 g RuS × 1 mole/133 g = 0.5 mole RuS

Since 4 RuS molecules combine with 9 O_2 molecules, given 1 mole of O_2 and 0.5 mole RuS, O_2 is the limiting reactant.

Ratio for the limiting reactant:

molar ratio: 9 O_2 / 4 RuS = 1 mole O_2 / x mole RuS

4 RuS / 9 O_2 = x mole RuS / 1 mole O_2

x mole RuS / 1 mole O_2 = 4 RuS / 9 O_2

x mole RuS = 4 RuS / 9 O_2 × 1 mole O_2

x = (4 / 9) mole RuS

Since RuS and H_2SO_4 have the same ratio, 4 / 9 H_2SO_4 is produced.

(4 / 9) mole H_2SO_4 × 98 g/mole = 44 g

96. A is correct.

Ca	S	O$_4$
+2	x	(4×-2)

The sum of charges in a neutral molecule is zero:

$2 + x + (4 \times -2) = 0$

$2 + x + (-8) = 0$

$x = -2 + 8$

oxidation number of S = +6

97. C is correct.

Balanced double replacement reaction: $C_6H_{12}O_6\,(s) + 6\,O_2\,(g) \rightarrow 6\,CO_2\,(g) + 6\,H_2O\,(g)$

98. B is correct.

$SO_3 + H_2O \rightarrow H_2SO_4$ is a synthesis reaction because two compounds join to produce one product.

$3\,CuSO_4 + 2\,Al \rightarrow Al_2(SO_4)_3 + 3\,Cu$ is a single-replacement reaction because only one element (i.e., SO_4) is transferred from one reactant to another.

$2\,NaHCO_3 \rightarrow Na_2CO_3 + CO_2 + H_2O$ is a decomposition reaction where a compound is broken down into its constituent elements.

$C_3H_8 + 5\,O_2 \rightarrow 3\,CO_2 + 4\,H_2O$ is a combustion reaction because a hydrocarbon (i.e., propane) reacts with oxygen to produce carbon dioxide and water.

99. A is correct.

The coefficients give the stoichiometric number (i.e., relative quantities) of molecules (or atoms) involved in the reaction.

100. D is correct.

Calculate the molecular mass of NO:

Molecular mass of NO = atomic mass of N + atomic mass of O

Molecular mass of NO = N: 14.01 g/mole + O: 16.00 g/mole

Molecular mass of NO = 30.01 g/mole

Calculate moles of NO:

Moles of NO = mass of NO / molecular mass of NO

Moles of NO = 17.0 g / 30.01 g/mole

Moles of NO = 0.57 mole

For STP conditions, 1 mole of gas has a volume of 22.4 L:

Volume of NO = moles of NO × 22.4 L/mole

Volume of NO = 0.57 mole × 22.4 L

Volume of NO = 12.7 L

===

Practice Set 6: Questions 101–120

===

101. B is correct.

Use the mnemonic OIL RIG: <u>O</u>xidation <u>I</u>s <u>L</u>oss, <u>R</u>eduction <u>I</u>s <u>G</u>ain (of electrons).

Oxidizing agents undergo reduction (gain electrons) in a redox reaction.

Cl^- has a full octet electron configuration and has a slight tendency to pick up more electrons.

Na^+ has a full octet electron configuration. Sodium metal is very reactive, and it is a strong reducing agent. As a result, the conjugate ion (Na^+) is going to be a weak oxidizing agent.

Cl_2 is already in a stable electron configuration, but it can be reduced to Cl^-.

Cl_2 is very reactive and a strong oxidizing agent.

102. B is correct.

Use the mnemonic OIL RIG: <u>O</u>xidation <u>I</u>s <u>L</u>oss, <u>R</u>eduction <u>I</u>s <u>G</u>ain (of electrons).

Oxidation is the loss of electrons, while reduction is the gain of electrons.

An oxidizing agent undergoes reduction, while a reducing agent undergoes oxidation.

If the substance loses electrons (being oxidized), another substance must be gaining electrons (being reduced).

The original substance is referred to as a reducing agent, even though it is being oxidized.

103. A is correct.

A molecular formula expresses the actual number of atoms of each element in a molecule.

An empirical formula is the simplest formula for a compound. For example, if the molecular formula of a compound is C_6H_{16}, the empirical formula is C_3H_8.

Elemental formula and atomic formula are not valid chemistry terms.

104. E is correct.

Use the mnemonic OIL RIG: <u>O</u>xidation <u>I</u>s <u>L</u>oss, <u>R</u>eduction <u>I</u>s <u>G</u>ain (of electrons).

Oxidation is the loss of electrons, while reduction is the gain of electrons.

An oxidizing agent undergoes reduction, while a reducing agent undergoes oxidation.

Because the forward reaction is spontaneous, the reverse reaction is not spontaneous.

$FeCl_2$ cannot reduce I_2; therefore, $FeCl_2$ is the weakest reducing agent.

NaCl is not the answer because Na and Cl ions do not participate in the redox reaction; each has an identical oxidation number on both sides of the reaction.

105. C is correct.

Use the atomic mass from the periodic table:

$$(C = 6 \times 12 \text{ grams}) + (H = 12 \times 1 \text{ gram}) + (O = 6 \times 16 \text{ grams}) = 180 \text{ g/mole}$$

$$180 \text{ g/mole} \times 3.5 \text{ moles} = 630 \text{ grams}$$

106. B is correct.

The number of oxygens on both sides of the reaction equation must be equal.

On the right side, there are 8 CO_2 molecules; since each CO_2 molecule contains 2 oxygens, multiply $8 \times 2 = 16$.

On the right side are 10 H_2O molecules; $10 \times 1 = 10$.

Add 16 and 10 to get the number of oxygens on the right side: $16 + 10 = 26$.

Since there are 26 oxygens on the right side, there should be 26 oxygens on the left side.

Each O_2 molecule contains 2 oxygens; $26 / 2 = 13$.

Therefore, coefficient 13 is needed to balance the equation.

$$2\, C_4H_{10}\,(g) + 13\, O_2\,(g) \rightarrow 8\, CO_2\,(g) + 10\, H_2O\,(g)$$

107. D is correct.

A mole is a unit of measurement used to express amounts of a chemical substance.

Avogadro's number is the number of constituent particles (i.e., often atoms or molecules) that equals approximately 6.022×10^{23}.

Avogadro's number equals the number of atoms in 1 mole (g atomic weight) of an element.

The number of molecules in a mole is 6.02×10^{23}, which is Avogadro's number.

108. A is correct.

The balanced equation is:

$$2\, RuS + 9/2\, O_2 + 2\, H_2O \rightarrow Ru_2O_3 + 2\, H_2SO_4$$

To avoid fractional coefficients, multiply coefficients by 2:

$$4\, RuS + 9\, O_2 + 4\, H_2O \rightarrow 2\, Ru_2O_3 + 4\, H_2SO_4$$

The sum of the coefficients:

$$4 + 9 + 4 + 2 + 4 = 23$$

109. C is correct.

The % by mass alone is not sufficient because it provides only the empirical formula.

The molecular weight alone is not sufficient because two different compounds might have similar molecular weights.

If molecular mass and % by mass are known, the empirical formula is multiplied by a factor to determine the molecular weight.

110. E is correct.

Balanced equation (synthesis) for reaction II:

$$4 \text{ Al } (s) + 3 \text{ Br}_2 (l) \rightarrow 2 \text{ Al}_2\text{Br}_3 (s)$$

111. B is correct.

First, calculate the molecular mass (MW) of $MgSO_4$:

MW of $MgSO_4$ = atomic mass of Mg + atomic mass of S + (4 × atomic mass of O)

MW of $MgSO_4$ = 24.3 g/mole + 32.1 g/mole + (4 × 16.0 g/mole)

MW of $MgSO_4$ = 120.4 g/mole

Then, use the molecular mass to determine the number of moles:

Moles of $MgSO_4$ = mass of $MgSO_4$ / MW of $MgSO_4$

Moles of $MgSO_4$ = 60.2 g / 120.4 g/mole

Moles of $MgSO_4$ = 0.5 mole

112. A is correct.

Grams of a compound often contain a very large number of atoms.

72.9 g of Mg × (1 mol/24.3 g) = 3 moles Mg

3 moles Mg × (6.02×10^{23} atoms/mol) = 1.81×10^{24} atoms

113. A is correct.

To obtain the percent mass composition of oxygen, calculate the mass of oxygen in the molecule, divide it by the molecular mass and multiply by 100%.

% mass oxygen = (mass of oxygen in molecule / molecular mass) × 100%

% mass oxygen = [(4 moles × 16.00 g/mole) / (303.39 g/mole)] × 100%

% mass oxygen = 0.211 × 100% = 21.1%

114. A is correct.

Balanced reaction:

$$4 \, RuS \, (s) + 9 \, O_2 + 4 \, H_2O \rightarrow 2 \, Ru_2O_3 \, (s) + 4 \, H_2SO_4$$

Set up a ratio and solve:

$$4 \, RuS \, / \, 2 \, Ru_2O_3 = 7 \text{ moles } RuS \, / \, x \text{ moles } Ru_2O_3$$

$$x \text{ moles } Ru_2O_3 = 4 \, RuS \times 7 \text{ moles } RuS \, / \, 2 \, Ru_2O_3$$

$$x \text{ moles } Ru_2O_3 = 7 \text{ moles } RuS \times 2 \, Ru_2O_3 \, / \, 4 \, RuS$$

$$x = 3.5 \text{ moles}$$

115. C is correct.

Oxidation numbers of S in compounds:

SO_4^{2-}	+6
$S_2O_3^{2-}$	+2
S^{2-}	−2

116. B is correct.

Li is in group IA, and its oxidation number is +1.

Therefore, in Li_2O_2, the oxidation number of oxygen is −1.

Oxygen has an oxidation number of −1 (like in peroxides, H_2O_2).

117. C is correct.

A double-replacement reaction occurs when parts of two ionic compounds (e.g., HBr and KOH) are exchanged to make two new compounds (H_2O and KBr).

$2 \, HI \rightarrow H_2 + I_2$ is a decomposition reaction where a compound is broken down into its constituent elements.

$SO_2 + H_2O \rightarrow H_2SO_4$ is a synthesis (or composition) reaction when two species combine to form a more complex chemical product.

$CuO + H_2 \rightarrow Cu + H_2O$ is a single-replacement reaction, which occurs when one element is transferred from one reactant to another element (i.e., the oxygen atom leaves the Cu and joins H_2).

118. B is correct.

Density = mass / volume

Density = (5 μg × 1g /1,000,000 μg) / (25 μL × 1 L / 1,000,000 μL)

Density = (5×10^{-6} g) / (2.5×10^{-5} L)

Density = 0.2 g/L

Perform the above calculation (i.e., density = mass/volume) for each answer choice to determine the lowest density.

119. E is correct.

Balanced double replacement reaction:

$$2 \, C_6H_{14} \, (g) + 19 \, O_2 \, (g) \rightarrow 12 \, CO_2 \, (g) + 14 \, H_2O \, (g)$$

120. A is correct.

A mole is a unit of measurement used to express amounts of a chemical substance.

The number of molecules in a mole is 6.02×10^{23}, which is Avogadro's number.

==

Practice Set 7: Questions 121–140

==

121. A is correct.

Use the mnemonic OIL RIG: Oxidation Is Loss, Reduction Is Gain (of electrons).

Oxidizing agents undergo reduction (gain electrons) in a redox reaction.

Cl^- has a full octet electron configuration and has little tendency to pick up more electrons.

Na^+ has a full octet electron configuration. Sodium metal is very reactive, and it is a strong reducing agent. As a result, the conjugate ion (Na^+) is going to be a weak oxidizing agent.

Cl_2 is already in a stable electron configuration, but it can be reduced to Cl^- quite easily. Cl_2 is very reactive and a strong oxidizing agent.

122. A is correct.

Use the mnemonic OIL RIG: Oxidation Is Loss, Reduction Is Gain (of electrons).

Oxidation is the loss of electrons, while reduction is the gain of electrons.

An oxidizing agent undergoes reduction, while a reducing agent undergoes oxidation.

Therefore, the substance that is reduced always gains electrons. It does not have to contain an element that increases in oxidation number.

It is not the reducing agent; the substance that is reduced is the oxidizing agent.

123. A is correct.

A molecular formula expresses the actual number of atoms of each element in a molecule.

An empirical formula is the simplest formula for a compound.

For example, if the molecular formula of a compound is C_6H_{16}, the empirical formula is C_3H_8.

Elemental formula and atomic formula are not chemistry terms.

124. E is correct.

Multiply the values in the bracket by the multiplier (i.e., subscript) outside the bracket.

For $C_6H_3(C_3H_7)_2(C_2H_5)$: $3 + (2 \times 7) + 5 = 22$ hydrogens

125. D is correct.

The *law of conservation of mass* states that for a closed system, the mass of the system must remain constant over time.

The law implies that during any chemical reaction, the mass of the reactants or starting materials must be equal to the mass of the products. From an understanding of this principle, a scientist could undertake quantitative studies of the transformations of substances.

126. E is correct.

The number of oxygens on both sides of the reaction equation must be equal.

On the right side, there are 8 CO_2 molecules; since each CO_2 molecule contains 2 oxygens, multiply $8 \times 2 = 16$.

On the right side are 10 H_2O molecules; $10 \times 1 = 10$.

Add 16 and 10 to get the number of oxygens on the right side: $16 + 10 = 26$.

Since there are 26 oxygens on the right side, there should be 26 oxygens on the left side.

Each O_2 molecule contains 2 oxygens; $26 / 2 = 13$.

Therefore, coefficient 13 is needed to balance the equation.

$$2\ C_4H_{10}\ (g) + 13\ O_2\ (g) \rightarrow 8\ CO_2\ (g) + 10\ H_2O\ (g)$$

127. C is correct.

Use the mnemonic OIL RIG: Oxidation Is Loss, Reduction Is Gain (of electrons).

Oxidation is the loss of electrons, while reduction is the gain of electrons.

An oxidizing agent undergoes reduction, while a reducing agent undergoes oxidation.

The oxidation number for F is 0 as a reactant and –1 as a product; a decrease in oxidation number means that F is reduced in this reaction.

128. B is correct.

Balanced equation (double replacement): $N_2H_4 + 2\ H_2O_2 \rightarrow N_2 + 4\ H_2O$

129. E is correct.

First, calculate the molecular mass (MW) of CH_4:

Molecular mass of CH_4 = atomic mass of C + (4 × atomic mass of H)

Molecular mass of CH_4 = 12.01 g/mole + (4 × 1.01 g/mole)

Molecular mass of CH_4 = 16.05 g/mole

Use this information to determine the number of moles:

Moles of CH_4 = mass of CH_4 / molecular mass of CH_4

Moles of CH_4 = 6.40 g / 16.05 g/mole

Moles of CH_4 = 0.4 mole

Then, multiply by Avogadro's number to obtain the number of molecules:

Number of CH_4 molecules = moles of CH_4 × Avogadro's number

Number of CH_4 molecules = 0.4 moles × 6.02×10^{23} molecules/mole

Number of CH_4 molecules = 2.40×10^{23} molecules

130. E is correct.

To determine the limiting reactant, divide the quantity by the coefficient of each reactant.

Al: 0.2 mole / 2 = 0.1

Fe_2O_3: = 0.4 mole / 1 = 0.4

Whichever reactant has a smaller number is the limiting reactant.

131. B is correct.

Start by balancing the number of nitrogen by adding 2 to NH_3:

__ H_2 + __ N_2 + → 2 NH_3

Now balance the hydrogen by adding 3 to H_2:

3 H_2 + __ N_2 → 2 NH_3

Balanced equation:

3 H_2 + N_2 → 2 NH_3

132. B is correct.

In order to calculate the number of atoms, start by calculating the molecular mass (MW) of fructose:

MW of fructose = (6 × atomic mass of C) + (12 × atomic mass of H) + (6 × atomic mass of O)

MW of fructose = (6 × 12.01 g/mole) + (12 × 1.01 g/mole) + (6 × 16.00 g/mole)

MW of fructose = 180.18 g/mole

Using the molecular mass, calculate the number of moles in the fructose sample:

Moles of fructose = mass of fructose / MW of fructose

Moles of fructose = 5.40 g / 180.18 g/mole

Moles of fructose = 0.03 mole

Because there are 6 oxygen atoms for each molecule of fructose:

Moles of oxygen = 6 × moles of fructose

Moles of oxygen = 6 × 0.03 mole

Moles of oxygen = 0.18 mole

One mole of atoms has 6.02×10^{23} atoms; this is Avogadro's Number (N_A).

Number of oxygen atoms = moles of oxygen × Avogadro's number

Number of oxygen atoms = 0.18 mole × (6.02×10^{23} atoms/mole)

Number of oxygen atoms = 1.08×10^{23} atoms

133. A is correct.

The molecular formula of iron (III) oxide is Fe_3O_4; therefore, only choices A and B are correct equations for this reaction.

$2\ Fe_2O_3 + 3\ C\ (s) \rightarrow 4\ Fe\ (l) + 3\ CO_2\ (g)$ is properly balanced because the coefficients are reduced to the lowest values.

134. E is correct.

In the balanced equation, the molar ratio of Sn^{4+} and Sn^{2+} are equal because the number of moles of Sn^{2+} (reactant) = the number of moles of Sn^{4+} (product).

135. A is correct.

Calculate the oxidation number for each species.

A: Mn = +7

B: Br = +5

C: C = +3

D: S = +6

E: K = +1

136. B is correct.

$2\ H_2O_2\ (s) \rightarrow 2\ H_2O\ (l) + O_2\ (g)$ is incorrectly classified. The decomposition part of the classification is correct because one complex compound (H_2O_2) is being broken down into two or more parts (H_2O and O_2).

However, it is a redox reaction because oxygen is being lost from H_2O_2, one of the three indicators of a redox reaction (i.e., electron loss/gain, hydrogen loss/gain, oxygen loss/gain). Additionally, one of the products is a single element (O_2), which is a clue that it is a redox reaction.

$AgNO_3\ (aq) + KOH\ (aq) \rightarrow KNO_3\ (aq) + AgOH\ (s)$ is correctly classified as a non-redox reaction because the reactants and products are compounds (no single elements), and there is no change in oxidation state (i.e., no gain or loss of electrons). It is a precipitation reaction because the chemical reaction occurs in an aqueous solution, and one of the products formed (AgOH) is insoluble, which makes it a precipitate.

$Pb(NO_3)_2\ (aq) + 2\ Na\ (s) \rightarrow Pb\ (s) + 2\ NaNO_3\ (aq)$ is correctly classified as a redox reaction because the nitrate (NO_3^-), which dissociates, is oxidized (loses electrons), the Na^+ is reduced (gains electrons). The two combine to form the ionic compound $NaNO_3$. It is a single-replacement reaction because one element is substituted for another element in a compound, making a new compound ($2NaNO_3$) and an element (Pb).

$HNO_3\ (aq) + LiOH\ (aq) \rightarrow LiNO_3\ (aq) + H_2O\ (l)$ is correctly classified as a non-redox reaction because the reactants and products are compounds (no single elements), and there is no change in oxidation state (i.e., no gain or loss of electrons). It is a double-replacement reaction because parts of two ionic compounds are exchanged to make two new compounds.

137. A is correct.

Formula mass is a synonym for molecular mass/molecular weight (MW).

Start by calculating the mass of the formula unit (CHCl):

$$CHCl = (12.01 \text{ g/mol} + 1.01 \text{ g/mol} + 35.45 \text{ g/mol}) = 48.47 \text{ g/mol}$$

Divide the molar mass by the formula unit mass:

$$194 \text{ g/mol} / 48.47 \text{ g/mol} = 4.0025$$

Round it to the closest whole number: 4

Multiply the formula unit by 4:

$$(CHCl)_4 = C_4H_4Cl_4$$

138. A is correct.

Moles of Al = mass of Al / atomic mass of Al

Moles of Al = (8.52 g) / (26.98 g/mol)

Moles of Al = 0.316 moles

139. E is correct.

Use the oxidation values of:

$$H = +1; O = -2; Cl = -1$$

H	Cr_2	O_4	Cl
1	$2x$	4×-2	-1

$$1 + 2x - 8 - 1 = 0$$

$$2x = 8$$

$$x = 4$$

In HCr_2O_4Cl, Cr has an oxidation number of +4.

140. B is correct.

The molar mass (or *molecular weight*) is the mass in grams of the atoms that make up a mole of a particular molecule.

The unit used to measure molar mass is grams per mole.

$$\text{Molar mass} = \text{grams} / \text{mole}$$

Notes for active learning

Thermochemistry – Detailed Explanations

===

Practice Set 1: Questions 1–20

===

1. A is correct.

A reaction's enthalpy is specified by ΔH (not ΔG). ΔH determines whether a reaction is exothermic (releases heat to surroundings) or endothermic (absorbs heat from surroundings).

To predict spontaneity of reaction, use the Gibbs free energy equation:

$$\Delta G = \Delta H° - T\Delta S$$

The reaction is spontaneous (i.e., exergonic) if ΔG is negative.

The reaction is nonspontaneous (i.e., endergonic) if ΔG is positive.

2. B is correct.

Heat is energy, and the energy from the heat is transferred to the gas molecules, which increases their movement (i.e., kinetic energy).

Temperature is a measure of the average kinetic energy of the molecules.

3. E is correct.

Heat capacity is the amount of heat required to increase the temperature of *the whole sample* by 1 °C.

Specific heat is the heat required to increase the temperature of *1 gram* of sample by 1 °C.

Heat = mass × specific heat × change in temperature:

$$q = m \times c \times \Delta T$$

$$q = 21.0 \text{ g} \times 0.382 \text{ J/g·°C} \times (68.5 \text{ °C} - 21.0 \text{ °C})$$

$$q = 21.0 \text{ g} \times 0.382 \text{ J/g·°C} \times (47.5 \text{ °C})$$

$$q = 381 \text{ J}$$

4. E is correct.

This is a theory/memorization question.

However, the problem can be solved by comparing the options for the most plausible.

Kinetic energy is correlated with temperature, so the formula that involves temperature would be a good choice.

Another approach is to analyze the units. Apply the units to the variables and evaluate them to obtain the answer.

For example, choices A and B have moles, pressure (Pa, bar or atm), and area (m^2), so it is not possible for them to result in energy (J) when multiplied.

Options C and D have molarity (moles/L) and volume (L or m^3); they cannot result in energy (J) when multiplied.

Option E is nRT: mol × (J/mol K) × K; units but J cancel, so this is a plausible formula for kinetic energy.

Additionally, for questions using formulas, dimensional analysis limits the answer choices on this type of question.

5. A is correct.

ΔH refers to enthalpy (or heat).

Endothermic reactions have heat as a reactant.

Exothermic reactions have heat as a product.

Exothermic reactions release heat and cause the temperature of the immediate surroundings to rise (i.e., a net loss of energy).

Endothermic reactions absorb heat and cool the surroundings (i.e., a net gain of energy).

Endothermic reactions absorb energy to break strong bonds to form a less stable state (i.e., positive enthalpy). Exothermic reactions release energy when forming stronger bonds to produce a more stable state (i.e., negative enthalpy).

6. D is correct.

Endothermic reactions absorb energy to break strong bonds to form a less stable state (i.e., positive enthalpy). Exothermic reactions release energy when forming stronger bonds to produce a more stable state (i.e., negative enthalpy).

The spontaneity of a reaction is determined by the calculation of ΔG using:

$$\Delta G = \Delta H - T\Delta S$$

The reaction is spontaneous when the value of ΔG is negative.

In this problem, the reaction is endothermic, which means that ΔH is positive.

The reaction decreases S, which means that ΔS value is negative.

Substituting these values into the equation:

$$\Delta G = \Delta H - T(-\Delta S)$$

$$\Delta G = \Delta H + T\Delta S, \text{ with } \Delta H \text{ and } \Delta S \text{ both positive values}$$

For this problem, the value of ΔG is positive.

The reaction does not occur because ΔG needs to be negative for a spontaneous reaction to proceed, the products are more stable than the reactants.

7. E is correct.

Chemical energy is potential energy stored in molecular bonds.

Electrical energy is the kinetic energy associated with the motion of electrons (e.g., in wires, circuits, lightning). The potential energy stored in a flashlight battery is electrical energy.

Heat is the spontaneous transfer of energy from a hot object to a cold one with no displacement or deformation of the objects (i.e., no work done).

8. A is correct.

Entropy is higher for less organized states (i.e., more random or disordered).

9. C is correct.

ΔH refers to enthalpy (or heat).

Endothermic reactions have heat as a reactant. Exothermic reactions have heat as a product.

Exothermic reactions release heat and cause the temperature of the immediate surroundings to rise (i.e., a net loss of energy). An endothermic process absorbs heat and cools the surroundings (i.e., a net gain of energy).

Endothermic reactions absorb energy to break strong bonds to form a less stable state (i.e., positive enthalpy).

Exothermic reactions release energy when forming stronger bonds to produce a more stable state (i.e., negative enthalpy).

The reaction is nonspontaneous (i.e., endergonic) if the products are less stable than the reactants and ΔG is positive.

The reaction is spontaneous (i.e., exergonic) if the products are more stable than the reactants and ΔG is negative.

10. E is correct.

This cannot be determined when the given $\Delta G = \Delta H - T\Delta S$, terms cancel.

The system is at equilibrium when $\Delta G = 0$, but in this question, both sides of the equation cancel:

$$\underbrace{X - RY}_{\Delta G \text{ term}} = \underbrace{X - RY}_{\substack{\Delta H - T\Delta S \\ \text{term}}}, \quad \text{so } 0 = 0$$

11. C is correct.

Gibbs free energy:

$$\Delta G = \Delta H - T\Delta S$$

Stable molecules are spontaneous because they have a negative (or relatively low) ΔG.

ΔG is most negative (most stable) when ΔH is the smallest and ΔS is largest.

12. A is correct.

ΔH refers to enthalpy (or heat).

Endothermic reactions have heat as a reactant.

Exothermic reactions have heat as a product.

The reaction is nonspontaneous (i.e., endergonic) if the products are less stable than the reactants and ΔG is positive.

The reaction is spontaneous (i.e., exergonic) if the products are more stable than the reactants and ΔG is negative.

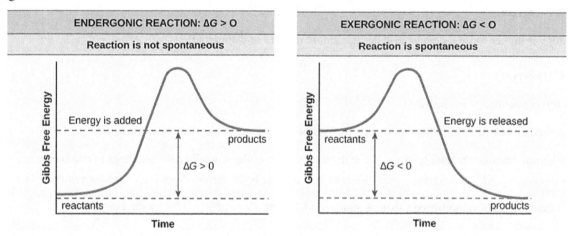

Exothermic reactions release heat and cause the temperature of the immediate surroundings to rise (i.e., a net loss of energy).

Endothermic reactions absorb heat and cool the surroundings (i.e., a net gain of energy).

Endothermic reactions absorb energy to break strong bonds to form a less stable state (i.e., positive enthalpy).

Exothermic reactions release energy when forming stronger bonds to produce a more stable state (i.e., negative enthalpy).

13. B is correct.

Gibbs free energy:

$$\Delta G = \Delta H - \text{T}\Delta S$$

Entropy (ΔS) determines the favorability of chemical reactions.

If entropy is large, the reaction is more likely to proceed.

14. D is correct.

In a chemical reaction, bonds within reactants are broken, and new bonds form to create products. Therefore, in bond dissociation problems,

ΔH reaction = sum of bond energy in reactants – the sum of bond energy in products

For $H_2C=CH_2 + H_2 \rightarrow CH_3–CH_3$:

$\Delta H_{reaction}$ = sum of bond energy in reactants – sum of bond energy in products

$\Delta H_{reaction}$ = [(C=C) + 4(C–H) + (H–H)] – [(C–C) + 6(C–H)]

$\Delta H_{reaction}$ = [612 kJ + (4 × 412 kJ) + 436 kJ] – [348 kJ + (6 × 412 kJ)]

$\Delta H_{reaction}$ = –124 kJ

Remember that this is the opposite of ΔH_f problems, where:

$\Delta H_{reaction}$ = (sum of ΔH_f products) – (sum of ΔH_f reactants)

15. C is correct.

Bond dissociation energy is the energy required to break a bond between two gaseous atoms and helps estimate the enthalpy change in a reaction.

16. B is correct.

The ΔH value is positive, which means the reaction is endothermic and absorbs energy from the surroundings.

ΔH value is expressed as the energy released/absorbed per mole (for species with a coefficient of 1).

If the coefficient is 2 then it is the energy transferred per 2 moles. For example, 2 moles of NO are produced from 1 mole for each reagent for the stated reaction.

Therefore, 43.2 kcal is produced when 2 moles of NO are produced.

17. A is correct.

State functions depend only on the initial and final states of the system and are independent of the paths taken to reach the final state.

Common examples of a state function in thermodynamics include internal energy, enthalpy, entropy, pressure, temperature, and volume.

Work and heat relate to the change in energy of a system when it moves from one state to another, depending on how a system changes between states.

18. B is correct.

The temperature (i.e., average kinetic energy) of a substance remains constant during a phase change (e.g., solid to liquid or liquid to gas).

For example, the heat (i.e., energy) breaks bonds between the ice molecules as they change phases into the liquid phase. Since the average kinetic energy of the molecules does not change at the moment of the phase change (i.e., melting), the temperature of the molecules does not change.

19. C is correct.

In a chemical reaction, bonds within reactants are broken, and new bonds form to create products.

Bond dissociation:

$\Delta H_{\text{reaction}}$ = (sum of bond energy in reactants) – (sum of bond energy in products)

$O=C=O + 3\ H_2 \rightarrow CH_3–O–H + H–O–H$

$\Delta H_{\text{reaction}}$ = (sum of bond energy in reactants) – (sum of bond energy in products)

$\Delta H_{\text{reaction}}$ = [2(C=O) + 3(H–H)] – [3(C–H) + (C–O) + (O–H) + 2(O–H)]

$\Delta H_{\text{reaction}}$ = [(2 × 743 kJ) + (3 × 436 kJ)] – [(3 × 412 kJ) + 360 kJ + 463 kJ + (2 × 463 kJ)]

$\Delta H_{\text{reaction}}$ = –191 kJ

This is the reverse of ΔH_f problems, where:

$\Delta H_{\text{reaction}}$ = (sum of $\Delta H_{f\ \text{product}}$) – (sum of $\Delta H_{f\ \text{reactant}}$)

20. C is correct.

Entropy indicates how a system is organized; increased entropy means less order.

The entropy of the universe is continuously increasing because a system moves towards more disorder unless energy is added to the system.

The second law of thermodynamics states that the entropy of interconnected systems, without the addition of energy, always increases.

==

Practice Set 2: Questions 21–40

==

21. A is correct.

Gibbs free energy: $\Delta G = \Delta H - T\Delta S$

The value that is the largest positive value is the most endothermic, while the value that is the largest negative value is the most exothermic.

ΔH refers to enthalpy (or heat).

Endothermic reactions have heat as a reactant. Exothermic reactions have heat as a product.

Exothermic reactions release heat and cause the temperature of the immediate surroundings to rise (i.e., a net loss of energy). An endothermic process absorbs heat and cools the surroundings (i.e., a net gain of energy).

Endothermic reactions absorb energy to break strong bonds to form a less stable state (i.e., positive enthalpy).

Exothermic reactions release energy when forming stronger bonds to produce a more stable state (i.e., negative enthalpy).

22. C is correct.

Potential Energy Stored energy and the energy of position (gravitational)	Kinetic Energy Energy of motion: motion of waves, electrons, atoms, molecules, and substances.
Chemical Energy Chemical energy is the energy stored in the bonds of atoms and molecules. Examples of stored chemical energy: biomass, petroleum, natural gas, propane, coal.	**Thermal Energy** Thermal energy is internal energy in substances; it is the vibration and movement of atoms and molecules within substances. Geothermal energy is an example of thermal energy.

23. A is correct.

By convention, the ΔG for an element in its standard state is 0.

24. B is correct.

Test strategy: when given a choice among similar explanations, carefully evaluate the most descriptive one.

However, check the statement because sometimes the most descriptive option is not accurate.

Here, the longest option is accurate, so that would be the best choice among all options.

25. C is correct.

ΔG is negative for spontaneous reactions.

26. C is correct.

An insulator reduces conduction (i.e., transfer of thermal energy through matter).

Air and vacuum are excellent insulators. Storm windows, which have air wedged between two glass panes, work by utilizing this conduction principle.

27. A is correct.

ΔH refers to enthalpy (or heat).

Endothermic reactions have heat as a reactant.

Exothermic reactions have heat as a product.

Exothermic reactions release heat and cause the temperature of the immediate surroundings to rise (i.e., a net loss of energy).

Endothermic reactions absorb heat and cool the surroundings (i.e., a net gain of energy).

The reaction is nonspontaneous (i.e., endergonic) if the products are less stable than the reactants and ΔG is positive. The reaction is spontaneous (i.e., exergonic) if the products are more stable than the reactants and ΔG is negative.

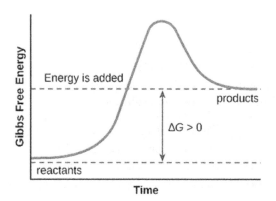 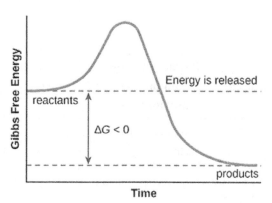

Endergonic reactions are nonspontaneous, and have products with more energy than the reactants

Exergonic reactions are spontaneous and have reactants with more energy than the products

28. D is correct.

Heat capacity is the amount of heat required to increase the temperature of *the whole sample* by 1 °C.

Specific heat is the heat required to increase the temperature of *1 gram* of sample by 1 °C.

q = mass × heat of condensation

q = 16 g × 1,380 J/g

q = 22,080 J

29. A is correct.

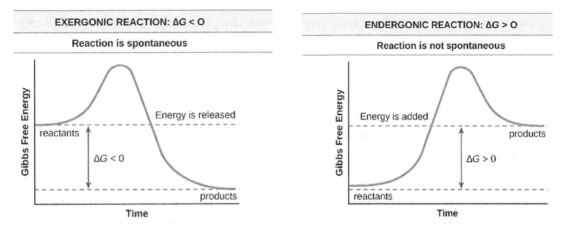

$\Delta G = \Delta H - T\Delta S$ refers to exergonic when ΔG is negative and endergonic when ΔG is positive.

If ΔS is 0, then $\Delta G = \Delta H$ because the $T\Delta S$ term cancels because it equals zero).

ΔH refers to exothermic when ΔH is negative and endothermic when ΔH is positive.

30. E is correct.

Heat of formation is the heat required or released upon creating one mole of the substance from its elements.

If the given reaction is reversed and divided by two, it would be the formation reaction of NH_3:

$\frac{1}{2} N_2 + 3/2 H_2 \rightarrow NH_3$

ΔH of this reaction would be reversed (i.e., plus to minus sign) and divided by two:

$\Delta H = -(92.4 \text{ kJ/mol}) / 2$

$\Delta H = -46.2 \text{ kJ/mol}$

31. B is correct.

Gibbs free energy:

$\Delta G = \Delta H - T\Delta S$

With a positive ΔH and negative ΔS, ΔG is positive, so the reaction is nonspontaneous.

32. C is correct.

Entropy (ΔS) measures the degree of disorder in a system.

When a change causes the components of a system to transform into a higher energy phase (solid $\rightarrow$ liquid, liquid $\rightarrow$ gas), the entropy of the system increases.

33. D is correct.

During phase changes, only the mass and heat of specific phase changes are required to calculate the energy released or absorbed.

Because water is turning into ice, the specific heat required is the heat of solidification.

34. B is correct.

Nuclear energy is considered energy because splitting or combining atoms results in enormous amounts of energy used in atomic bombs and nuclear power plants.

Potential Energy Stored energy and the energy of position (gravitational)	Kinetic Energy Energy of motion: motion of waves, electrons, atoms, molecules, and substances.
Nuclear Energy Nuclear energy is the energy stored in the nucleus of an atom. It is the energy that holds the nucleus together. The nucleus of the uranium atom is an example of nuclear energy.	**Electrical Energy** Electrical energy is the movement of electrons. Lightning and electricity are examples of electrical energy.

35. D is correct.

Entropy is a measurement of disorder.

Gases contain more energy than liquids of the same element and have a higher degree of randomness. Entropy is the unavailable energy that cannot be converted into mechanical work.

36. A is correct.

There are two different methods available to solve this problem. Both methods are based on the definition of ΔH formation as energy consumed/released when one mole of the molecule is produced from its elemental atoms.

Method 1.

Rearrange and add the equations to create the formation reaction of C_2H_5OH.

Information provided by the problem:

$$C_2H_5OH + 3\ O_2 \rightarrow 2\ CO_2 + 3\ H_2O \qquad \Delta H = 327 \text{ kcal}$$

$$H_2O \rightarrow H_2 + \tfrac{1}{2}\ O_2 \qquad \Delta H = 68.3 \text{ kcal}$$

$$C + O_2 \rightarrow CO_2 \qquad \Delta H = -94.1 \text{ kcal}$$

Place C_2H_5OH on the product side. For the other two reactions, arrange them so the elements are on the left and the molecules are on the right (remember that the goal is to create a formation reaction: elements forming a molecule).

When reversing the direction of a reaction, change the positive/negative sign of ΔH.

Multiplying the whole reaction by a coefficient would multiply ΔH by the same ratio.

$$2\ CO_2 + 3\ H_2O \rightarrow C_2H_5OH + 3\ O_2 \qquad \Delta H = -327\ \text{kcal}$$

$$3\ H_2 + 3/2\ O_2 \rightarrow 3\ H_2O \qquad \Delta H = -204.9\ \text{kcal}$$

$$\underline{2\ C + 2\ O_2 \rightarrow 2\ CO_2 \qquad\qquad \Delta H = -188.2\ \text{kcal}}$$

$$2\ C + 3\ H_2 + 1/2\ O_2 \rightarrow C_2H_5OH \qquad \Delta H = -720.1\ \text{kcal}$$

ΔH formation of C_2H_5OH is -720.1 kcal.

Method 2.

The heat of formation (ΔH_f) data can be used to calculate ΔH of a reaction:

$$\Delta H_{reaction} = \text{sum of } \Delta H_{f\ product} - \text{sum of } \Delta H_{f\ reactant}$$

Apply the formula on the first equation:

$$C_2H_5OH + 3\ O_2 \rightarrow 2\ CO_2 + 3\ H_2O \quad \Delta H = 327\ \text{kcal}$$

$$\Delta H_{reaction} = \text{sum of } \Delta H_{f\ product} - \text{sum of } \Delta H_{f\ reactant}$$

$$327\ \text{kcal} = [2(\Delta H_f\ CO_2) + 3(\Delta H_f\ H_2O)] - [(\Delta H_f\ C_2H_5OH) + 3(\Delta H_f\ O_2)]$$

Oxygen (O_2) is an element, so $\Delta H_f = 0$

For CO_2 and H_2O, ΔH_f information can be obtained from the other 2 reactions:

$$H_2O \rightarrow H_2 + \tfrac{1}{2}\ O_2 \qquad \Delta H = 68.3\ \text{kcal}$$

$$C + O_2 \rightarrow CO_2 \qquad \Delta H = -94.1\ \text{kcal}$$

The CO_2 reaction is already a formation reaction – creating one mole of a molecule from its elements. Therefore, ΔH_f of $CO_2 = -94.1$ kcal

If the H_2O reaction is reversed, it will be a formation reaction.

$$H_2 + \tfrac{1}{2}\ O_2 \rightarrow H_2O \qquad \Delta H = -68.3\ \text{kcal}$$

Using those values, calculate ΔH_f of C_2H_5OH:

$$327\ \text{kcal} = [2(\Delta H_f\ CO_2) + 3(\Delta H_f\ H_2O)] - [(\Delta H_f\ C_2H_5OH) + 3(\Delta H_f\ O_2)]$$

$$327\ \text{kcal} = [2(-94.1\ \text{kcal}) + 3(-68.3\ \text{kcal})] - [(\Delta H_f\ C_2H_5OH) + 3(0\ \text{kcal})]$$

$$327\ \text{kcal} = [(-188.2\ \text{kcal}) + (-204.9\ \text{kcal})] - (\Delta H_f\ C_2H_5OH)$$

$$327\ \text{kcal} = (-393.1\ \text{kcal}) - (\Delta H_f\ C_2H_5OH)$$

$$\Delta H_f\ C_2H_5OH = -720.1\ \text{kcal}$$

37. B is correct.

$$S + O_2 \rightarrow SO_2 + 69.8\ \text{kcal}$$

The reaction indicates that 69.8 kcal of energy is released (i.e., on the product side).

All species in this reaction have the coefficient of 1, which means that ΔH is calculated by reacting 1 mole of each reagent to create 1 mole of product.

The atomic mass of sulfur is 32.1 g, which means that this reaction uses 32.1 g of sulfur.

---

38. D is correct.

The disorder of the system is decreasing as more complex and ordered molecules are forming from the reaction of H_2 and O_2 gases.

3 moles of gas are combining to form 2 moles of gas.

Additionally, two molecules are joining to form one molecule.

39. E is correct.

In thermodynamics, an isolated system is enclosed by rigid, immovable walls through which neither matter nor energy passes.

Temperature is a measure of energy and is not a form of energy. Therefore, the temperature cannot be exchanged between the system and the surroundings.

A closed system can exchange energy (as heat or work) with its surroundings, but not matter.

An isolated system cannot exchange energy (as heat or work) or matter with the surroundings.

An open system can exchange energy and matter with the surroundings.

40. E is correct.

State functions depend only on the initial and final states of the system and are independent of the paths taken to reach the final state.

Common examples of a state function in thermodynamics include internal energy, enthalpy, entropy, pressure, temperature, and volume.

==

Practice Set 3: Questions 41–60

==

41. A is correct.

When comparing various fuels in varying forms (e.g., solid, liquid or gas), it's easiest to compare energy released in terms of mass, because matter has mass regardless of its state (as opposed to volume, which is convenient for a liquid or gas, but not for a solid).

Moles are more complicated because they have to be converted to mass before a direct comparison can be made between the fuels.

42. C is correct.

Convection is heat carried by fluids (i.e., liquids and gases). Heat is prevented from leaving the system through convection when the lid is placed on the cup.

43. B is correct.

For a reaction to be spontaneous, the entropy ($\Delta S_{system} + \Delta S_{surrounding}$) has to be positive (i.e., greater than 0).

44. E is correct.

Explosion of gases: chemical $\rightarrow$ heat energy

Heat converting into steam, which in turn moves a turbine: heat $\rightarrow$ mechanical energy

Generator creates electricity: mechanical $\rightarrow$ electrical energy

Potential Energy **Stored energy and the energy of position (gravitational)**	**Kinetic Energy** **Energy of motion: motion of waves, electrons, atoms, molecules, and substances.**
Chemical Energy Chemical energy is the energy stored in the bonds of atoms and molecules. Examples of stored chemical energy: biomass, petroleum, natural gas, propane, coal. **Nuclear Energy** Nuclear energy is the energy stored in the nucleus of an atom. It is the energy that holds the nucleus. The nucleus of the uranium atom is an example of nuclear energy.	**Radiant Energy** Radiant energy is electromagnetic energy that travels in transverse waves. Radiant energy includes visible light, x-rays, gamma rays, and radio waves. Solar energy is an example of radiant energy. **Thermal Energy** Thermal energy is internal energy in substances; it is the vibration and movement of atoms and molecules within substances. Geothermal energy is an example of thermal energy.

Stored Mechanical Energy	Motion
Stored mechanical energy is energy stored in objects by the application of a force. Compressed springs and stretched rubber bands are examples of stored mechanical energy.	The movement of objects or substances from one place to another is motion. Wind and hydropower are examples of motion.
Gravitational Energy	**Sound**
Gravitational Energy is the energy of place or position. Water in a reservoir behind a hydropower dam is an example of gravitational potential energy. When the water is released to spin the turbines, it becomes kinetic energy.	Sound is the movement of energy through substances in longitudinal (compression/rarefaction) waves.
	Electrical Energy
	Electrical energy is the movement of electrons. Lightning and electricity are examples of electrical energy.

45. A is correct. The relationship between enthalpy (ΔH) and internal energy (ΔE):

Enthalpy = Internal energy + work (for gases, work = PV)

$$\Delta H = \Delta E + \Delta(PV)$$

Solving for ΔE:

$$\Delta E = \Delta H - \Delta(PV)$$

According to ideal gas law:

$$PV = nRT$$

Substitute ideal gas law to the previous equation:

$$\Delta E = \Delta H - \Delta(nRT)$$

R and T are constant, which leaves Δn as the variable.

The reaction is C_2H_2 (g) + $2H_2$ (g) → C_2H_6 (g).

There are three gas molecules on the left and one on the right, which means $\Delta n = 1 - 3 = -2$.

In this problem, the temperature is not provided. However, the presence of degree symbols ($\Delta G°$, $\Delta H°$, $\Delta S°$) indicates that those are standard values, which are measured at 25 °C or 298.15 K.

Double-check the units before performing calculations; ΔH is in kilojoules, while the gas constant (R) is 8.314 J/mol K. Convert ΔH to joules before calculating.

Use the Δn and T values to calculate ΔE:

$$\Delta E = \Delta H - \Delta nRT$$

$$\Delta E = -311,500 \text{ J/mol} - (-2 \times 8.314 \text{ J/mol K} \times 298.15 \text{ K})$$

$$\Delta E = -306,542 \text{ J} \approx -306.5 \text{ kJ}$$

46. D is correct.

A closed system can exchange energy (as heat or work) but not matter with its surroundings.

An isolated system cannot exchange energy (as heat or work) or matter with the surroundings.

An open system can exchange energy and matter with the surroundings.

47. C is correct.

An increase in entropy indicates an increase in disorder; find a reaction that creates a higher number of molecules on the product side compared to the reactant side.

48. C is correct.

When enthalpy is calculated, work done by gases is not taken into account; however, the energy of the reaction includes the work done by gases.

The largest difference between the energy of the reaction and enthalpy would be for reactions with the largest difference in the number of gas molecules between the product and reactants.

49. B is correct.

ΔH refers to enthalpy (or heat).

Endothermic reactions have heat as a reactant.

Exothermic reactions have heat as a product.

Endothermic reactions absorb energy to break strong bonds to form a less stable state (i.e., positive enthalpy). Exothermic reactions release energy when forming stronger bonds to produce a more stable state (i.e., negative enthalpy).

The reaction is nonspontaneous (i.e., endergonic) if the products are less stable than reactants and ΔG is positive.

The reaction is spontaneous (i.e., exergonic) if the products are more stable than the reactants and ΔG is negative.

50. B is correct.

All thermodynamic functions in $\Delta G = \Delta H - T\Delta S$ refer to the system.

51. C is correct.

To predict spontaneity of reaction, use the Gibbs free energy equation:

$$\Delta G = \Delta H° - T\Delta S$$

The reaction is spontaneous if ΔG is negative.

Substitute the given values to the equation:

$$\Delta G = -113.4 \text{ kJ/mol} - [T \times (-145.7 \text{ J/K mol})]$$

$$\Delta G = -113.4 \text{ kJ/mol} + (T \times 145.7 \text{ J/K mol})$$

Important: note that the units are not identical, $\Delta H°$ is in kJ and $\Delta S°$ is in J. Convert kJ to J (1 kJ = 1,000 J):

$$\Delta G = -113,400 \text{ J/mol} + (T \times 145.7 \text{ J/K mol})$$

It can be predicted that the value of ΔG would be negative if the value of T is small.

If T goes higher, ΔG approaches a positive value, and the reaction would be nonspontaneous.

52. E is correct.

Enthalpy: U + PV

53. C is correct.

State functions depend only on the initial and final states of the system and are independent of the paths taken to reach the final state.

Common examples of a state function in thermodynamics include internal energy, enthalpy, entropy, pressure, temperature, and volume.

An extensive property is a property that changes when the size of the sample changes. Examples include mass, volume, length, and charge.

Entropy has no absolute zero value, and a substance at zero Kelvin has zero entropy (i.e., no motion).

54. C is correct.

In an exothermic reaction, the bonds formed are stronger than the bonds broken.

55. D is correct.

Calculate the value of $\Delta H_f = \Delta H_{f \text{ product}} - \Delta H_{f \text{ reactant}}$

$$\Delta H_f = -436.8 \text{ kJ mol}^{-1} - (-391.2 \text{ kJ mol}^{-1})$$

$$\Delta H_f = -45.6 \text{ kJ mol}^{-1}$$

To determine spontaneity, calculate the Gibbs free energy:

$$\Delta G = \Delta H° - T\Delta S$$

The reaction is spontaneous if ΔG is negative:

$$\Delta G = -45.6 \text{ kJ} - T\Delta S$$

Typical ΔS values are around 100–200 J.

If the ΔH value is –45.6 kJ or –45,600 J, the value of ΔG would still be negative unless $T\Delta S$ is less than –45,600.

Therefore, the reaction would be spontaneous over a broad range of temperatures.

56. E is correct.

Enthalpy is an extensive property: it varies with the quantity of the matter.

$$\text{Reaction 1}: \quad P_4 + 6\,Cl_2 \rightarrow 4\,PCl_3 \qquad \Delta H = -1{,}289 \text{ kJ}$$

$$\text{Reaction 2}: 3\,P_4 + 18\,Cl_2 \rightarrow 12\,PCl_3 \qquad \Delta H = ?$$

The only difference between reactions 1 and 2 is the coefficients.

In reaction 2, coefficients are three times reaction 1. Reaction 2 consumes three times as many reactants as reaction 1; reaction 2 releases three times as much energy as reaction 1.

$$\Delta H = 3 \times -1{,}289 \text{ kJ} = -3{,}867 \text{ kJ}$$

57. C is correct.

ΔH refers to enthalpy (or heat).

Endothermic reactions have heat as a reactant.

Exothermic reactions have heat as a product.

Endothermic reactions absorb energy to break strong bonds to form a less stable state (i.e., positive enthalpy). Since endothermic reactions consume energy, there should be heat on the reactants' side.

Exothermic reactions release energy when forming stronger bonds to produce a more stable state (i.e., negative enthalpy).

The reaction is nonspontaneous (i.e., endergonic) if the products are less stable than the reactants and ΔG is positive.

The reaction is spontaneous (i.e., exergonic) if the products are more stable than the reactants and ΔG is negative.

58. D is correct.

According to the second law of thermodynamics, the entropy gain of the universe must be positive (i.e., increased disorder).

The entropy of a system, however, can be negative if the surroundings experience an increase in entropy greater than the negative entropy change experienced by the system.

59. E is correct.

In thermodynamics, an isolated system is enclosed by rigid, immovable walls through which neither matter nor energy passes.

Bell jars are designed not to let air or other materials in or out. Insulated means that heat cannot get in or out.

Evaporation indicates a phase change (vaporization) from liquid to gaseous phase; however, it does not imply that the matter has escaped its system.

A closed system can exchange energy (as heat or work) with its surroundings, but not matter.

An isolated system cannot exchange energy (as heat or work) or matter with the surroundings.

An open system can exchange energy and matter with the surroundings.

60. E is correct.

The *law of conservation of energy* states that the energy (e.g., potential or kinetic) of an isolated system remains constant and is conserved. Energy can be neither created nor destroyed but is transformed from one form to another. For instance, chemical energy can be converted to kinetic energy in the explosion of a firecracker.

The *law of conservation of mass* states that for any system closed to transfers of matter and energy, the mass of the system must remain constant over time.

Law of Definite Proportions (Proust's law) states that a chemical compound contains the same proportion of elements by mass. The law of definite proportions forms the basis of stoichiometry (i.e., from the known amounts of separate reactants, the amount of the product can be calculated).

Avogadro's law is an experimental gas law relating the volume of a gas to the amount of substance of gas present. It states that equal volumes of gases at the same temperature and pressure have the same number of molecules.

Boyle's law is an experimental gas law that describes how the pressure of a gas tends to increase as the volume of a gas decreases. It states that the absolute pressure exerted by a given mass of an ideal gas is inversely proportional to the volume it occupies if the temperature and amount of gas remain unchanged within a closed system.

==

Practice Set 4: Questions 61–80

==

61. C is correct.

ΔH refers to enthalpy (or heat).

Endothermic reactions have heat as a reactant.

Exothermic reactions have heat as a product.

Exothermic reactions release heat and cause the temperature of the immediate surroundings to rise (i.e., a net loss of energy), while an endothermic process absorbs heat and cools the surroundings (i.e., a net gain of energy).

Endothermic reactions absorb energy to break strong bonds to form a less stable state (i.e., positive enthalpy). Exothermic reactions release energy when forming stronger bonds to produce a more stable state (i.e., negative enthalpy).

The reaction is nonspontaneous (i.e., endergonic) if the products are less stable than the reactants and ΔG is positive.

The reaction is spontaneous (i.e., exergonic) if the products are more stable than the reactants and ΔG is negative.

62. C is correct.

Temperature is a measure of the average kinetic energy of the molecules.

Kinetic energy is proportional to temperature.

In most thermodynamic equations, temperatures are expressed in Kelvin, so convert the Celsius temperatures to Kelvin:

$$25 \text{ °C} + 273.15 = 298.15 \text{ K}$$

$$50 \text{ °C} + 273.15 = 323.15 \text{ K}$$

Calculate the kinetic energy using simple proportions:

$$KE = (323.15 \text{ K} / 298.15 \text{ K}) \times 500 \text{ J}$$

$$KE = 540 \text{ J}$$

63. C is correct.

The equation relates to the change in entropy (ΔS) at different temperatures.

Entropy increases at higher temperatures (i.e., increased kinetic energy).

---

64. B is correct.

A change is exothermic when energy is released to the surroundings.

Energy loss occurs when a substance changes to a more rigid phase (e.g., gas → liquid or liquid → solid).

65. A is correct.

Potential Energy Stored energy and the energy of position (gravitational)	Kinetic Energy Energy of motion: motion of waves, electrons, atoms, molecules, and substances.
Chemical Energy	**Electrical Energy**
Chemical energy is the energy stored in the bonds of atoms and molecules. Examples of stored chemical energy: biomass, petroleum, natural gas, propane, coal.	Electrical energy is the movement of electrons. Lightning and electricity are examples of electrical energy.

66. B is correct.

At equilibrium, there is no potential, and therefore neither direction of the reaction is favored.

Reaction potential (E) indicates the tendency of a reaction to be spontaneous in either direction.

Because the solution is in equilibrium, no further reactions occur in either direction. Therefore, $E = 0$.

67. D is correct.

The transformation from solid to gas is accompanied by an increase in entropy (i.e., disorder).

The molecules of gas have greater kinetic energy than the molecules of solids.

68. C is correct.

The reaction is nonspontaneous (i.e., endergonic) if the products are less stable than the reactants and ΔG is positive.

The reaction is spontaneous (i.e., exergonic) if the products are more stable than the reactants and ΔG is negative.

69. E is correct.

The spontaneity of a reaction is determined by evaluating the Gibbs free energy, or ΔG.

$\Delta G = \Delta H - \text{T}\Delta S$

A reaction is spontaneous if $\Delta G < 0$ (or ΔG is negative).

For ΔG to be negative, ΔH has to be less than $\text{T}\Delta S$.

A reaction is nonspontaneous if $\Delta G > 0$ (or ΔG is positive and ΔH is greater than $\text{T}\Delta S$).

70. B is correct.

The reaction is nonspontaneous (i.e., endergonic) if the products are less stable than the reactants and ΔG is positive.

The reaction is spontaneous (i.e., exergonic) if the products are more stable than the reactants and ΔG is negative.

Entropy (ΔS) is determined by the relative number of molecules (compounds):

The reactants have one molecule (with a coefficient total of 2).

The products have two molecules (with a coefficient total of 7).

The number of molecules has increased from reactants to products. Therefore, the entropy (i.e., disorder) has increased.

71. D is correct.

ΔG indicates the energy made available to do non P–V work.

This is a useful quantity for processes like batteries and living cells that do not have expanding gases.

72. C is correct.

The internal energy of a system is the energy contained within the system, excluding the kinetic energy of motion of the system as a whole and the potential energy of the system as a whole due to external force fields.

The internal energy of a system can be changed by transfers of matter and by work and heat transfer. When impermeable container walls prevent matter transfer, the system is said to be closed.

The First law of thermodynamics states that the increase in internal energy is equal to the heat added plus the work done on the system by its surroundings. If the container walls do not pass energy or matter, the system is isolated, and its internal energy remains constant.

Bond energy is internal potential energy (PE), while thermal energy is internal kinetic energy (KE).

73. D is correct.

Endothermic reactions absorb energy to break strong bonds to form a less stable state (i.e., positive enthalpy).

Exothermic reactions release energy when forming stronger bonds to produce a more stable state (i.e., negative enthalpy).

To predict spontaneity of reaction, use the Gibbs free energy equation:

$$\Delta G = \Delta H^\circ - T\Delta S$$

The reaction is nonspontaneous (i.e., endergonic) if the products are less stable than the reactants and ΔG is positive.

The reaction is spontaneous (i.e., exergonic) if the products are more stable than the reactants and ΔG is negative.

It is the total amount of energy that is determinative. Some bonds are stronger than others, so there is a net gain or net loss of energy when formed.

74. E is correct.

The starting material is ice at −25 °C, which means that its temperature must be raised to 0 °C before it melts.

The first term in the heat calculation is:

specific heat of ice (A) × mass × change of temperature

75. C is correct.

The heat of formation of water vapor is similar to the highly exothermic process for the heat of combustion of hydrogen.

Heat is required to convert H_2O (*l*) into vapor; therefore, the heat of formation of water vapor must be less exothermic than of H_2O (*l*).

Solid → liquid → gas: endothermic because the reaction absorbs energy to break strong bonds to form a less stable state (i.e., positive enthalpy).

Gas → liquid → solid: exothermic because the reaction releases energy during the formation of stronger bonds to produce the more stable state (i.e., negative enthalpy).

76. E is correct.

77. B is correct.

Einstein established the equivalence of mass and energy, whereby the energy equivalent of mass is given by $E = mc^2$.

Energy can be converted to mass, and mass can be converted to energy as long as the total energy, including mass-energy, is conserved.

In nuclear fission, the mass of the parent nucleus is greater than the sum of the masses of the daughter nuclei. This deficit in mass is converted into energy.

78. D is correct.

A closed system can exchange energy (as heat or work) but not matter with its surroundings.

An isolated system cannot exchange energy (as heat or work) or matter with the surroundings.

An open system can exchange energy and matter with the surroundings.

79. A is correct.

Gases have the highest entropy (i.e., disorder or randomness).

Solutions have higher entropy than pure phases since the molecules are more scattered.

The liquid phase has less entropy than the aqueous phase.

80. A is correct.

In a chemical reaction, bonds within reactants are broken, and new bonds form.

Therefore, in bond dissociation problems:

$\Delta H_{reaction}$ = sum of bond energy in reactants – the sum of bond energy in products.

This is the opposite of ΔH_f problems, where:

$\Delta H_{reaction}$ = sum of ΔH_f product – sum of ΔH_f reactant

For $N_2 + 3\ H_2 \rightarrow 2\ NH_3$

$\Delta H_{reaction}$ = sum of bond energy in reactants – sum of bond energy in products

$\Delta H_{reaction} = [(N\equiv N) + 3(H–H)] – [2 \times 3(N–H)]$

$\Delta H_{reaction} = [(946\ kJ) + (3 \times 436\ kJ) – (2 \times (3 \times 389\ kJ)]$

$\Delta H_{reaction} = –80\ kJ$

Notes for active learning

Kinetics and Equilibrium – Detailed Explanations

==

Practice Set 1: Questions 1–20

==

1. A is correct.

General formula for the equilibrium constant of a reaction:

$$aA + bB \leftrightarrow cC + dD$$

$$K_{eq} = ([C]^c \times [D]^d) / ([A]^a \times [B]^b)$$

For the reaction above: $K_{eq} = [C] / [A]^2 \times [B]^3$

2. B is correct.

Two gases having the same temperature have the same average molecular kinetic energy.

Therefore,

$$(\tfrac{1}{2})m_A v_A^2 = (\tfrac{1}{2})m_B v_B^2$$

$$m_A v_A^2 = m_B v_B^2$$

$$(m_B / m_A) = (v_A / v_B)^2$$

$$(v_A / v_B) = 2$$

$$(m_B / m_A) = 2^2 = 4$$

The only gases listed with a mass ratio of about 4 to 1 are iron (55.8 g/mol) and nitrogen (14 g/mol).

3. A is correct.

The slowest reaction would be the reaction with the highest activation energy and the lowest temperature.

4. A is correct.

The rate law is calculated experimentally by comparing trials and determining how changes in the initial concentrations of the reactants affect the rate of the reaction.

$$\text{rate} = k[A]^x \cdot [B]^y$$

where k is the rate constant, and the exponents x and y are the partial reaction orders (i.e., determined experimentally). They are not equal to the stoichiometric coefficients.

To determine the order of reactant A, find two experiments where the concentrations of B are identical and concentrations of A are different.

==

Use data from experiments 2 and 3 to calculate order of A:

$\text{Rate}_3 / \text{Rate}_2 = ([A]_3 / [A]_2)^{\text{order of A}}$

$0.500 / 0.500 = (0.060 / 0.030)^{\text{order of A}}$

$1 = (2)^{\text{order of A}}$

Order of A = 0

5. C is correct.

The balanced equation: $C_2H_6O + 3\,O_2 \rightarrow 2\,CO_2 + 3\,H_2O$

The rate of reaction/consumption is proportional to the coefficients.

The rate of carbon dioxide production is twice the rate of ethanol consumption: $2 \times 4.0\ \text{M s}^{-1} = 8.0\ \text{M s}^{-1}$

6. D is correct.

General formula for the equilibrium constant of a reaction:

$aA + bB \leftrightarrow cC + dD$

$K_{eq} = ([C]^c \times [D]^d) / ([A]^a \times [B]^b)$

For equilibrium constant calculation, only include species in aqueous or gas phases.

$K_{eq} = 1 / [Cl_2]$

7. B is correct.

$K_{eq} = [\text{products}] / [\text{reactants}]$

If the K_{eq} is less than 1 (e.g., 6.3×10^{-14}), then the numerator (i.e., products) is smaller than the denominator (i.e., reactants), and fewer products have formed relative to the reactants.

If the reaction favors reactants compared to products, the equilibrium lies to the left.

8. A is correct.

The higher the energy of activation, the slower the reaction is. The height of this barrier is independent of the determination of spontaneous (ΔG is negative with products more stable than reactants) or nonspontaneous reactions (ΔG is positive with products less stable than reactants).

9. E is correct.

The main function of catalysts is lowering the reaction's activation energy (energy barrier), thus increasing its rate (k).

Temperature is a measure of the average kinetic energy (i.e., $KE = \frac{1}{2}mv^2$) of the molecules.

Except for zero-order chemical reactions, increased concentration of reactants increases the probability that the reactants collide with sufficient energy and orientation to overcome the energy of activation barrier and proceed toward products.

10. D is correct.

Every reaction has activation energy: the amount of energy required by the reactant to start reacting.

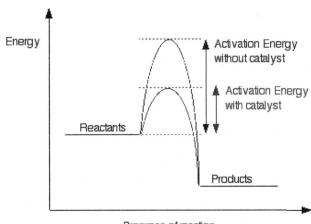

On the graph, activation energy can be estimated by calculating the distance between the initial energy level (i.e., reactant) and the peak (i.e., transition state). Activation energy is *not* the difference between energy levels of the initial and final state (reactant and product).

A catalyst provides an alternative pathway for the reaction to proceed to product formation. It lowers the energy of activation (i.e., relative energy between reactants and transition state) and, therefore, speeds the reaction rate.

Catalysts do not affect the Gibbs free energy (ΔG: stability of products vs. reactants) or the enthalpy (ΔH: bond breaking in reactants or bond making in products).

11. D is correct.

The activation energy is the energy barrier that must be overcome for the transformation of reactants into products.

A reaction with higher activation energy (energy barrier) has a decreased rate (k).

12. B is correct.

The activation energy is the energy barrier that must be overcome to transform reactants into products.

A reaction with lower activation energy (energy barrier) has an increased rate (k).

13. D is correct.

The activation energy is the energy barrier that must be overcome to transform reactants into products.

A reaction with higher activation energy (energy barrier) has a decreased rate (k).

If the graphs are on the same scale, the height of the activation energy (R to the highest peak on the graph) is greatest in graph d.

14. B is correct.

If the graphs are on the same scale, the height of the activation energy (R to the highest peak on the graph) is the smallest in graph b.

The main function of catalysts is lowering the reaction's activation energy (energy barrier), thus increasing its rate (k).

Although catalysts do decrease the amount of energy required to reach the rate-limiting transition state, they do *not* decrease the relative energy of the products and reactants. Therefore, a catalyst does not affect ΔG.

A catalyst provides an alternative pathway for the reaction to proceed to product formation. It lowers the energy of activation (i.e., relative energy between reactants and transition state) and, therefore, speeds the reaction rate.

Catalysts do not affect the Gibbs free energy (ΔG: stability of products vs. reactants) or the enthalpy (ΔH: bond breaking in reactants or bond making in products).

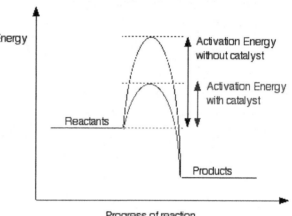

Catalysts have no effect on ΔG (relative levels of R and P). Graph D shows an endergonic reaction with products less stable than reactants. The other graphs show an exergonic reaction with the same relative difference between the more stable and less stable reactants; only the activation energy is different.

15. B is correct.

A reaction is at equilibrium when the forward reaction rate equals the rate of the reverse reaction. Achieving equilibrium is common for reactions. Catalysts, by definition, are regenerated during a reaction and are not consumed by the reaction.

16. D is correct.

The molecules must collide with sufficient energy, frequency of collision, and proper orientation to overcome the activation energy barrier.

17. B is correct.

A catalyst lowers the energy of activation, which increases the rate of the reaction.

The primary function of catalysts is lowering the reaction's activation energy (energy barrier), thus increasing its rate (k).

Although catalysts do decrease the amount of energy required to reach the rate-limiting transition state, they do *not* decrease the relative energy of the products and reactants. Therefore, a catalyst does not affect ΔG.

18. E is correct.

When changing the conditions of a reaction, Le Châtelier's principle states that the position of equilibrium shifts to counteract the change. If the concentration of reactants or products changes, the position of the equilibrium changes. Adding reactants or removing products shifts the equilibrium to the right.

According to Le Châtelier's Principle, adding or removing a solid at equilibrium does not affect equilibrium. However, solid NaOH dissociates into Na^+ and OH^- when placed in an aqueous solution. The OH^- combines with H^+ to form water. Reducing H^+ as a product drives the reaction to the right.

19. A is correct.

When changing the conditions of a reaction, Le Châtelier's principle states that the position of equilibrium shifts to counteract the change. If the reaction temperature, pressure, or volume changes, the position of equilibrium changes.

Adding reactants or removing products shifts the equilibrium to the right.

Heat (i.e., energy related to temperature) is a reactant, and increasing its value drives the reaction toward product formation.

Decreasing the pressure on the reaction vessel drives the reaction toward reactants (i.e., to the left). The relative molar concentration of the reactants (i.e., $2 + 6$) is greater than the products (i.e., $4 + 3$).

The other choices listed drive the reaction toward reactants (i.e., to the left).

20. C is correct.

First, calculate the concentration/molarity of each reactant:

H_2: 0.20 moles / 4.00 L = 0.05 M

X_2: 0.20 moles / 4.00 L = 0.05 M

HX: 0.800 moles / 4.00L = 0.20 M

Use the concentrations to calculate Q:

$Q = [HX]^2 / [H_2] \cdot [X_2]$

$Q = (0.20)^2 / (0.05 \times 0.05)$

$Q = 16$

Because $Q < K_c$ (i.e., $16 < 24.4$), the reaction shifts to the right.

For Q to increase and match K_c, the numerator (i.e., $[HX]^2$), which is the product (right side) of the equation, needs to be increased.

===

Practice Set 2: Questions 21–40

===

21. E is correct.

General formula for the equilibrium constant of a reaction:

$aA + bB \leftrightarrow cC + dD$

$K_{eq} = ([C]^c \times [D]^d) / ([A]^a \times [B]^b)$

For equilibrium constant (K_c) calculation, only include species in aqueous or gas phases:

$K_c = [CO_2] [H_2O]^2 / [CH_4]$

22. A is correct.

The rate law is calculated by determining how changes in concentrations of the reactants affect the initial rate of the reaction.

rate = $k[A]^{x} \cdot [B]^{y}$

where k is the rate constant, and the exponents x and y are the partial reaction orders. They are not equal to the stoichiometric coefficients.

Whenever the fast (i.e., second) step follows the slow (i.e., first) step, the fast step is assumed to reach equilibrium, and the equilibrium concentrations are used for the rate law of the slow step.

23. E is correct.

The main function of catalysts is lowering the reaction's activation energy (energy barrier), thus increasing its rate (k).

Although catalysts do decrease the amount of energy required to reach the rate-limiting transition state, they do *not* decrease the relative energy of the products and reactants. Therefore, a catalyst does not affect ΔG.

A catalyst provides an alternative pathway for the reaction to proceed to product formation. It lowers the energy of activation (i.e., relative energy between reactants and transition state) and speeds the reaction rate.

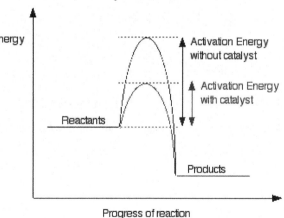

Catalysts do not affect the Gibbs free energy (Δ*G*: stability of products vs. reactants) or the enthalpy (Δ*H*: bond breaking in reactants or bond making in products).

24. B is correct.

In equilibrium reactions, equilibrium varies between conditions, so any proportion of product/reactant mass is possible.

The question refers to the state *after* a reaction reached equilibrium.

For a reaction that favors the formation of products ($K_{eq} > 1$), the amount of product would be more than reactants. For a reaction that favors reactants ($K_{eq} < 1$), the amount of product would be less than reactants.

When $K_{eq} = 1$, the amounts of products and reactants would be equal.

25. C is correct.

The rate law is calculated experimentally by comparing trials and determining how changes in the initial concentrations of the reactants affect the rate of the reaction.

$$\text{rate} = k[A]^x \cdot [B]^y$$

where k is the rate constant, and the exponents x and y are the partial reaction orders (i.e., determined experimentally). They are not equal to the stoichiometric coefficients.

Rate laws cannot be determined from the balanced equation (i.e., used to determine equilibrium) unless the reaction occurs in a single step.

From the data, when the concentration doubles, the rate quadrupled. Since 4 is 2^2, the rate law is second order, therefore the rate $= k[H_2]^2$.

26. D is correct.

Lowering temperature shifts the equilibrium to the right because heat is a product (lowering the temperature is similar to removing the product). Increasing temperature (i.e., heating the system) has the opposite effect and shifts the equilibrium to the left (i.e., toward products).

Removal of H_2 (i.e., reactant) shifts the equilibrium to the left.

The addition of NH_3 (i.e., products) shifts the equilibrium to the left.

A catalyst lowers the energy of activation but does not affect the position of the equilibrium.

27. B is correct.

The order of the reaction has to be determined experimentally.

It is possible for a reaction's order to be identical to its coefficients. For example, in a single-step reaction or during the slow step of a multi-step reaction, the coefficients correlate to the rate law.

28. A is correct.

General formula for the equilibrium constant of a reaction:

$$a A + b B \leftrightarrow c C + d D$$

$$K_{eq} = ([C]^c \times [D]^d) / ([A]^a \times [B]^b)$$

For equilibrium constant calculation, only include species in aqueous or gas phases:

$$K_{eq} = 1 / [CO] \times [H_2]^2$$

29. E is correct.

Two peaks in this reaction indicate two energy-requiring steps with one intermediate (i.e., C) and each peak (i.e., B and D) as an activated complex (i.e., transition states).

The activated complex (i.e., transition state) is undergoing bond breaking/bond making events.

30. A is correct.

The activation energy for the slow step of a reaction is the distance from the starting material (or an intermediate) to the activated complex (i.e., transition state) with the absolute highest energy (i.e., highest point on the graph).

31. C is correct.

The activation energy for the slow step of a reverse reaction is the distance from an intermediate (or product) to the activated complex (i.e., transition state) with the absolute highest energy (i.e., highest point on the graph).

The slow step may not have the greatest magnitude for activation energy (e.g., E→ D on the graph).

32. B is correct.

The change in energy (or ΔH) is the difference between the energy of the reactants and products.

33. A is correct.

As the temperature (i.e., average kinetic energy) increases, the particles move faster (i.e., increased kinetic energy) and collide more frequently per unit time. This increases the reaction rate.

34. A is correct.

When changing the conditions of a reaction, Le Châtelier's principle states that equilibrium shifts to counteract the change. If the reaction temperature, pressure, or volume is changed, the position of equilibrium changes.

According to Le Châtelier's principle, endothermic ($+\Delta H$) reactions increase the formation of products at higher temperatures.

Since both sides of the reaction have 2 gas molecules, a change in pressure does not affect equilibrium.

35. B is correct.

All chemical reactions eventually reach equilibrium, the state at which the reactants and products are present in concentrations that have no further tendency to change with time.

Therefore, the rate of production of each of the products (i.e., forward reaction) equals the rate of their consumption by the reverse reaction.

36. E is correct.

Chemical equilibrium refers to a dynamic process whereby the *rate* at which a reactant molecule is being transformed into a product is the same as the *rate* for a product molecule to be transformed into a reactant.

All chemical reactions eventually reach equilibrium, the state at which the reactants and products are present in concentrations that have no further tendency to change with time.

37. A is correct.

Expression for equilibrium constant: $K = [H_2O]^2 \cdot [Cl_2]^2 / [HCl]^4 \cdot [O_2]$

Solve for $[Cl_2]$:

$$[Cl_2]^2 = (K \times [HCl]^4 \cdot [O_2]) / [H_2O]^2$$

$$[Cl_2]^2 = [46.0 \times (0.150)^4 \times 0.395] / (0.625)^2$$

$$[Cl_2]^2 = 0.0235$$

$$[Cl_2] = 0.153 \text{ M}$$

38. A is correct.

To determine the amount of each compound at equilibrium, consider the chemical reaction written in the form:

$$a\text{A} + b\text{B} \leftrightarrow c\text{C} + d\text{D}$$

The equilibrium constant (K_c) is defined as:

$$K_c = ([\text{C}]^c \times [\text{D}]^d) / ([\text{A}]^a \times [\text{B}]^b)$$

or $K_c = [\text{products}] / [\text{reactants}]$

If the K_{eq} is greater than 1, the numerator (i.e., products) is larger than the denominator (i.e., reactants), and more products have formed relative to the reactants.

If the reaction favors products compared to reactants, the equilibrium lies to the right.

39. D is correct.

Gases (as opposed to liquids and solids) are most sensitive to changes in pressure. There are three moles of hydrogen gas as a reactant and three moles of water vapor as a product.

Therefore, changes in pressure result in proportionate changes to the forward and reverse reactions.

40. E is correct.

When changing the conditions of a reaction, Le Châtelier's principle states that the position of equilibrium shifts to counteract the change. If the reaction temperature, pressure, volume, or concentration changes, the position of the equilibrium changes.

CO is a reactant, and increasing its concentration shifts the equilibrium toward product formation (i.e., to the right).

==

Practice Set 3: Questions 41–60

==

41. A is correct.

General formula for the equilibrium constant of a reaction:

$$a\text{A} + b\text{B} \leftrightarrow c\text{C} + d\text{D}$$

$$K_{eq} = ([\text{C}]^c \times [\text{D}]^d) / ([\text{A}]^a \times [\text{B}]^b)$$

For equilibrium constant calculation, only include species in aqueous or gas phases:

$$K_{eq} = [\text{NO}]^4 \times [\text{H}_2\text{O}]^6 / [\text{NH}_3]^4 \times [\text{O}_2]^5$$

42. E is correct.

The high levels of CO_2 cause a person to hyperventilate to reduce the number of CO_2 (reactant).

Hyperventilating has the effect of driving the reaction toward products.

Removing products (i.e., HCO_3^- and H^+) or intermediates (H_2CO_3) drives the reaction toward products.

43. B is correct.

The units of the rate constants:

Zero order reaction: M sec^{-1}

First order reaction: sec^{-1}

Second order reaction: L mole^{-1} sec^{-1}

44. E is correct.

A reaction proceeds when the reactant(s) have sufficient energy to overcome activation energy and proceed to products. Increasing the temperature increases the molecule's kinetic energy, increases the frequency of collision, and increases the probability that the reactants will overcome the barrier (energy of activation) to form products.

45. B is correct.

Low activation energy increases the rate because the energy required for the reaction to proceed is lower.

High temperature increases the rate because faster-moving molecules have a greater probability of collision, facilitating the reaction.

Combined, lower activation energy and a higher temperature results in the highest relative rate for the reaction.

46. C is correct.

The rate law is calculated by comparing trials and determining how changes in the initial concentrations of the reactants affect the rate of the reaction.

$$\text{rate} = k[A]^x \cdot [B]^y$$

where k is the rate constant, and the exponents x and y are the partial reaction orders (i.e., determined experimentally). They are not equal to the stoichiometric coefficients.

Start by identifying two reactions where XO concentration is constant, and O_2 is different: experiments 1 and 2. When the concentration of O_2 is doubled, the rate is doubled. This indicates that the order of the reaction for O_2 is 1.

Now, determine the order for XO.

Find 2 reactions where the concentration of O_2 is constant and XO is different: experiments 2 and 3. When the concentration of XO is tripled, the rate is multiplied by a factor of 9. $3^2 = 9$, which means the order of the reaction with respect to XO is 2.

Therefore, the expression of rate law is:

$$\text{rate} = k[XO]^2 \cdot [O_2]$$

47. B is correct.

General formula for the equilibrium constant of a reaction:

$$a\text{A} + b\text{B} \leftrightarrow c\text{C} + d\text{D}$$

$$K_{eq} = ([C]^c \times [D]^d) / ([A]^a \times [B]^b)$$

For equilibrium constant calculation, only include species in aqueous or gas phases.

$$K_{eq} = 1 / [CO_2]$$

48. D is correct.

$$K_{eq} = [\text{products}] / [\text{reactants}]$$

If the numerator (i.e., products) is smaller than the denominator (i.e., reactants), and fewer products have formed relative to the reactants. Therefore, K_{eq} is less than 1.

49. C is correct.

The activation energy is the energy barrier for a reaction to proceed. The forward reaction is measured from the reactants to the highest energy level in the reaction.

A reaction mechanism with low energy of activation proceeds faster than a reaction with high energy of activation.

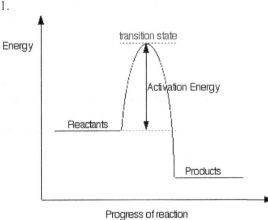

50. E is correct.

As the average kinetic energy (i.e., $KE = \frac{1}{2}mv^2$) increases, the particles move faster and collide more frequently per unit time and possess greater energy when they collide. This increases the reaction rate. Hence the reaction rate of most reactions increases with increasing temperature.

For reversible reactions, it is common for the rate law to depend on the concentration of the products. The overall rate is negative (except for autocatalytic reactions).

As the concentration of products decreases, their collision frequency decreases, and the reverse reaction rate decreases, therefore, the overall rate of the reaction increases.

The primary function of catalysts is lowering the reaction's activation energy (energy barrier), thus increasing its rate (k).

51. C is correct.

General formula for the equilibrium constant of a reaction:

$$a\text{A} + b\text{B} \leftrightarrow c\text{C} + d\text{D}$$

$$K_{eq} = ([\text{C}]^c \times [\text{D}]^d) / ([\text{A}]^a \times [\text{B}]^b)$$

For the ionization equilibrium (K_i) constant calculation, only include species in aqueous or gas phases (which, in this case, is all):

$$K_i = [\text{H}^+]\cdot[\text{HS}^-] / [\text{H}_2\text{S}]$$

52. B is correct.

As the temperature (i.e., average kinetic energy) increases, the particles move faster and collide more frequently per unit time and possess greater energy when they collide. This increases the reaction rate. Hence the reaction rate of most reactions increases with increasing temperature.

53. B is correct.

Reactions are spontaneous when Gibbs free energy (ΔG) is negative, and the reaction is described as exergonic; the products are more stable than the reactants.

Reactions are nonspontaneous when Gibbs free energy (ΔG) is positive, and the reaction is described as endergonic; the products are less stable than the reactants.

54. C is correct.

When changing the conditions of a reaction, Le Châtelier's principle states that equilibrium shifts to counteract the change. If the reaction temperature, pressure, or volume is changed, the position of equilibrium changes.

There are 2 moles on each side, so changes in pressure (or volume) do not shift the equilibrium.

55. A is correct.

All of the choices are possible events of chemical reactions, but they do not need to happen for a reaction to occur, except that reactant particles must collide (i.e., make contact) with each other.

56. D is correct.

The enthalpy or heat of reaction (ΔH) is not a function of temperature.

Since the reaction is exothermic, Le Châtelier's principle states that increasing the temperature decreases the forward reaction.

57. E is correct.

When changing the conditions of a reaction, Le Châtelier's principle states that the position of equilibrium shifts to counteract the change. If the reaction's concentration changes, the position of the equilibrium changes.

According to Le Châtelier's Principle, adding or removing a solid at equilibrium does not affect equilibrium. However, adding solid $KC_2H_3O_2$ is equivalent to adding K^+ and $C_2H_3O_2^-$ (i.e., product) as it dissociates in an aqueous solution. The increased concentration of products shifts the equilibrium toward reactants (i.e., to the left).

Increasing the pH decreases the $[H^+]$ and therefore would shift the reaction to products (i.e., right).

58. E is correct.

When changing the conditions of a reaction, Le Châtelier's principle states that the position of equilibrium shifts to counteract the change. If the reaction temperature, pressure, or volume is changed, the position of equilibrium changes.

The $K_{eq} = 2.8 \times 10^{-21}$ which indicates a low concentration of products compared to reactants:

$$K_{eq} = [products] / [reactants]$$

59. E is correct.

Chemical equilibrium refers to a dynamic process whereby the rate at which a reactant molecule is being transformed into a product is the same as the rate for a product molecule to be transformed into a reactant. The rate of the forward reaction is equal to the rate of the reverse reaction.

60. C is correct.

All chemical reactions eventually reach equilibrium, the state at which the reactants and products are present in concentrations that have no further tendency to change with time.

Catalysts speed up reactions and therefore increase the rate at which equilibrium is reached.

However, they never alter the thermodynamics of a reaction and therefore do not change free energy ΔG.

Catalysts do not alter the equilibrium constant.

===

Practice Set 4: Questions 61–80

===

61. E is correct.

To determine the order with respect to W, compare the data for trials 2 and 4.

The concentrations of X and Y do not change when comparing trials 2 and 4, but the concentration of W changes from 0.015 to 0.03, which corresponds to an increase by a factor of 2. However, the rate did not increase (0.08 remains 0.08); therefore, the order with respect to W is 0.

The order for X is determined by comparing the data for trials 1 and 3.

The concentrations for W and Z are constant, and the concentration for X increases by a factor of 3 (from 0.05 to 0.15). The rate of the reaction increased by a factor of 9 (0.04 to 0.36). Since $3^2 = 9$, the order of the reaction with respect to X is two.

The order with respect to Y is found by comparing the data from trials 1 and 5. [W] and [X] do not change, but [Y] goes up by a factor of 4. The rate from trials 1 to 5 goes up by a factor of 2. Since $4^{1/2} = 2$, the order of the reaction with respect to Y is ½.

The overall order is found by the sum of the orders for X, Y, and Z: $0 + 2 + ½ = 2½$.

62. B is correct.

Since the order with respect to W is zero, the rate of formation of Z does not depend on the concentration of W.

63. D is correct.

The rate law is calculated by comparing trials and determining how changes in the initial concentrations of the reactants affect the rate of the reaction.

$$\text{rate} = k[A]^x \cdot [B]^y$$

where k is the rate constant, and the exponents x and y are the partial reaction orders (i.e., determined experimentally). They are not equal to the stoichiometric coefficients.

Use the partial orders determined in question **61** to express the rate law:

$$\text{rate} = k[W]^0[X]^2[Z]^{½}$$

Substitute the values for trial #1:

$$0.04 = k\,[0.01]^0[0.05]^2[0.04]^{½}$$

$$0.04 = (k)\cdot(1)\cdot(2.5 \times 10^{-3})\cdot(2 \times 10^{-1})$$

$$k = 80$$

---

64. A is correct.

When changing the conditions of a reaction, Le Châtelier's principle states that the position of the equilibrium shifts to counteract the change. If the reaction temperature, pressure, or volume changes, the position of the equilibrium changes.

Decreasing the temperature of the system favors the production of more heat. It shifts the reaction towards the exothermic side. Because $\Delta H > 0$ (i.e., heat is a reactant), the reaction is endothermic. Therefore, decreasing the temperature shifts the equilibrium toward the reactants.

65. D is correct.

General formula for the equilibrium constant of a reaction:

$$a\text{A} + b\text{B} \leftrightarrow c\text{C} + d\text{D}$$

$$K_{eq} = ([\text{C}]^c \times [\text{D}]^d) / ([\text{A}]^a \times [\text{B}]^b)$$

For the equilibrium constant (K_i), only include species in aqueous phases:

$$K_i = [\text{H}^+] \times [\text{H}_2\text{PO}_4^-] / [\text{H}_3\text{PO}_4]$$

66. C is correct.

Catalysts speed up reactions and therefore increase the rate at which equilibrium is reached. However, catalysts do not alter the thermodynamics of a reaction and therefore do not change free energy ΔG.

A catalyst provides an alternative pathway for the reaction to proceed to product formation. It lowers the energy of activation (i.e., relative energy of reactants and transition state) and speeds the reaction rate.

Catalysts do not affect the Gibbs free energy (ΔG: stability of products vs. reactants) or the enthalpy (ΔH: bond breaking in reactants or bond making in products).

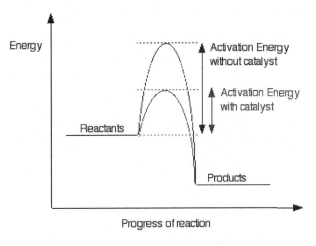

67. D is correct.

General formula for the equilibrium constant of a reaction:

$$a\text{A} + b\text{B} \leftrightarrow c\text{C} + d\text{D}$$

$$K_{eq} = ([\text{C}]^c \times [\text{D}]^d) / ([\text{A}]^a \times [\text{B}]^b)$$

For the ionization equilibrium constant (K_i) calculation, only include species in aqueous or gas phases (which, in this case, is all):

$$K_i = [\text{H}^+] \cdot [\text{HSO}_3^-] / [\text{H}_2\text{SO}_3]$$

68. D is correct.

$$K_{eq} = [\text{products}] / [\text{reactants}]$$

If the K_{eq} is greater than 1, then the numerator (i.e., products) is greater than the denominator (i.e., reactants), and more products have formed relative to the reactants.

69. B is correct.

The activation energy is the energy barrier for a reaction to proceed. For the forward reaction, it is measured from the energy of the reactants to the highest energy level in the reaction.

70. A is correct.

Lowering the temperature decreases the reaction rate because the molecules involved in the reaction move and collide more slowly, so the reaction occurs at a slower speed.

Increasing the concentration of reactants due to the increased probability that the reactants collide with sufficient energy and orientation to overcome the energy of the activation barrier and proceed toward products. The primary function of catalysts is lowering the reaction's activation energy (energy barrier), thus increasing its rate (k).

71. B is correct.

The activation energy is the energy barrier that must be overcome to transform reactant(s) into product(s). The molecules must collide with the proper orientation and energy (i.e., kinetic energy is the average temperature) to overcome the energy of the activation barrier.

72. B is correct.

The activation energy is the energy barrier that must be overcome to transform reactant(s) into product(s). The molecules must collide with the proper orientation and energy (i.e., kinetic energy is the average temperature) to overcome the energy of the activation barrier.

Solids describe molecules with limited relative motion. Liquids are molecules with more relative motion than solids, while gases exhibit the most relative motion.

73. B is correct.

The rate law is calculated experimentally by comparing trials and determining how changes in the initial concentrations of the reactants affect the rate of the reaction.

$$\text{rate} = k[\text{A}]^x \cdot [\text{B}]^y$$

where k is the rate constant, and the exponents x and y are the partial reaction orders (i.e., determined experimentally). They are not equal to the stoichiometric coefficients.

Comparing Trials 1 and 3, [A] increased by a factor of 3, as did the reaction rate; thus, the reaction is first order with respect to A.

Comparing Trials 1 and 2, [B] increased by a factor of 4 and the reaction rate increased by a factor of $16 = 4^2$. Thus, the reaction is second order with respect to B. Therefore, the rate = $k[\text{A}] \cdot [\text{B}]^2$

74. D is correct.

The molecules must collide with sufficient energy, frequency, and proper orientation to overcome the barrier of activation energy.

75. D is correct.

When changing the conditions of a reaction, Le Châtelier's principle states that equilibrium shifts to counteract the change. If the reaction temperature, pressure, or volume is changed, the position of equilibrium changes.

Removing NH_3 and adding N_2 shifts the equilibrium to the right.

Removing N_2 or adding NH_3 shifts the equilibrium to the left.

76. B is correct.

Increasing $[HC_2H_3O_2]$ and increasing $[H^+]$ affects the equilibrium because both species are part of the equilibrium equation.

$$K_{eq} = [products] / [reactants]$$

$$K_{eq} = [C_2H_3O_2^-] / [HC_2H_3O_2] \cdot [H^+]$$

Adding solid $NaC_2H_3O_2$: in aqueous solutions, $NaC_2H_3O_2$ dissociates into Na^+ (*aq*) and $C_2H_3O_2^-$ (*aq*). Therefore, the overall concentration of $C_2H_3O_2^-$ (*aq*) increases, and it affects the equilibrium.

Adding solid $NaNO_3$: this is a salt and does not react with reactant or product molecules; therefore, it will not affect the equilibrium.

Adding solid NaOH: the reactant is an acid because it dissociates into hydrogen ions and anions in aqueous solutions. It reacts with bases (e.g., NaOH) to create water and salt. Thus, NaOH reduces the concentration of the reactant, which affects the equilibrium.

Increasing $[HC_2H_3O_2]$ and $[H^+]$ increases reactants and drives the reaction toward product formation.

77. D is correct.

Calculate the value of the equilibrium expression: hydrogen iodide concentration decreases for equilibrium to be reached.

$$K = [products] / [reactants]$$

$$K = [HI]^2 / [H_2] \cdot [I_2]$$

To determine the state of a reaction, calculate its reaction quotient (*Q*). *Q* has the same calculation method as equilibrium constant (*K*); the difference is *K* has to be calculated at the point of equilibrium, whereas *Q* can be calculated at any time.

If $Q > K$, reaction shifts towards reactants (more reactants, less products)

If $Q = K$, reaction is at equilibrium

If $Q < K$, reaction shifts towards products (more products, less reactants)

continued...

$$Q = [HI]^2 / [H_2] \cdot [I_2]$$

$$Q = (3)^2 / 0.4 \times 0.6$$

$$Q = 37.5$$

Because $Q > K$ (i.e., (37.5 > 35), the reaction shifts toward reactants (larger denominator), and the amount of product (HI) decreases.

78. C is correct.

Decreasing the concentration of reactants (or decreasing pressure) decreases the reaction rate because there is a decreased probability that any two reactant molecules collide with sufficient energy to overcome the energy of activation and form products.

79. B is correct.

Chemical equilibrium expression is dependent on the stoichiometry of the reaction, more specifically, the coefficients of each species involved in the reaction.

The mechanism refers to the pathway of product formation (e.g., S_N1 or S_N2) but does not affect the relative energies of the reactants and products (i.e., a position for the equilibrium).

The rate refers to the time needed to achieve equilibrium but does not affect the equilibrium position.

80. E is correct.

$$K = [products] / [reactants]$$

$$K = [CH_3OH] / [H_2O]^2 \cdot [CO]$$

None of the stated changes tend to decrease the *magnitude* of the equilibrium constant K.

For exothermic reactions ($\Delta H < 0$), decreasing the temperature increases the magnitude of K.

Changes in volume, pressure, or concentration do not affect the magnitude of K.

==

Practice Set 5: Questions 81–100

==

81. E is correct.

When changing the conditions of a reaction, Le Châtelier's principle states that the position of equilibrium shifts to counteract the change. If the reaction's concentration changes, the position of the equilibrium changes.

Adding reactants or removing products shifts the equilibrium to the right (i.e., toward product formation).

82. A is correct.

When two or more reactions are combined to create a new reaction, the K_c of the resulting reaction is a product of K_c values of the individual reactions.

When a reaction is reversed, the new $K_c = 1$ / old K_c.

Start by analyzing the reactions provided the problem:

      (1) $2 NO + Cl_2 \leftrightarrow 2 NOCl$ $K_c = 3.2 \times 10^3$

      (2) $2 NO_2 \leftrightarrow 2 NO + O_2$ $K_c = 15.5$

      (3) $NOCl + \frac{1}{2} O_2 \leftrightarrow NO_2 + \frac{1}{2} Cl_2$ $K_c = ?$

Reaction 3 can be created by combining reactions 1 and 2.

For reaction 1, notice that on reaction 3, NOCl is located on the left. To match reaction 3, reverse reaction 1:

      $2 NOCl \leftrightarrow 2 NO + Cl_2$

The K_c of this reversed reaction is:

      $K_c = 1 / (3.2 \times 10^{-3}) = 312.5$

Reaction 2 needs to be reversed to match the position of NO in reaction 3:

      $2 NO + O_2 \leftrightarrow 2 NO_2$

The new K_c will be:

      $K_c = 1 / 15.5 = 0.0645$

Since K of the resulting reaction is the product of K_c from both reactions, add those reactions:

      $2 NOCl \leftrightarrow 2 NO + Cl_2$

      $2 NO + O_2 \leftrightarrow 2 NO_2$

      ──────────────────────────

      $2 NOCl + O_2 \leftrightarrow 2 NO_2 + Cl_2$

      $K_c = 312.5 \times 0.064 = 20.16$

Divide the equation by 2:

$$NOCl + \tfrac{1}{2} O_2 \leftrightarrow NO_2 + \tfrac{1}{2} Cl_2$$

The new K_c is the square root of initial K_c:

$$K_c = \sqrt{20.16} = 4.49$$

83. C is correct.

Increased pressure or increased concentration of reactants increases the probability that the reactants collide with sufficient energy and orientation to overcome the energy of activation barrier and proceed toward products.

84. C is correct.

The reaction rate is the speed at which reactants are consumed, or a product is formed.

The rate of a chemical reaction at a constant temperature depends only on the concentrations of the substances that influence the rate. The reactants influence the rate of reaction, but occasionally products can influence the rate of the reaction.

85. A is correct.

General formula for the equilibrium constant of a reaction:

$$aA + bB \leftrightarrow cC + dD$$

$$K_{eq} = ([C]^c \times [D]^d) / ([A]^a \times [B]^b)$$

For the reaction above:

$$K_{eq} = [B] \times [C]^3 / [A]^2$$

86. B is correct.

The rate law is calculated by comparing trials and determining how changes in the initial concentrations of the reactants affect the rate of the reaction.

$$\text{rate} = k[A]^x \cdot [B]^y$$

where k is the rate constant, and the exponents x and y are the partial reaction orders (i.e., determined experimentally). They are not equal to the stoichiometric coefficients.

87. C is correct.

Rates of formation/consumption of species in a reaction are proportional to their coefficients.

If the rate of a species is known, the rate of other species can be calculated using simple proportions:

$$\text{coefficient }_{NOBr} / \text{coefficient }_{Br2} = \text{rate }_{\text{formation NOBr}} / \text{rate }_{\text{consumption Br2}}$$

$$\text{rate }_{\text{consumption Br2}} = \text{rate }_{\text{formation NOBr}} / (\text{coefficient }_{NOBr} / \text{coefficient }_{Br2})$$

$$\text{rate }_{\text{consumption Br2}} = 4.50 \times 10^{-4} \text{ mol L}^{-1} \text{ s}^{-1} / (2 / 1)$$

$$\text{rate }_{\text{consumption Br2}} = 2.25 \times 10^{-4} \text{ mol L}^{-1} \text{ s}^{-1}$$

88. C is correct.

The activation energy is the energy barrier that must be overcome to transform reactant(s) into product(s).

A reaction with lower activation energy (energy barrier) has an increased rate (k).

89. A is correct.

General formula for the equilibrium constant of a reaction:

$aA + bB \leftrightarrow cC + dD$

$K_{eq} = ([C]^c \times [D]^d) / ([A]^a \times [B]^b)$

For the reaction above:

$K_{eq} = [C]^2 \cdot [D] / [A] \cdot [B]^2$

90. D is correct.

When changing the conditions of a reaction, Le Châtelier's principle states that the position of the equilibrium shifts to counteract the change.

If the reaction's water concentration (i.e., product) is changed, equilibrium changes toward reactants.

91. D is correct.

For the general equation:

$aA + bB \leftrightarrow cC + dD$

The equilibrium constant is defined as:

$K_{eq} = ([C]^c \times [D]^d) / ([A]^a \times [B]^b)$

or

$K_{eq} = [products] / [reactants]$

If K_{eq} is less than 1 (e.g., 4.3×10^{-17}), then the numerator (i.e., products) is smaller than the denominator (i.e., reactants), and fewer products have formed relative to the reactants.

If the reaction favors reactants compared to products, the equilibrium lies to the left.

92. E is correct.

The primary function of catalysts is lowering the reaction's activation energy (energy barrier), thus increasing its rate (k).

The activation energy is the energy barrier for a reaction to proceed. For the forward reaction, it is measured from the energy of the reactants to the highest energy level in the reaction.

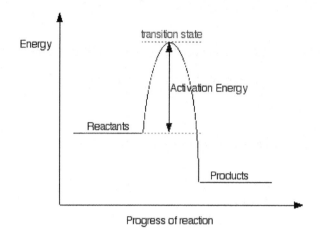

The transition state signifies "bond making" and "bond making" events in the transformation of the reactant(s) into product(s).

93. C is correct.

Use the equilibrium concentrations to calculate K_c:

$$K_c = [N_2] \cdot [H_2]^3 / [NH_3]^2$$

$$K_c = 0.04 \times (0.12)^3 / (0.4)^2$$

$$K_c = 4.3 \times 10^{-4}$$

94. A is correct.

Increasing the concentration of reactants (or increasing pressure) increases the reaction rate because there is an increased probability that any two reactant molecules collide with sufficient energy to overcome the energy of activation and form products.

95. D is correct.

When changing the conditions of a reaction, Le Châtelier's principle states that equilibrium will shift to counteract the change. If the reaction temperature, pressure, or volume is changed, the position of the equilibrium changes.

Add the molar coefficients in the reaction (i.e., 3 for the reactants and 1 for the product).

In general, decreasing the pressure tends to favor the side of the reaction that has a higher molar coefficient sum (i.e., toward the reactants in this example).

Decreasing the volume favors the side of the reaction that has a smaller molar coefficient sum (i.e., products in this example). Therefore, it would increase the yield of methanol.

Decreasing the temperature of the system favors the production of more heat. It shifts the reaction towards the exothermic side. Because $\Delta H < 0$, the reaction is exothermic in the forward direction (towards product). Therefore, decreasing the temperature increases the yield of methanol.

96. E is correct.

Decreasing the pH: increases the concentration of H^+, which shifts the equilibrium to the left (i.e., increases the chlorine concentration)

Adding HCl: increases the concentration of H^+, which shifts the equilibrium to the left.

Adding HClO: increases the concentration of HClO, which shifts the equilibrium to the left.

Adding NaClO: increases the concentration of ClO^-, which shifts the equilibrium to the left.

97. A is correct.

The diffusion rate is proportional to the speed of the gas.

Since lighter gases travel faster than heavier gases, H_2 diffuses at a faster rate than O_2.

At the same temperature, two gases have the same average molecular kinetic energy:

$\frac{1}{2}m_H v_H^2 = \frac{1}{2}m_O v_O^2$

$(v_H / v_O)^2 = m_O / m_H$

Since $m_O / m_H = 16$,

$(v_H / v_O)^2 = 16$

$v_H / v_O = 4{:}1$ ratio

98. E is correct.

The primary function of catalysts is lowering the reaction's activation energy (energy barrier), thus increasing its rate (k).

Although catalysts do decrease the amount of energy required to reach the rate-limiting transition state, they do *not* decrease the relative energy of the products and reactants. Therefore, a catalyst does not affect ΔG.

Catalysts speed up reactions and therefore increase the rate at which equilibrium is reached.

However, catalysts never alter the thermodynamics of a reaction and therefore do not change free energy ΔG.

99. E is correct.

Balanced reaction:

$3\,A + 2\,B \rightarrow 4\,C$

Moles of A used in the reaction = 1.4 moles – 0.9 moles = 0.5 moles

Moles of C formed = (coefficient C / coefficient A) × moles of A used

Moles of C formed = (4 / 3) × 0.5 moles = 0.67 moles

100. D is correct.

When two or more reactions are combined to create a new reaction, the K_c of the resulting reaction is a product of K_c values of the individual reactions.

When a reaction is reversed, new K_c = 1 / old K_c.

$$PCl_3 + Cl_2 \leftrightarrow PCl_5 \qquad K_c = K_{c1}$$

$$2\,NO + Cl_2 \leftrightarrow 2\,NOCl \qquad K_c = K_{c2}$$

Reverse the first reaction:

$$PCl_5 \leftrightarrow PCl_3 + Cl_2 \qquad K_c = 1 / K_{c1}$$

Add the two reactions to create a third reaction:

$$PCl_5 \leftrightarrow PCl_3 + Cl_2 \qquad K_c = 1 / K_{c1}$$

$$2\,NO + Cl_2 \leftrightarrow 2\,NOCl \qquad K_c = K_{c2}$$

———————————————————————

$$PCl_5 + 2\,NO \leftrightarrow PCl_3 + 2\,NOCl$$

$$K = [(1 / K_1) \times K_2] = K_2 / K_1$$

==

Practice Set 6: Questions 101–120

==

101. C is correct.

Mass action expression is related to the equilibrium constant expression.

General formula for the equilibrium constant of a reaction:

aA + bB ↔ cC + dD

$K_{eq} = ([C]^c \times [D]^d) / ([A]^a \times [B]^b)$

To determine the mass action expression, only include species in aqueous or gas phases:

$K_c = [CrCl_3]^4 / [CCl_4]^3$

102. A is correct.

When changing the conditions of a reaction, Le Châtelier's principle states that the position of the equilibrium shifts to counteract the change. If the reaction's concentration is changed, the position of equilibrium changes.

Adding KNO_2 increases the amount of K+ and NO_2^- (product). Therefore, the reaction shifts toward the reactants (i.e., to the left).

103. E is correct.

Rate expressions are determined experimentally from the rate data for a reaction.

The calculation is not similar to that for the equilibrium constant.

Since rate data are not given for this reaction, the rate expression cannot be determined.

104. D is correct.

The rate law is calculated experimentally by comparing trials and determining how changes in the initial concentrations of the reactants affect the rate of the reaction.

rate = $k[A]^x \cdot [B]^y$

where k is the rate constant, and the exponents x and y are the partial reaction orders (i.e., determined experimentally). They are not equal to the stoichiometric coefficients.

To write a complete rate law, the order of each reactant is required. Determine the order of reactant D by using 2 experiments where the concentrations of E are identical, and concentrations of D are different: experiments 1 and 2.

Use data from these experiments to calculate order of D:

Rate$_2$ / Rate$_1$ = ([D]$_2$ / [D]$_1$)$^{\text{order of D}}$

$(0.000500) / (0.000250) = (0.200 / 0.100)^{\text{order of D}}$

$2 = (2)^{\text{order of D}}$

Order of D = 1

Use the same procedure to calculate order of E.

Data from experiments 1 and 3:

$$Rate_3 / Rate_1 = ([E]_3 / [E]_1)^{order\ of\ E}$$

$$(0.001000) / (0.000250) = (0.500 / 0.250)^{order\ of\ E}$$

$$4 = (2)^{order\ of\ E}$$

Order of E = 2

Rate = $k[D] \cdot [E]^2$

105. D is correct.

The rate of formation/consumption of species in a reaction is proportional to their coefficients.

If the rate of a species is known, the rate of other species can be calculated using simple proportions.

Because the rate of formation of a species is already known, there is no need to calculate the rate.

(The extra information on the problem is just there to distract test-takers).

$$rate_{consumption\ H2} / rate_{formation\ NO2} = coefficient_{H2} / coefficient_{NO2}$$

$$rate_{consumption\ H2} = (coefficient_{H2} / coefficient_{NO2}) \times rate_{formation\ NO2}$$

$$rate_{consumption\ H2} = (4 / 2) \times 2.8 \times 10^{-4}\ M/min$$

$$rate_{consumption\ H2} = 5.6 \times 10^{-4}\ M/min$$

106. C is correct.

General formula for the equilibrium constant of a reaction:

$$aA + bB \leftrightarrow cC + dD$$

$$K_{eq} = ([C]^c \times [D]^d) / ([A]^a \times [B]^b)$$

For equilibrium constant calculation, only include species in aqueous or gas phases (which, in this problem, is all of them):

$$K_{eq} = [CO_2]^2 / [CO]^2 \times [O_2]$$

107. B is correct.

Reactions proceed faster at higher temperatures.

Catalysts lower the energy of activation (i.e., energy barrier) for the reaction, and the reaction rate increases.

108. D is correct.

For the general equation:

$$aA + bB \leftrightarrow cC + dD$$

The equilibrium constant is defined as:

$$K_{eq} = ([C]^c \times [D]^d) / ([A]^a \times [B]^b)$$

or

$$K_{eq} = [products] / [reactants]$$

If K_{eq} is less than 1 (e.g., 6.3×10^{-14}), then the numerator (i.e., products) is smaller than the denominator (i.e., reactants), and fewer products have formed relative to the reactants.

If the reaction favors reactants compared to products, the equilibrium lies to the left.

109. E is correct.

Typically, the forward reaction's constant is denoted as k_1 and the reverse is denoted as k_{-1}.

The overall constant of step 1 would be k_1 / k_{-1}.

110. D is correct.

A catalyst lowers the activation energy (E_a) of a reaction.

The primary function of catalysts is lowering the reaction's activation energy (energy barrier), thus increasing its rate (k).

Although catalysts do decrease the amount of energy required to reach the rate-limiting transition state, they do *not* decrease the relative energy of the products and reactants.

A catalyst provides an alternative pathway for the reaction to proceed to product formation. It lowers the energy of activation (i.e., relative energy between reactants and transition state) and speeds the reaction rate.

Catalysts do not affect the Gibbs free energy (ΔG: stability of products vs. reactants) or the enthalpy (ΔH: bond breaking in reactants or bond making in products).

111. D is correct.

When changing the conditions of a reaction, Le Châtelier's principle states that the position of equilibrium shifts to counteract the change. If the reaction's concentration changes, the position of equilibrium changes.

Adding reactants or removing products shifts the equilibrium to the right.

112. C is correct.

Increased pressure or increased concentration of reactants increases the probability that the reactants collide with sufficient energy and orientation to overcome the energy of activation barrier and proceed toward products.

113. B is correct.

Increasing the concentration of reactants (or increasing pressure) increases the reaction rate because there is an increased probability that any two reactant molecules collide with sufficient energy to overcome the energy of activation and form products.

---

114. B is correct.

The number of collisions does not have to be large for a reaction to occur. However, a minimum threshold of collisions needs to be reached before a reaction can happen. Reactions tend to happen more readily at higher temperatures because higher temperatures result in a larger number of collisions.

Some reactions only break bonds but do not form new bonds (e.g., radical cleavage of chlorine in the ozone decomposition process). Some reactions only form bonds and do not break bonds (e.g., $Na^+ + Cl^- \rightarrow NaCl$).

It is not necessary that all chemical bonds in the reactants must break for a reaction to occur.

115. C is correct.

When changing the reaction conditions, Le Châtelier's principle states that equilibrium shifts to counteract the change. If the reaction temperature, pressure, or volume is changed, the position of equilibrium changes.

A catalyst lowers the reaction's activation energy (energy barrier) and therefore increases the rate (k) of the reaction. A catalyst does not affect the relative energy of the reactants and products and does not affect the equilibrium.

116. E is correct.

All chemical reactions eventually reach equilibrium, the state at which the reactants and products are present in concentrations that have no further tendency to change with time.

When the concentration in the reactant side of a reaction is reduced, some product transforms (i.e., the reaction is reversed) into reactant to compensate for the reduction of reactants. This results in a decrease in the concentration from the product side of the reaction.

117. A is correct.

Activated complexes (i.e., transition state for molecules undergoing bond breaking and bond making) decompose rapidly, have specific geometries that are extremely reactive. Activated complexes (as opposed to intermediates) cannot be isolated.

118. A is correct.

All chemical reactions eventually reach equilibrium, the state at which the reactants and products are present in concentrations that have no further tendency to change with time. Therefore, the rate in the forward and reverse directions is equal.

119. A is correct.

Note the following balanced reaction and calculation:

$2\,NH_3 \rightarrow N_2 + 3H_2$

$K = [\text{products}] / [\text{reactants}]$

$K = [N_2] \cdot [H_2]^3 / [NH_3]^2$

$K = (0.3)^3 \cdot (0.2) / (0.1)^2$

120. C is correct.

Usually, to solve this kind of problem, determine the order of the reactants (X, Y, Z).

However, look at the answer choices; they pertain to the order of X. Therefore, calculate the order of X.

Find 2 experiments where the concentrations of other reactants (Y and Z) are constant: experiments 3 and 4.

Notice that when the concentration of X is reduced from 0.600 M to 0.200 M (i.e., a factor of 3), the rate decreases by a factor of 3 (from 15 to 5). Therefore, the reaction is first order with respect to X.

Notes for active learning

Solution Chemistry – Detailed Explanations

===

Practice Set 1: Questions 1–20

===

1. E is correct.

Solubility is proportional to pressure. Use simple proportions to compare solubility in different pressures:

$P_1 / S_1 = P_2 / S_2$

$S_2 = P_2 / (P_1 / S_1)$

$S_2 = 4.5 \text{ atm} / (1.00 \text{ atm} / 1.90 \text{ cc}/100 \text{ mL})$

$S_2 = 8.55 \text{ cc} / 100 \text{ mL}$

2. E is correct.

The correct interpretation of molarity is moles of solute per liter of solvent.

3. A is correct.

NaCl is a charged, ionic salt (Na^+ and Cl^-) and therefore is extremely water-soluble.

Hexanol hydrogen bonds with water due to the hydroxyl group, but the long hydrocarbon chain reduces its solubility, and the chains interact via hydrophobic interactions forming micelles.

Aluminum hydroxide has poor solubility in water and requires the addition of Brønsted-Lowry acids to dissolve completely.

4. A is correct.

Spectator ions appear on both sides of the ionic equation.

Ionic equation of the reaction:

$Ba^{2+}(aq) + 2\,Cl^-(aq) + 2\,K^+(aq) + CrO_4^-(aq) \rightarrow BaCrO_4(s) + 2\,K^+(aq) + 2\,Cl^-(aq)$

5. D is correct.

Higher pressure exerts more pressure on the solution, which allows more molecules to dissolve in the solvent.

Lower temperature increases gas solubility.

As the temperature increases, solvent and solute molecules move faster, and it is more difficult for the solvent molecules to bond with the solute.

6. A is correct.

Solutes dissolve in the solvent to form a solution.

7. B is correct.

When the compound is dissolved in water, the attached hydrate/water crystals dissociate and become part of the water solvent.

Therefore, the only ions left are 1 Co^{2+} and 2 NO_3^-.

8. D is correct.

The mass of a solution is calculated by taking the mass of the solution and container and subtracting the mass of the container.

9. A is correct.

An insoluble compound is incapable of being dissolved (especially with reference to water).

Hydroxide salts of Group I elements are soluble, while hydroxide salts of Group II elements (Ca, Sr and Ba) are slightly soluble.

Salts containing nitrate ions (NO_3^-) are generally soluble.

Most sulfate salts are soluble. Important exceptions: $BaSO_4$, $PbSO_4$, Ag_2SO_4 and $SrSO_4$.

Hydroxide salts of transition metals and Al^{3+} are insoluble.

Thus, $Fe(OH)_3$, $Al(OH)_3$ and $Co(OH)_2$ are not soluble.

10. C is correct.

% v/v solution = (volume of acetone / volume of solution) × 100%

% v/v solution = {25 mL / (25 mL + 75 mL)} × 100%

% v/v solution = 25%

11. C is correct.

The van der Waals forces in the hydrocarbons are relatively weak, while hydrogen bonds between H_2O molecules are strong intermolecular bonds.

Bonds between polar H_2O and nonpolar octane are weaker than the bonds between two polar H_2O molecules (i.e., hydrogen bonds).

12. A is correct.

To find the K_{sp}, take the molarities (or concentrations) of the products (cC and dD) and multiply them.

For K_{sp}, only aqueous species are included in the calculation.

If any of the products have coefficients, raise the product to that coefficient power and multiply the concentration by that coefficient:

$K_{sp} = [C]^c \cdot [D]^d$

$K_{sp} = [Cu^{2+}]^3 \cdot [PO_4^{3-}]^2$

The reactant (aA) is solid and is not included in the K_{sp} equation.

Solids are not included when calculating equilibrium constant expressions because their concentrations do not change the expression. Any change in their concentrations is insignificant and is thus omitted.

13. C is correct.

The *like dissolves like* rule applies when a solvent is miscible with a solute that has similar properties.

A polar solute (e.g., ethanol) is miscible with a polar solvent (e.g., water).

14. C is correct.

In a net ionic equation, substances that do not dissociate in aqueous solutions are written in their molecular form (not broken down into ions).

Gases, liquids, and solids are written in molecular form.

Some solutions do not dissociate into ions in water or do so in very little amounts (e.g., weak acids such as HF, CH_3COOH). They are also written in molecular form.

15. B is correct.

A supersaturated solution contains more of the dissolved material than could be dissolved by the solvent under normal conditions. Increased heat allows for a solution to become supersaturated. The term also refers to the vapor of a compound with a higher partial pressure than the vapor pressure of that compound.

A supersaturated solution forms a precipitate if seed crystals are added as the solution reaches a lower energy state when solutes precipitate from the solution.

16. E is correct.

Mass % = mass of solute / mass of solution

Rearrange that equation to solve for mass of solution:

mass of solution = mass of solute / mass %

mass of solution = 122 g / 7.50%

mass of solution = 122 g / 0.075

mass of solution = 1,627 g

17. B is correct.

Determine moles of NH_3:

Moles of NH_3 = mass of NH_3 / molecular weight of NH_3

Moles of NH_3 = 15.0 g / (14.01 g/mol + 3 × 1.01 g/mol)

Moles of NH_3 = 15.0 g / (17.04 g/mol)

Moles of NH_3 = 0.88 mol

Determine the volume of the solution (solvent + solute):

Volume of solution = mass / density

Volume of solution = (250 g + 15 g) / 0.974 g/mL

Volume of solution = 272.1 mL

Convert volume to liters:

Volume = 272.1 mL × 0.001 L / mL

Volume = 0.2721 L

Divide moles by the volume to calculate molarity:

Molarity = moles / liter of solution

Molarity = 0.88 mol / 0.2721 L = 3.23 M

18. A is correct.

Solutions with the highest concentration of ions have the highest boiling point.

Calculate the concentration of ions in each solution:

0.2 M $Al(NO_3)_3$ = 0. 2 M × 4 ions = 0.8 M

0. 2 M $MgCl_2$ = 0. 2 M × 3 ions = 0.6 M

0. 2 M glucose = 0. 2 M × 1 ion = 0.2 M (glucose does not dissociate into ions in solution)

0. 2 M Na_2SO_4 = 0. 2 M × 3 ions = 0.6 M

Water = 0 M

19. A is correct.

Calculate the number of ions in each option:

A: $Li_3PO_4 \rightarrow 3\ Li^+ + PO_4^{3-}$	(4 ions)
B: $Ca(NO_3)_2 \rightarrow Ca^{2+} + 2\ NO_3^-$	(3 ions)
C: $MgSO_4 \rightarrow Mg^{2+} + SO_4^{2-}$	(2 ions)
D: $(NH_4)_2SO_4 \rightarrow 2\ NH_4^+ + SO_4^{2-}$	(3 ions)
E: $(NH_4)_4Fe(CN)_6 \rightarrow 4\ NH_4^+ + Fe(CN)_6^{4-}$	(5 ions)

20. D is correct.

An electrolyte is a substance that produces an electrically conducting solution when dissolved in a polar solvent (e.g., water).

The dissolved electrolyte separates into positively-charged cations and negatively-charged anions.

Strong electrolytes dissociate entirely (or almost completely) because the resulting ions are stable in the solution.

==

Practice Set 2: Questions 21–40

==

21. B is correct.

Note that carbonate salts help remove 'hardness' in water.

Generally, SO_4, NO_3 and Cl salts tend to be soluble in water, while CO_3 salts are less soluble.

22. A is correct.

A *spectator ion* exists in the same form on both the reactant and product sides of a chemical reaction.

Balanced equation:

$$Pb(NO_3)_2 \ (aq) + H_2SO_4 \ (aq) \rightarrow PbSO_4 \ (s) + 2 \ HNO_3 \ (aq)$$

23. C is correct.

A polar solute is miscible with a polar solvent.

Ascorbic acid is polar and is therefore miscible in water.

24. E is correct.

Apply the formula for the molar concentration of a solution:

$$M_1V_1 = M_2V_2$$

Substitute the given volume and molar concentrations of HCl, solve for the final volume of HCl of the resulting dilution:

$$V_2 = [M_1V_1] / (M_2)$$

$$V_2 = [(2.00 \ M \ HCl) \times (0.125 \ L \ HCl)] / (0.400 \ M \ HCl)$$

$$V_2 = 0.625 \ L \ HCl$$

Solve for the volume of water needed to be added to the initial volume of HCl in order to obtain the final diluted volume of HCl:

$$V_{H_2O} = V_2 - V_1$$

$$V_{H_2O} = (0.625 \ L \ HCl) - (0.125 \ L \ HCl)$$

$$V_{H_2O} = 0.500 \ L = 500 \ mL$$

25. A is correct.

The *–ate* ending indicates the species with more oxygen than species ending in *–ite*.

However, it does not indicate a specific number of oxygen molecules.

_____

26. C is correct.

Solubility is the property of a solid, liquid, or gaseous substance (i.e., solute); it dissolves in a solid, liquid, or gaseous solvent to form a solution (i.e., solute in the solvent).

The solubility of a substance depends on the physical and chemical properties of the solute and solvent and the temperature, pressure, and pH of the solution.

Gaseous solutes (e.g., oxygen) exhibit complex behavior with temperature. As the temperature rises, gases usually become less soluble in water but more soluble in organic solvents.

27. A is correct.

A saturated solution contains the maximum amount of dissolved material in the solvent under normal conditions. Increased heat allows for a solution to become supersaturated. The term refers to the vapor of a compound with a higher partial pressure than the vapor pressure of that compound.

A saturated solution forms a precipitate as more solute is added to the solution.

28. D is correct.

Immiscible refers to the property (of solutions) of when two or more substances (e.g., oil and water) are mixed and eventually separate into two layers.

Miscible is when two liquids are mixed but do not necessarily interact chemically.

In contrast to miscibility, *soluble* means the substance (solid, liquid, or gas) can *dissolve* in another solid, liquid, or gas. In other words, a substance *dissolves* when it becomes incorporated into another substance. In contrast to miscibility, solubility involves a *saturation point*, at which a substance cannot dissolve any further and a mass, the *precipitate*, begins to form.

29. D is correct.

In this case, hydration means dissolving in water rather than reacting with water.
Therefore, the compound dissociates into its ions (without involving water in the actual chemical reaction).

30. A is correct.

Depending on the solubility of a solute, there are three possible results:

 1) a dilute solution has less solute than the maximum amount that it is able to dissolve;

 2) a saturated solution has exactly the same amount as its solubility;

 3) a precipitate forms if there is more solute than is able to be dissolved, so the excess solute separates from the solution (i.e., crystallization).

Precipitation lowers the concentration of the solute to the saturation level to increase the stability of the solution.

Gas solubility is inversely proportional to temperature and proportional to pressure.

31. E is correct.

Like dissolves like means that polar substances tend to dissolve in polar solvents and nonpolar substances in nonpolar solvents.

Molecules that can form hydrogen bonds with water are soluble.

Salts are ionic compounds.

The anion and cation bond with the polar molecule of water and are soluble.

32. B is correct.

Like dissolves like means that polar substances tend to dissolve in polar solvents and nonpolar substances in nonpolar solvents.

Methanol (CH_3OH) is a polar molecule. Therefore, methanol is soluble in water because it can make a hydrogen bond with water.

33. B is correct.

To find the K_{sp}, take the molarities or concentrations of the products (cC and dD) and multiply them. For K_{sp}, only aqueous species are included in the calculation.

If any of the products have coefficients, raise the product to that coefficient power and multiply the concentration by that coefficient:

$$K_{sp} = [C]^c \cdot [D]^d$$
$$K_{sp} = [Au^{3+}] \cdot [Cl^-]^3$$

The reactant (aA) is solid and is not included in the K_{sp} equation.

Solids are not included when calculating equilibrium constant expressions because their concentrations do not change the expression. Any change in their concentrations is insignificant and is thus omitted.

34. B is correct.

Break down the molecules into their constituent ions: $CaCO_3 + 2\,H^+ + 2\,NO_3^- \rightarrow Ca^{2+} + 2\,NO_3^- + CO_2 + H_2O$

Remove species that appear on both sides of the reaction: $CaCO_3 + 2\,H^+ \rightarrow Ca^{2+} + CO_2 + H_2O$

35. D is correct.

According to the problem, there are 0.950 moles of nitrate ion in a $Fe(NO_3)_3$ solution. Because there are three nitrates (NO_3) ions for each $Fe(NO_3)_3$ molecule, the moles of $Fe(NO_3)_3$ can be calculated:

(1 mole / 3 mole) × 0.950 moles = 0.317 moles

Calculate the volume of solution:

Volume of solution = moles of solute / molarity

Volume of solution = 0.317 moles / 0.550 mol/L

Volume of solution = 0.576 L

Convert the volume into milliliters: 0.576 L × 1,000 mL/L = 576 mL

36. A is correct.

mEq represents the amount in milligrams of a solute equal to 1/1,000 of its gram equivalent weight, taking into account the valency of the ion. Millimolar (mM) = mEq / valence.

Divide the given concentration of Ca^{2+} by 2:

$[Ca^{2+}]$ = [48 mEq Ca^{2+}] / 2 = 24 mM Ca^{2+}

$[Ca^{2+}]$ = 24 mM Ca^{2+}

Divide the concentration of Ca^{2+} by 1,000 to obtain the concentration of Ca^{2+} in units of molarity:

$[Ca^{2+}]$ = 24 mM Ca^{2+} × [(1 M) / (1,000 mM)]

$[Ca^{2+}]$ = 0.024 M Ca^{2+}

37. A is correct.

Start by calculating the number of moles:

Moles of LiOH = mass of LiOH / molar mass of LiOH

Moles of LiOH = 36.0 g / (24.0 g/mol)

Moles of LiOH = 1.50 moles

Divide moles by volume to calculate molarity:

Molarity = moles / volume

Molarity = 1.50 moles / (975 mL × 0.001 L/mL)

Molarity = 1.54 M

38. C is correct.

The glucose content is 8.50 % (m/v). This means that the solute (glucose) is measured in grams, but the solution volume is measured in milliliters.

Therefore, the mass and volume of the solution are interchangeable (1 g = 1 mL).

% (m/v) of glucose = mass of glucose / volume of solution

volume of solution = mass of glucose / % (m/v) of glucose

volume of solution = 60 g / 8.50%

volume of solution = 706 mL

39. D is correct.

When AgCl dissociates, equal amounts of Ag^+ and Cl^- are produced.

If the concentration of Cl^- is B, this must be the concentration of Ag^+.

Therefore, B can be the concentration of Ag.

The concentration of silver ion can be determined by dividing the K_{sp} by the concentration of chloride ion:

$$K_{sp} = [Ag^+] \cdot [Cl^-].$$

Therefore, the concentration of Ag can be A/B moles/liter.

40. B is correct.

Strong electrolytes dissociate entirely (or almost completely) in water.

Strong acids and bases dissociate almost completely, but weak acids and bases dissociate only slightly.

===

Practice Set 3: Questions 41–60

===

41. E is correct.

Solute-solute and solvent-solvent attractions are important in establishing bonding amongst themselves, both for solvent and solute.

Once they are mixed in a solution, the solute-solvent attraction becomes the major attraction force, but the other two forces (solute-solute and solvent-solvent attractions) are still present.

42. B is correct.

Since the empirical formula for magnesium iodide is MgI_2, two moles of dissolved I^- result from each mole of dissolved MgI_2.

Therefore, if $[MgI_2] = 0.40$ M, then $[I^-] = 2(0.40$ M$) = 0.80$ M.

43. B is correct.

Ammonia forms hydrogen bonds, and SO_2 is polar.

44. A is correct.

The chlorite ion (chlorine dioxide anion) is ClO_2^-.

Chlorite is a compound that contains this group, with chlorine in an oxidation state of +3.

formula	Cl^-	ClO^-	ClO_2^-	ClO_3^-	ClO_4^-
Anion name	chloride	hypochlorite	chlorite	chlorate	perchlorate
Oxidation state	−1	+1	+3	+5	+7

45. D is correct.

Water softeners cannot remove ions from the water without replacing them.

The process replaces the ions that cause scaling (precipitation) with non-reactive ions.

46. B is correct.

These compounds are rarely encountered in most chemistry problems and are part of the long list of exceptions to the solubility rule.

Salts containing nitrate ions (NO_3^-) are generally soluble.

47. A is correct.

Hydration involves the interaction of water molecules with the solute. The water molecules exchange bonding relationships with the solute, whereby water-water bonds break and water-solute bonds form.

When an ion is hydrated, it is surrounded and bonded by water molecules. The average number of water molecules bonding to an ion is known as its hydration number.

Hydration numbers can vary but often are either 4 or 6.

48. E is correct.

Solvation describes the process whereby the solvent surrounds the solute molecules.

49. D is correct.

Miscible refers to the property (of solutions) when two or more substances (e.g., water and alcohol) are mixed without separating.

50. C is correct.

The *like dissolves like* rule applies when a solvent is miscible with a solute that has similar properties.

A nonpolar solute is immiscible with a polar solvent.

Retinol (vitamin A)

51. B is correct.

Only aqueous (*aq*) species are broken down into their ions for ionic equations.

Solids, liquids, and gases stay the same.

52. E is correct.

For K_{sp}, only aqueous species are included in the calculation.

The decomposition of $PbCl_2$:

$$PbCl_2 \rightarrow Pb^{2+} + 2\ Cl^-$$

$$K_{sp} = [Pb^{2+}] \cdot [Cl^-]^2$$

When x moles of $PbCl_2$ fully dissociate, x moles of Pb and $2x$ moles of Cl^- are produced:

$$K_{sp} = (x) \cdot (2x)^2$$

$$K_{sp} = 4x^3$$

53. B is correct.

Strong electrolytes dissociate completely (or almost entirely) in water.

Strong acids and bases dissociate nearly completely (i.e., form stable anions), but weak acids and bases dissociate only slightly (i.e., form unstable anions).

54. A is correct.

A *spectator ion* exists in the same form on both the reactant and product sides of a chemical reaction.

The balanced equation for potassium hydroxide and nitric acid:

$$KOH + HNO_3 \rightarrow KNO_3 + H_2O$$

Ionic equation:

$$K^+ \, {}^-OH + H^+ \, NO_3^- \rightarrow K^+ \, NO_3^- + H_2O$$

55. A is correct.

When a solution is diluted, the moles of solute (n) is constant.

However, the molarity and volume changes because n = MV:

$$n_1 = n_2$$

$$M_1V_1 = M_2V_2$$

$$V_2 = (M_1V_1) / M_2$$

$$V_2 = (0.20 \text{ M} \times 6.0 \text{ L}) / 14 \text{ M}$$

$$V_2 = 0.086 \text{ L}$$

Convert to milliliters:

$$0.086 \text{ L} \times (1{,}000 \text{ mL} / \text{L}) = 86 \text{ mL}$$

56. B is correct.

When a solution is diluted, the moles of solute are constant.

Use this formula to calculate the new molarity:

$$M_1V_1 = M_2V_2$$

$$160 \text{ mL} \times 4.50 \text{ M} = 595 \text{ mL} \times M_2$$

$$M_2 = (160 \text{ mL} / 595 \text{ mL}) \times 4.50 \text{ M}$$

$$M_2 = 1.21 \text{ M}$$

57. B is correct.

Like dissolves like means that polar substances dissolve in polar solvents and nonpolar substances in nonpolar solvents.

Since benzene is nonpolar, look for a nonpolar substance.

Silver chloride is ionic, while CH_2Cl_2, H_2S and SO_2 are polar.

58. B is correct.

Molarity can vary based on temperature because it involves the volume of the solution.

At different temperatures, the volume of water varies slightly, which affects the molarity.

59. D is correct.

Concentration:

solute / volume

For example: 10 g / 1 liter = 10 g/liter

2 g / 1 liter = 2 g/liter

60. E is correct.

The NaOH content is 5.0% (w/v). It means that the solute (NaOH) is measured in grams, but the solution volume is measured in milliliters.

Therefore, in this problem, the mass and volume of the solution are interchangeable (1 g = 1 mL).

mass of NaOH = % NaOH × volume of solution

mass of NaOH = 5.0% × 75.0 mL

mass of NaOH = 3.75 g

===

Practice Set 4: Questions 61–80

===

61. D is correct.

Adding NaCl increases [Cl⁻] in solution (i.e., common ion effect), which increases the precipitation of $PbCl_2$ because the ion product increases. This increase causes lead chloride to precipitate and the concentration of free chloride in solution to decrease.

62. A is correct.

AgCl has a stronger tendency to form than $PbCl_2$ because of its smaller K_{sp}.

Therefore, as AgCl forms, an equivalent amount of $PbCl_2$ dissolves.

63. A is correct.

For the dissolution of $PbCl_2$:

$$PbCl_2 \rightarrow Pb^{2+} + 2\ Cl^-$$

$$K_{sp} = [Pb^{2+}]\cdot[2\ Cl^-]^2 = 10^{-5}$$

$$K_{sp} = (x)\cdot(2x)^2 = 10^{-5}$$

$$4x^3 = 10^{-5}$$

$$x \approx 0.014$$

$$[Cl^-] = 2(0.014)$$

$$[Cl^-] = 0.028$$

For the dissolution of AgCl:

$$AgCl \rightarrow Ag^+ + Cl^-$$

$$K_{sp} = [Ag^+]\cdot[Cl^-]$$

$$K_{sp} = (x)\cdot(x)$$

$$10^{-10} = x^2$$

$$x = 10^{-5}$$

$$[Cl^-] = 10^{-5}$$

64. A is correct.

The intermolecular bonding (i.e., van der Waals) in alkanes is similar.

65. C is correct.

Ideally, dilute solutions are so dilute that solute molecules do not interact.

Therefore, the mole fraction of the solvent approaches one.

66. D is correct.

Henry's law states that, at a constant temperature, the amount of gas that dissolves in a volume of liquid is directly proportional to the partial pressure of that gas in equilibrium with that liquid.

Tyndall effect is light scattering by particles in a colloid (or particles in a very fine suspension).

A colloidal suspension contains microscopically dispersed insoluble particles (i.e., colloid) suspended throughout the liquid. The colloid particles are larger than those of the solution but not large enough to precipitate due to gravity.

67. C is correct.

Miscible refers to the property (of solutions) when two or more substances (e.g., water and alcohol) are mixed without separating.

68. B is correct.

Start by calculating moles of glucose:

Moles of glucose = mass of glucose / molar mass of glucose

Moles of glucose = 10.0 g / (180.0 g/mol)

Moles of glucose = 0.0555 mol

Then, divide moles by volume to calculate molarity:

Molarity of glucose = moles of glucose / volume of glucose

Molarity of glucose = 0.0555 mol / (100 mL × 0.001 L/mL)

Molarity = 0.555 M

69. E is correct.

An insoluble compound is incapable of being dissolved (especially regarding water).

Hydroxide salts of Group I elements are soluble.

Hydroxide salts of Group II elements (Ca, Sr and Ba) are slightly soluble.

Hydroxide salts of transition metals and Al^{3+} are insoluble. Thus, $Fe(OH)_3$, $Al(OH)_3$, $Co(OH)_2$ are not soluble.

Most sulfate salts are soluble. Important exceptions are $BaSO_4$, $PbSO_4$, Ag_2SO_4 and $SrSO_4$.

Salts containing Cl^-, Br^- and I^- are generally soluble. Exceptions are halide salts of Ag^+, Pb^{2+} and $(Hg_2)^{2+}$. $AgCl$, $PbBr_2$ and Hg_2Cl_2 are insoluble.

Chromates are frequently insoluble. Examples: $PbCrO_4$, $BaCrO_4$

70. C is correct.

The overall reaction is:

$$Zn\ (s) + 2\ HCl\ (aq) \rightarrow ZnCl_2\ (aq) + H_2\ (g)$$

Because HCl and $ZnCl_2$ are ionic, they dissociate into ions in aqueous solutions:

$$Zn\ (s) + 2\ H^+\ (aq) + 2\ Cl^-\ (aq) \rightarrow Zn^{2+}\ (aq) + 2\ Cl^-\ (aq) + H_2\ (g)$$

The common ion (Cl^-) cancels to yield the net ionic reaction:

$$Zn\ (s) + 2\ H^+\ (aq) \rightarrow Zn^{2+}\ (aq) + H_2\ (g)$$

71. E is correct.

The _like dissolves like_ rule applies when a solvent is miscible with a solute that has similar properties.

A polar solute (e.g., ketone, alcohols, or carboxylic acids) is miscible with a polar (water) solvent.

The three molecules are polar and are therefore soluble in water.

72. A is correct.

Calculation of the solubility constant (K_{sp}) is similar to the equilibrium constant.

The concentration of each species is raised to the power of their coefficients and multiplied with each other.

For K_{sp}, only aqueous species are included in the calculation.

$$CaF_2\ (s) \rightarrow Ca^{2+}\ (aq) + 2F^-\ (aq)$$

The concentration of fluorine ions can be determined using the concentration of calcium ions:

$$[F^-] = 2 \times [Ca^{2+}]$$

$$[F^-] = 2 \times 0.00021\ M$$

$$[F^-] = 4.2 \times 10^{-4}\ M$$

Then, K_{sp} can be determined:

$$K_{sp} = [Ca^{2+}] \cdot [F^-]^2$$

$$K_{sp} = (2.1 \times 10^{-4}\ M) \times (4.2 \times 10^{-4}\ M)^2$$

$$K_{sp} = 3.7 \times 10^{-11}$$

73. E is correct.

To determine the number of equivalents:

$$(4\ moles\ /\ liter) \times (3\ equivalents\ /\ mole) \times (1/3\ liter)$$

$$= 4\ equivalents$$

74. A is correct.

An electrolyte is a substance that produces an electrically conducting solution when dissolved in a polar solvent (e.g., water).

The dissolved electrolyte separates into positively-charged cations and negatively-charged anions.

Electrolytes conduct electricity.

75. A is correct.

This is a frequently encountered insoluble salt in chemistry problems.

Most sulfate salts are soluble. Important exceptions are $BaSO_4$, $PbSO_4$, Ag_2SO_4 and $SrSO_4$.

76. C is correct.

When the ion product is equal to or greater than K_{sp}, precipitation of the salt occurs.

If the ion product value is less than K_{sp}, precipitation does not occur.

77. C is correct.

Gas solubility is inversely proportional to temperature, so a low temperature increases solubility.

The high pressure of O_2 above the solution increases the pressure of the solution (which improves solubility) and the amount of O_2 available for dissolving.

78. C is correct.

Molality is the number of solute moles dissolved in 1,000 grams of solvent.

The mass of the solution is 1,000 g + mass of solute.

Mass of solute (CH_3OH) = moles CH_3OH × molecular mass of CH_3OH

Mass of solute (CH_3OH) = 8.60 moles × [12.01 g/mol + (4 × 1.01 g/mol) + 16 g/mol]

Mass of solute (CH_3OH) = 8.60 moles × (32.05 g/mol)

Mass of solute (CH_3OH) = 275.63 g

Total mass of solution: 1,000 g + 275.63 g = 1,275.63 g

Volume of solution = mass / density

Volume of solution = 1,275.63 g / 0.94 g/mL

Volume of solution = 1,357.05 mL

Divide moles by volume to calculate molarity:

Molarity = number of moles / volume

Molarity = 8.60 moles / 1,357.05 mL

Molarity = 6.34 M

79. E is correct.

By definition, a 15.0% aqueous solution of KI contains 15% KI and the remainder (100 % – 15% = 85%) is water.

If there are 100 g of KI solution, it has 15 g of KI and 85 g of water.

The answer choice of 15 g KI / 100 g water is incorrect because 100 g is the mass of the solution; the actual mass of water is 85 g.

80. D is correct.

Molarity equals moles of solute divided by liters of solution.

$$1 \text{ cm}^3 = 1 \text{ mL}$$

Molarity:

$$(0.75 \text{ mol}) / (0.075 \text{ L}) = 10 \text{ M}$$

===

Practice Set 5: Questions 81–107

===

81. D is correct.

Mass % of solute = (mass of solute / mass of seawater) × 100%

Mass % of solute = (1.35 g / 25.88 g) × 100%

Mass % of solute = 5.22%

82. A is correct.

Some solutions are colored (e.g., Kool-Aid powder dissolved in water).

The color of chemicals is a physical property of chemicals from (most commonly) the excitation of electrons due to absorption of energy by the chemical. The observer sees not the absorbed color but the wavelength that is reflected.

Most simple inorganic (e.g., sodium chloride) and organic compounds (e.g., ethanol) are colorless.

Transition metal compounds are often colored because of transitions of electrons between *d*-orbitals of different energy.

Organic compounds tend to be colored when there is extensive conjugation (i.e., alternating double and single bonds), causing the energy gap between the HOMO (i.e., highest occupied molecular orbital) and LUMO (i.e., lowest unoccupied molecular orbital) to decrease, bringing the absorption band from the UV to the visible region. Color is due to the energy absorbed by the compound when an electron transitions from the HOMO to the LUMO.

A physical change is a change in physical properties, such as melting, the transition to gas, changes to crystal form, color, volume, shape, size, and density.

83. E is correct.

As the name suggests, the solubility product constant (K_{sp}) is a product of the solubility of each ion in an ionic compound.

Start by writing the dissociation equation of AgCl in a solution:

$$AgCl\ (s) \rightarrow Ag^+\ (aq) + Cl^-\ (aq)$$

The solubility of AgCl is provided in the problem: 1.3×10^{-4} mol/L. This is the maximum amount of AgCl that can be dissolved in water in standard conditions.

There are 1.3×10^{-4} mol AgCl dissolved in one liter of water, which means that there are 1.3×10^{-4} mol of Ag^+ ions and 1.3×10^{-4} mol of Cl^- ions.

K_{sp} is calculated using the same method as the equilibrium constant: ion concentration to the power of ion coefficient. Only aqueous species are included in the calculation.

continued...

For AgCl:

$K_{sp} = [Ag] \cdot [Cl]$

$K_{sp} = (1.3 \times 10^{-4}) \cdot (1.3 \times 10^{-4})$

$K_{sp} = 1.7 \times 10^{-8}$

84. E is correct.

Concentration:

solute / volume

For example:

10 g / 1 liter = 10 g/liter

5 g / 0.5 liter = 10 g/liter

85. B is correct.

The reaction is endothermic because the temperature of the solution drops as the reaction absorbs heat from the environment, and $\Delta H°$ is positive.

The solution is unsaturated at 1 molar, so dissolving more salt at standard conditions is spontaneous, and $\Delta G°$ is negative.

86. E is correct.

The heat of a solution is the enthalpy change (or energy absorbed as heat at constant pressure) when a solution forms. For solution formation, solvent-solvent bonds and solute-solute bonds must be broken while solute-solvent bonds are formed.

The breaking of bonds absorbs energy, while the formation of bonds releases energy.

If the heat of the solution is negative, energy is released.

87. E is correct.

The reaction is endothermic if the bonds formed have lower energy than the bonds broken.

From $\Delta G = \Delta H - T\Delta S$, the entropy of the system increases if the reaction is spontaneous.

88. B is correct.

Hydrogen bonding in H_2O is stronger than van der Waals forces in the nonpolar hydrocarbon of benzene.

89. D is correct.

Balanced reaction:

$BaCl_2 + K_2CrO_4 \rightarrow BaCrO_4 + 2\ KCl$

Most Cl and K compounds are soluble; therefore, KCl is more likely to be soluble than $BaCrO_4$.

90. C is correct.

Sodium carbonate reacts with calcium and magnesium ions (responsible for the 'hard water' phenomena) and forms precipitates of calcium carbonate and magnesium carbonate. This reduces the hardness of the water.

91. C is correct.

Calculate moles of $NaHCO_3$:

Moles of $NaHCO_3$ = mass $NaHCO_3$ / molar mass $NaHCO_3$

Moles of $NaHCO_3$ = 0.400 g / 84.0 g/mol

Moles of $NaHCO_3$ = 4.76×10^{-3} mol

In this reaction, $NaHCO_3$ and HCl have coefficients of 1.

Therefore, moles $NaHCO_3$ = moles HCl = 4.76×10^{-3} mol.

Use this to calculate volume of HCl:

volume of HCl = moles / molarity

volume of HCl = 4.76×10^{-3} mol / 0.25 M

volume of HCl = 0.019 L

Convert volume to milliliters:

0.019 L × (1,000 mL/L) = 19.0 mL

92. D is correct.

Calculation of concentration in ppm (parts per million) is similar to percentage calculation; the only difference is the multiplication factor is 10^6 ppm instead of 100%.

Concentration of contaminant = (mass of contaminant / mass of solution) × 10^6 ppm

Concentration of contaminant = $[(8.8 \times 10^{-3}$ g) / (5,246 g)] × 10^6 ppm

Concentration of contaminant = 1.68 ppm

93. A is correct.

The bicarbonate ion is the conjugate base of carbonic acid: H_2CO_3

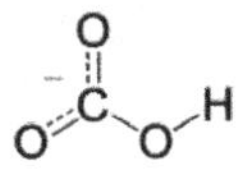

The bicarbonate ion carries a –1 formal charge and is the conjugate base of carbonic acid (H_2CO_3); it is the conjugate acid of the carbonate ion (CO_2^{-3}).

The equilibrium reactions are shown below:

$CO_3^{2-} + 2 H_2O \leftrightarrow HCO_3^- + H_2O + OH^- \leftrightarrow H_2CO_3 + 2 OH^-$

$H_2CO_3 + 2 H_2O \leftrightarrow HCO_3^- + H_3O^+ + H_2O \leftrightarrow CO_3^{2-} + 2 H_3O^+$

94. E is correct.

Solutes dissolve in the solvent to form a solution.

95. E is correct.

Hydration involves the interaction of water molecules with the solute. The water molecules exchange bonding relationships with the solute, whereby water-water bonds break and water-solute bonds form.

96. C is correct.

Hydration involves the interaction of water molecules with the solute. The water molecules exchange bonding relationships with the solute whereby water-water bonds break and water-solute bonds form.

Hydrates are not new compounds but rather compound molecules surrounded by water crystals.

A colloidal suspension contains microscopically dispersed insoluble particles (i.e., colloid) suspended throughout the liquid. The colloid particles are larger than those of the solution but not large enough to precipitate due to gravity.

97. A is correct.

Reaction:

$$2\ AgNO_3\ (aq) + K_2CrO_4\ (aq) \rightarrow 2\ KNO_3\ (aq) + AgCrO_4\ (s)$$

Salts containing nitrate ions (NO_3^-) are generally soluble.

98. E is correct.

The *like dissolves like* rule applies when a solvent is miscible with a solute that has similar properties.

A polar solute is miscible with a polar solvent, and the rule applies for the nonpolar solute / solvent.

The three molecules are nonpolar and are therefore miscible with the nonpolar liquid bromine. The attractive forces between each of the solutes and the liquid bromine solution are van der Waals.

99. C is correct.

Molecules can have covalent bonds in one situation and ionic bonds in another.

Hydrogen chloride (HCl) is a gas in which hydrogen and chlorine are covalently bound.

However, if HCl is bubbled into the water, it ionizes completely to yield H^+ and Cl^- of a hydrochloric acid solution. Hydrochloric acid contains two nonmetals and is a covalently bonded molecule.

Hydrochloric acid is a strong acid and dissociates into ions when placed in water (i.e., strong electrolyte).

100. D is correct.

The addition of solute to a liquid lowers the vapor pressure of the liquid because some water molecules are bonding to the solute. This results in fewer water molecules available for vaporization and therefore reduces the vapor pressure.

101. A is correct.

Calculation of the solubility constant (K_{sp}) is similar to the equilibrium constant.

The concentration of each species is raised to the power of their coefficients and multiplied with each other.

For K_{sp}, only aqueous species are included in the calculation.

Barium hydroxide dissolves reversibly via the reaction:

$$Ba(OH)_2\,(s) \leftrightarrow Ba^{2+}\,(aq) + 2\,OH^-\,(aq)$$

Therefore, the equilibrium expression is:

$$K_{sp} = [Ba^{2+}]\cdot[OH^-]^2$$

102. A is correct.

Start by calculating the moles of H_3PO_4 required:

moles H_3PO_4 = molarity of solution × volume

moles H_3PO_4 = 0.2 M × (200 mL × 0.001 L/mL)

moles H_3PO_4 = 0.04 mole

Then, use this information to calculate mass:

Mass of H_3PO_4 = moles × molar mass

Mass of H_3PO_4 = 0.04 moles × [(3 × 1.01 g/mol) + 30.97 g/mol + (4 × 16 g/mol)]

Mass of H_3PO_4 = 0.04 moles × (3.03 g/mol + 30.97 g/mol + 64 g/mol)

Mass of H_3PO_4 = 3.9 g

103. D is correct.

Ionic product constant of water:

$$K_w = [H_3O^+]\cdot[OH^-]$$

Rearrange the equation to solve for [OH^-]:

$$[OH^-] = K_w / [H_3O^+]$$

$$[OH^-] = [1 \times 10^{-14}] / [1 \times 10^{-8}]$$

$$[OH^-] = 1 \times 10^{-6}$$

104. B is correct.

Molarity of NaCl = moles of NaCl / volume of NaCl

Molarity of NaCl = 4.50 moles / 1.50 L

Molarity of NaCl = 3.0 M

105. B is correct.

Upon reaction with an acid, sulfite (SO_3^{2-}) compounds release SO_2 gas:

$$2\ HNO_3\ (aq) + Na_2SO_3\ (aq) \rightarrow 2\ NaNO_3\ (aq) + H_2O\ (l) + SO_2\ (g)$$

Because Na and NO_3 ions are present on both sides, they are considered spectator ions and can be removed from the net ionic equation:

$$2\ H^+\ (aq) + SO_3^{2-}\ (aq) \rightarrow H_2O\ (l) + SO_2\ (g)$$

106. D is correct.

All of the above units are used in chemistry, but % (m/v) and % (m/m) are used more often (molarity and molality, respectively).

107. D is correct.

An electrolyte is a substance that produces an electrically conducting solution when dissolved in a polar solvent (e.g., water).

The dissolved electrolyte separates into positively-charged cations and negatively-charged anions.

Electrolytes conduct electricity.

Notes for active learning

Notes for active learning

Acids and Bases – Detailed Explanations

===

Practice Set 1: Questions 1–20

===

1. A is correct.

An acid dissociates a proton to form the conjugate base, while a conjugate base accepts a proton to form the acid.

2. B is correct.

$$pH = -\log[H^+]$$

$$pH = -\log(0.10)$$

$$pH = 1$$

3. A is correct.

Salt is a combination of an acid and a base.

Its pH will be determined by the acid and base that created the salt.

Combinations include:

Weak acid and strong base: salt is basic

Strong acid and weak base: salt is acidic

Strong acid and strong base: salt is neutral

Weak acid and weak base: could be anything (acidic, basic, or neutral)

Examples of strong acids and bases that commonly appear in chemistry problems:

Strong acids: HCl, HBr, HI, H_2SO_4, $HClO_4$ and HNO_3

Strong bases: NaOH, KOH, $Ca(OH)_2$ and $Ba(OH)_2$

The hydroxides of Group I and II metals are considered strong bases. Examples include LiOH (lithium hydroxide), NaOH (sodium hydroxide), KOH (potassium hydroxide), RbOH (rubidium hydroxide), CsOH (cesium hydroxide), $Ca(OH)_2$ (calcium hydroxide), $Sr(OH)_2$ (strontium hydroxide) and $Ba(OH)_2$ (barium hydroxide).

4. C is correct.

The Brønsted-Lowry acid–base theory focuses on the ability to accept and donate protons (H^+).

A Brønsted-Lowry acid is a term for a substance that donates a proton (H^+) in an acid–base reaction, while a Brønsted-Lowry base is a substance that accepts a proton.

5. C is correct.

A solution's conductivity is correlated to the number of ions present in a solution. The bulb is shining brightly implies that the solution is an excellent conductor; meaning that the solution has a high concentration of ions.

6. C is correct.

An amphoteric compound can react as an acid (i.e., donates protons) as well as a base (i.e., accepts protons).

One type of amphoteric species is amphiprotic molecules, which can either accept or donate a proton (H^+). Examples of amphiprotic molecules include amino acids (i.e., an amine and carboxylic acid group) and self-ionizable compounds such as water.

7. E is correct.

Strong acids dissociate protons into the aqueous solution. The resulting anion is stable, which accounts for the ~100% ionization of the strong acid.

Weak acids do not completely (or appreciably) dissociate protons into the aqueous solution. The resulting anion is unstable, which accounts for a small ionization of the weak acid.

Each of the listed strong acids forms an anion stabilized by resonance.

$HClO_4$ (perchloric acid) has a pK_a of –10.

H_2SO_4 (sulfuric acid) is a diprotic acid and has a pK_a of –3 and 1.99.

HNO_3 (nitrous acid) has a pK_a of –1.4.

8. E is correct.

To find the pH of strong acid and strong base solutions (where $[H^+] > 10^{-6}$), use the equation:

$$pH = -\log[H_3O^+]$$

Given that $[H^+] < 10^{-6}$, the pH = 4.

Note that this approach only applies to strong acids and strong bases.

9. B is correct.

The self-ionization (autoionization) of water is an ionization reaction in pure water or an aqueous solution, in which a water molecule, H_2O, deprotonates (loses the nucleus of one of its hydrogen atoms) to become a hydroxide ion (^-OH).

10. B is correct.

An electrolyte is a substance that dissociates into cations (i.e., positive ions) and anions (i.e., negative ions) when placed in solution.

NH_3 is a weak acid with a pK_a of 38. Therefore, it does not readily dissociate into H^+ and $^-NH_2$.

HCO_3^- (bicarbonate) has a pK_a of about 10.3 and is considered a weak acid because it does not dissociate completely. Therefore, it does not readily dissociate into H^+ and CO_3^{2-}.

HCN (nitrile) is a weak acid with a pK_a of 9.3.

Therefore, it does not readily dissociate into H^+ and ^-CN (i.e., cyanide).

11. B is correct.

If the $[H_3O^+] = [^-OH]$, it has a pH of 7, and the solution is neutral.

12. E is correct.

The Arrhenius acid–base theory states that acids produce H^+ ions (protons) in H_2O solution and bases produce ^-OH ions (hydroxide) in H_2O solution.

13. C is correct.

Brønsted-Lowry acid–base focuses on the ability to accept and donate protons (H^+).

A Brønsted-Lowry acid is a term for a substance that donates a proton in an acid–base reaction, while a Brønsted-Lowry base is a substance that accepts a proton.

The definition is expressed in terms of an equilibrium expression

acid + base ↔ conjugate base + conjugate acid.

14. B is correct.

According to the Brønsted-Lowry acid–base theory:

An acid (reactant) dissociates a proton to become the conjugate base (product).

A base (reactant) gains a proton to become the conjugate acid (product).

The definition is expressed in terms of an equilibrium expression

acid + base ↔ conjugate base + conjugate acid.

15. C is correct.

The pH scale has a range from 1 to 14, with 7 being a neutral pH.

Acidic solutions have a pH below 7, while basic solutions have a pH above 7.

The pH scale is a log scale where 7 has a 50% deprotonated and 50% protonated (neutral) species (1:1 ratio).

At a pH of 6, there are 1 deprotonated : 10 protonated species. The ratio is 1:10.

At a pH of 5, there are 1 deprotonated : 100 protonated species. The ratio is 1:100.

A pH change of 1 unit changes the ratio by 10×.

Lower pH (< 7) results in more protonated species (e.g., cation), while an increase in pH (> 7) results in more deprotonated species (e.g., anion).

16. A is correct.

pI is the symbol for isoelectric point: pH where a protein ion has zero net charges.

To calculate pI of amino acids with 2 pK_a values, take the average of the pK_a's:

> pI = (pK_{a1} + pK_{a2}) / 2
>
> pI = (2.2 + 4.2) / 2
>
> pI = 3.2

17. D is correct.

Acidic solutions contain hydronium ions (H_3O^+). These ions are in the aqueous form because they are dissolved in water.

Although chemists often write H^+ (*aq*), referring to a single hydrogen nucleus (a proton), it exists as a hydronium ion (H_3O^+).

18. C is correct.

The equivalence point is the point at which chemically equivalent quantities of acid and base have been mixed. The moles of acid are equivalent to the moles of base.

The endpoint (related to but not the same as the equivalence point) refers to the point at which the indicator changes color in a colorimetric titration. The endpoint can be found by an indicator, such as phenolphthalein (i.e., it turns colorless in acidic solutions and pink in basic solutions).

A buffer is an aqueous solution that consists of a weak acid and its conjugate base, or vice versa. Buffered solutions resist changes in pH and are often used to keep the pH at a nearly constant value in many chemical applications. It does this by readily absorbing or releasing protons (H^+) and ^-OH.

When an acid is added to the solution, the buffer releases ^-OH and accepts H^+ ions from the acid. When a base is added, the buffer accepts ^-OH ions from the base and releases protons (H^+).

Using the Henderson-Hasselbalch equation: pH = pK_a + log([A^-] / [HA])

where [HA] = concentration of the weak acid, in units of molarity; [A^-] = concentration of the conjugate base, in units of molarity.

> pK_a = –log(K_a)

where K_a = acid dissociation constant.

From the question, the concentration of the conjugate base equals the concentration of the weak acid. This equates to the following expression:

> [HA] = [A^-]

Rearranging: [A^-] / [HA], like in the Henderson-Hasselbalch equation:

> [A^-] / [HA] = 1

Substituting the value into the Henderson-Hasselbalch equation:

$$pH = pK_a + \log(1)$$

$$pH = pK_a + 0$$

$$pH = pK_a$$

When pH = pK$_a$, the titration is in the buffering region.

19. D is correct.

The ionic product constant of water:

$$K_w = [H_3O^+]\cdot[^-OH]$$

$$[^-OH] = K_w / [H_3O^+]$$

$$[^-OH] = [1 \times 10^{-14}] / [7.5 \times 10^{-9}]$$

$$[^-OH] = 1.3 \times 10^{-6}$$

20. C is correct.

Consider the K_a presented in the question – which species is more acidic or basic than NH$_3$?

NH$_4^+$ is correct because it is NH$_3$ after absorbing one proton. The concentration of H$^+$ in water is proportional to K_a. NH$_4^+$ is the conjugate acid of NH$_3$.

A: H$^+$ is acidic.

B: NH$_2^-$ is NH$_3$ with one less proton. If a base loses a proton, it would be an even stronger base with a higher proton affinity, so it is not weaker than NH$_3$.

D: Water is neutral.

E: NaNH$_2$ is the neutral species of NH$_2^-$.

===

Practice Set 2: Questions 21–40

===

21. E is correct.

Acid as a reactant produces a conjugate base, while a base as a reactant produces a conjugate acid.

The conjugate base of a chemical species is that species after H^+ has dissociated.

Therefore, the conjugate base of HSO_4^- is SO_4^{2-}.

The conjugate base of H_3O^+ is H_2O.

22. A is correct.

Start by calculating the moles of $Ca(OH)_2$:

Moles of $Ca(OH)_2$ = molarity $Ca(OH)_2$ × volume of $Ca(OH)_2$

Moles of $Ca(OH)_2$ = 0.1 M × (30 mL × 0.001 L/mL)

Moles of $Ca(OH)_2$ = 0.003 mol

Use the coefficients from the reaction equation to determine moles of HNO_3:

Moles of HNO_3 = (coefficient of HNO_3) / [coefficient $Ca(OH)_2$ × moles of $Ca(OH)_2$]

Moles of HNO_3 = (2 / 1) × 0.003 mol

Moles of HNO_3 = 0.006 mol

Divide moles by molarity to calculate volume:

Volume of HNO_3 = moles of HNO_3 / molarity of HNO_3

Volume of HNO_3 = 0.006 mol / 0.2 M

Volume of HNO_3 = 0.03 L

Convert volume to milliliters:

0.03 L × 1000 mL / L = 30 mL

23. D is correct.

Acidic solutions have a pH less than 7 due to a higher concentration of H^+ ions relative to ^-OH ions.

Basic solutions have a pH greater than 7 due to a higher concentration of ^-OH ions relative to H^+ ions.

24. D is correct.

An amphoteric compound can react as an acid (i.e., donates protons) as well as a base (i.e., accepts protons).

Examples of amphoteric molecules include amino acids (i.e., an amine and carboxylic acid group) and self-ionizable compounds such as water.

---

25. B is correct.

Strong acids (i.e., reactants) proceed towards products.

$$K_a = [\text{products}] / [\text{reactants}]$$

The molecule with the largest K_a is the strongest acid.

$$pK_a = -\log K_a$$

The molecule with the smallest pK_a is the strongest acid.

Strong acids dissociate a proton to produce the weakest conjugate base (i.e., most stable anion).

Weak acids dissociate a proton to produce the strongest conjugate base (i.e., least stable anion).

26. B is correct.

Balanced reaction:

$$H_3PO_4 + 3\ LIOH = Li_3PO_4 + 3\ H_2O$$

In the neutralization of acids and bases, the result is salt and water.

Phosphoric acid and lithium hydroxide react, so the resulting compounds are lithium phosphate and water.

27. B is correct.

The stability of the compound determines base strength. If the compound is unstable in its present state, it seeks a bonding partner (e.g., H+ or another atom) by donating its electrons for the new bond formation.

The 8 strong bases are: LiOH (lithium hydroxide), NaOH (sodium hydroxide), KOH (potassium hydroxide), $Ca(OH)_2$ (calcium hydroxide), RbOH (rubidium hydroxide), $Sr(OH)_2$, (strontium hydroxide), CsOH (cesium hydroxide) and $Ba(OH)_2$ (barium hydroxide).

28. D is correct.

The formula for pH:

$$pH = -\log[H^+]$$

Rearrange to solve for $[H^+]$:

$$[H^+] = 10^{-pH}$$

$$[H^+] = 10^{-2}\ M = 0.01\ M$$

29. C is correct.

Ionization	Dissociation
The process that produces new charged particles.	The separation of charged particles that already exist in a compound.
Involves polar covalent compounds or metals.	Involves ionic compounds.
Involves covalent bonds between atoms	Involves ionic bonds in compounds
Produces charged particles.	Produces either charged particles or electrically neutral particles.
Irreversible	Reversible
Example: $HCl \rightarrow H^+ + Cl^-$ $Mg \rightarrow Mg^{2+} + 2\ e^-$	Example: $PbBr_2 \rightarrow Pb^{2+} + 2\ Br^-$

30. C is correct.

A base is a chemical substance with a pH greater than 7 and feels slippery because it dissolves the fatty acids and oils from the skin and reduces the friction between skin cells.

Under acidic conditions, litmus paper is red, and under basic conditions, it is blue.

Many bitter-tasting foods are alkaline because bitter compounds often contain amine groups, which are weak bases.

Acids are known to have a sour taste (e.g., lemon juice) because the sour taste receptors on the tongue detect the dissolved hydrogen (H^+) ions.

31. E is correct.

The 7 strong acids are HCl (hydrochloric acid), HNO_3 (nitric acid), H_2SO_4 (sulfuric acid), HBr (hydrobromic acid), HI (hydroiodic acid), $HClO_3$ (chloric acid), and $HClO_4$ (perchloric acid).

The 8 strong bases are: LiOH (lithium hydroxide), NaOH (sodium hydroxide), KOH (potassium hydroxide), $Ca(OH)_2$ (calcium hydroxide), RbOH (rubidium hydroxide), $Sr(OH)_2$, (strontium hydroxide), CsOH (cesium hydroxide) and $Ba(OH)_2$ (barium hydroxide).

32. C is correct.

The Arrhenius acid–base theory states that acids produce H^+ ions in H_2O solution and bases produce ^-OH ions in H_2O solution.

The Brønsted-Lowry acid–base theory focuses on the ability to accept and donate protons (H^+).

A Brønsted-Lowry acid is a term for a substance that donates a proton in an acid–base reaction, while a Brønsted-Lowry base is a substance that accepts a proton.

Lewis acids are defined as electron-pair acceptors, whereas Lewis bases are electron-pair donors.

33. A is correct.

By the Brønsted-Lowry acid–base theory: an acid (reactant) dissociates a proton to become the conjugate base (product), a base (reactant) gains a proton to become the conjugate acid (product).

The definition is expressed in terms of an equilibrium expression: acid + base $\leftrightarrow$ conjugate base + conjugate acid.

34. D is correct.

With polyprotic acids (i.e., more than one H^+ present), the pK_a indicates the pH at which the H^+ is deprotonated. If the pH goes above the first pK_a, one proton dissociates, and so on.

In this example, the pH is above the first and second pK_a, so two acid groups are deprotonated while the third acidic proton is unaffected.

35. E is correct.

It is important to identify the acid that is active in the reaction.

The parent acid is defined as the most protonated form of the buffer. The number of dissociating protons an acid can donate depends on the charge of its conjugate base.

$Ba_2P_2O_7$ is given as one of the products in the reaction. Because barium is a group 2B metal, it has a stable oxidation state of +2. Because two barium cations are present in the product, the charge of P_2O_7 ion (the conjugate base in the reaction) must be –4.

Therefore, the fully protonated form of this conjugate must be $H_4P_2O_7$, which is a tetraprotic acid because it has 4 protons that can dissociate.

36. C is correct.

The greater the concentration of H_3O^+, the more acidic the solution is.

37. B is correct.

A triprotic acid has three protons that can dissociate.

38. B is correct.

KCl and NaI are salts. Two salts react if one of the products precipitates.

In this example, the products (KI and NaCl) are soluble in water, so they do not react.

39. B is correct.

Sodium acetate is a basic compound because acetate is the conjugate base of acetic acid, a weak acid ("the conjugate base of a weak acid acts as a base in the water").

The addition of a base to any solution, even a buffered solution, increases the pH.

40. E is correct.

An acid anhydride is a compound that has two acyl groups bonded to the same oxygen atom.

Anhydride means *without water* and is formed via dehydration (i.e., removal of H_2O).

==

Practice Set 3: Questions 41–60

==

41. A is correct.

The Brønsted-Lowry acid–base theory focuses on the ability to accept and donate protons (H⁺).

A Brønsted-Lowry acid is a term for a substance that donates a proton (H⁺) in an acid–base reaction, while a Brønsted-Lowry base is a substance that accepts a proton.

42. A is correct.

Strong acids (i.e., reactants) proceed towards products.

K_a = [products] / [reactants]

The molecule with the largest K_a is the strongest acid.

$pK_a = -\log K_a$

The molecule with the smallest pK_a is the strongest acid.

Strong acids dissociate a proton to produce the weakest conjugate base (i.e., most stable anion).

Weak acids dissociate a proton to produce the strongest conjugate base (i.e., least stable anion).

43. D is correct.

Learn the ions involved in boiler scale formations: CO_3^{2-} and the metal ions.

44. E is correct.

A buffer is an aqueous solution that consists of a weak acid and its conjugate base, or vice versa.

Buffered solutions resist changes in pH and are often used to keep the pH at a nearly constant value in many chemical applications. It does this by readily absorbing or releasing protons (H⁺) and ⁻OH.

When an acid is added to the solution, the buffer releases ⁻OH and accepts H⁺ ions from the acid.

To create a buffer solution, there needs to be a pair of a weak acid/base and its conjugate, or a salt that contains an ion from the weak acid/base.

Since the problem indicates that sulfoxylic acid (H_2SO_2), which has a pK_a 7.97, needs to be in the mixture, the other component would be an HSO_2^- ion (bisulfoxylate) of $NaHSO_2$.

45. C is correct.

Acidic salt is a salt that still contains H⁺ in its anion. It is formed when a polyprotic acid is partially neutralized, leaving at least 1 H⁺.

For example: $H_3PO_4 + 2\,KOH \rightarrow K_2HPO_4 + 2\,H_2O$: (partial neutralization, K_2HPO_4 is acidic salt)

While: $H_3PO_4 + 3\,KOH \rightarrow K_3PO_4 + 3\,H_2O$: (complete neutralization, K_3PO_4 is not acidic salt)

46. E is correct.

An acid anhydride is a compound that has two acyl groups bonded to the same oxygen atom.

Anhydride means *without water* and is formed via dehydration (i.e., removal of H_2O).

47. A is correct.

By the Brønsted-Lowry definition, an acid donates protons, while a base accepts protons.

On the product side of the reaction, H_2O acts as a base (i.e., the conjugate base of H_3O^+), and HCl acts as an acid (i.e., the conjugate acid of Cl^-).

48. D is correct.

The ratio of the conjugate base to the acid must be determined from the pH of the solution and the pK_a of the acidic component in the reaction.

In the reaction, $H_2PO_4^-$ acts as the acid, and HPO_4^{2-} acts as the base, so the pK_a of $H_2PO_4^-$ should be used in the equation.

Substitute the given values into the Henderson-Hasselbalch equation:

$$pH = pK_a + \log[\text{salt} / \text{acid}]$$

$$7.35 = 6.87 + \log[\text{salt} / \text{acid}]$$

Since $H_2PO_4^-$ is acting as the acid, subtract 6.87 from both sides:

$$0.48 = \log[\text{salt} / \text{acid}]$$

The log base is 10, so the inverse log gives:

$$10^{0.48} = (\text{salt} / \text{acid})$$

$$(\text{salt} / \text{acid}) = 3.02$$

The ratio between the conjugate base or salt and the acid is 3.02 / 1.

49. B is correct.

An electrolyte is a substance that dissociates into cations (i.e., positive ions) and anions (i.e., negative ions) when placed in solution. The light bulb is dimly lit, indicating that the solution contains only a low concentration (i.e., partial ionization) of the ions.

An electrolyte produces an electrically conducting solution when dissolved in a polar solvent (e.g., water). The dissolved ions disperse uniformly through the solvent. If an electrical potential (i.e., voltage) is applied to such a solution, the cations of the solution migrate towards the electrode (i.e., an abundance of electrons).

In contrast, anions migrate towards the electrode (i.e., a deficit of electrons).

50. B is correct.

An acid is a chemical substance with a pH less than 7, producing H^+ ions in water. An acid can be neutralized by a base (i.e., a substance with a pH above 7) to form a salt.

Acids are known to have a sour taste (e.g., lemon juice) because the sour taste receptors on the tongue detect the dissolved hydrogen (H^+) ions.

However, acids are not known to have a slippery feel; this is characteristic of bases. Bases feel slippery because they dissolve the fatty acids and oils from the skin and reduce the friction between the skin cells.

51. C is correct.

The Arrhenius acid–base theory states that acids produce H^+ ions (protons) in H_2O solution and bases produce OH^- ions (hydroxide) in H_2O solution.

The Brønsted-Lowry acid–base theory focuses on the ability to accept and donate protons (H^+).

A Brønsted-Lowry acid is a term for a substance that donates a proton in an acid–base reaction, while a Brønsted-Lowry base is a substance that accepts a proton.

52. A is correct.

By the Brønsted-Lowry acid–base theory:

An acid (reactant) dissociates a proton to become the conjugate base (product).

A base (reactant) gains a proton to become the conjugate acid (product).

The definition is expressed in terms of an equilibrium expression:

acid + base ↔ conjugate base + conjugate acid.

53. A is correct.

By the Brønsted-Lowry acid–base theory:

An acid (reactant) dissociates a proton to become the conjugate base (product).

A base (reactant) gains a proton to become the conjugate acid (product).

The definition is expressed in terms of an equilibrium expression:

acid + base ↔ conjugate base + conjugate acid.

HCl dissociates completely and therefore is a strong acid.

Strong acids produce weak (i.e., stable) conjugate bases.

54. D is correct.

A diprotic acid has two protons that can dissociate.

55. A is correct.

Protons (H^+) migrate between amino acid and solvent, depending on the pH of solvent and pK_a of functional groups on the amino acid.

Carboxylic acid groups can donate protons, while amine groups can receive protons.

For the carboxylic acid group:

   If the pH of solution $< pK_a$: group is protonated and neutral

   If the pH of solution $> pK_a$: group is deprotonated and negative

For the amine group:

   If the pH of solution $< pK_a$: group is protonated and positive

   If the pH of solution $> pK_a$: group is deprotonated and neutral

56. E is correct.

Acidic solutions contain hydronium ions (H_3O^+). These ions are in the aqueous form because they are dissolved in water. Although chemists often write H^+ (*aq*), referring to a single hydrogen nucleus (a proton), it exists as the hydronium ion (H_3O^+).

57. A is correct.

With a K_a of 10^{-5}, the pH of a 1 M solution of the carboxylic acid, $CH_3CH_2CH2CO_2H$, would be 5 and is a weak acid.

Only $CH_3CH_2CH_2CO_2H$ can be considered a weak acid because it yields a (relatively) unstable anion.

58. C is correct.

Condition 1: pH = 4

   $[H_3O^+] = 10^{-pH} = 10^{-4}$

Condition 2: pH = 7

   $[H_3O^+] = 10^{-pH} = 10^{-7}$

Ratio of $[H_3O^+]$ in condition 1 and condition 2:

   $10^{-4} : 10^{-7} = 1{,}000 : 1$

Solution with a pH of 4 has 1,000 times greater $[H^+]$ than a solution with a pH of 7.

Note: $[H_3O^+]$ is equivalent to $[H^+]$

59. B is correct.

The pH of a buffer is calculated using the Henderson-Hasselbalch equation:

$$pH = pK_a + \log([\text{conjugate base}] / [\text{conjugate acid}])$$

When [acid] = [base], the fraction is 1.

Log 1 = 0,

$$pH = pK_a + 0$$

If the K_a of the acid is 4.6×10^{-4} (between 10^{-4} and 10^{-3}), the pK_a (and therefore the pH) is between 3 and 4.

60. C is correct.

$K_w = [H^+] \cdot [^-OH]$ is the definition of the ionization constant for water.

===

Practice Set 4: Questions 61–80

===

61. A is correct.

Weak acids react with a strong base, converted to its weak conjugate base, creating an overall basic solution.

Strong acids do react with a strong base, creating a neutral solution.

Weak acids partially dissociate when dissolved in water; unlike a strong acid, they do not readily form ions.

Strong acids are much more corrosive than weak acids.

62. C is correct.

K_w is the water ionization constant (or *water autoprotolysis constant*).

It can be determined experimentally and equals 1.011×10^{-14} at 25 °C (1.00×10^{-14} is used).

63. D is correct.

All options are correct descriptions of the reaction, but the correct choice is the most descriptive.

64. A is correct.

The Brønsted-Lowry acid–base theory focuses on the ability to accept and donate protons (H^+).

A Brønsted-Lowry acid is a term for a substance that donates a proton in an acid–base reaction, while a Brønsted-Lowry base is a substance that accepts a proton.

65. A is correct.

Buffered solutions resist changes in pH and are often used to keep the pH at a nearly constant value in many chemical applications. It does this by readily absorbing or releasing protons (H^+) and ^-OH.

H_2SO_4 is a strong acid. Weak acids and their salts are good buffers.

A buffer is an aqueous solution that consists of a weak acid and its conjugate base, or vice versa.

When an acid is added to the solution, the buffer releases ^-OH and accepts H^+ ions from the acid.

When a base is added, the buffer accepts ^-OH ions from the base and releases protons (H^+).

66. E is correct.

The main use of litmus paper is to test whether a solution is acidic or basic.

Litmus paper can be used to test for water-soluble gases that affect acidity or alkalinity; the gas dissolves in the water, and the resulting solution colors the litmus paper. For example, alkaline ammonia gas causes the litmus paper to change from red to blue.

Blue litmus paper turns red under acidic conditions, and red litmus paper turns blue under basic or alkaline conditions, with the color change occurring over the pH range 4.5-8.3 at 25 °C (77 °F).

Neutral litmus paper is purple.

Litmus can be prepared as an aqueous solution that functions similarly. Under acidic conditions, the solution is red, and under basic conditions, the solution is blue.

The properties listed above (turning litmus paper blue, bitter taste, slippery feel, and neutralizing acids) are true of bases.

An acidic solution has opposite qualities.

It has a pH lower than 7 and therefore turns litmus paper red.

It neutralizes bases, tastes sour, and does not feel slippery.

67. A is correct.

An electrolyte is a substance that dissociates into cations (i.e., positive ions) and anions (i.e., negative ions) when placed in solution. An electrolyte produces an electrically conducting solution when dissolved in a polar solvent (e.g., water). The dissolved ions disperse uniformly through the solvent. If an electrical potential (i.e., voltage) is applied to such a solution, the cations of the solution migrate towards the electrode (i.e., an abundance of electrons). In contrast, the anions migrate towards the electrode (i.e., a deficit of electrons).

An acid is a substance that ionizes when dissolved in suitable ionizing solvents such as water.

If a high proportion of the solute dissociates to form free ions, it is a strong electrolyte.

If most of the solute does not dissociate, it is a weak electrolyte.

The more free ions present, the better the solution conducts electricity.

68. A is correct.

The 7 strong acids are HCl (hydrochloric acid), HNO_3 (nitric acid), H_2SO_4 (sulfuric acid), HBr (hydrobromic acid), HI (hydroiodic acid), $HClO_3$ (chloric acid), and $HClO_4$ (perchloric acid).

69. B is correct.

The activity series determines if a metal displaces another metal in the solution.

The reaction can only occur if the added metal is above (i.e., activity series) the metal currently bonded with the anion.

An activity series ranks substances in their order of relative reactivity.

For example, magnesium metal can displace hydrogen ions from solution, so it is more reactive than elemental hydrogen:

$$Mg\ (s) + 2\ H^+\ (aq) \rightarrow H_2\ (g) + Mg^{2+}\ (aq)$$

Zinc can displace hydrogen ions from solution, so zinc is more reactive than elemental hydrogen:

$$Zn\ (s) + 2\ H^+\ (aq) \rightarrow H_2\ (g) + Zn^{2+}\ (aq)$$

Magnesium metal can displace zinc ions from solution:

$$Mg\ (s) + Zn^{2+}\ (aq) \rightarrow Zn\ (s) + Mg^{2+}\ (aq)$$

The metal activity series with the most active (i.e., most strongly reducing) metals appear at the top, and the least active metals near the bottom.

Li: $2\ Li\ (s) + 2\ H_2O\ (l) \rightarrow LiOH\ (aq) + H_2\ (g)$

K: $2\ K\ (s) + 2\ H_2O\ (l) \rightarrow 2\ KOH\ (aq) + H_2\ (g)$

Ca: $Ca\ (s) + 2\ H_2O\ (l) \rightarrow Ca(OH)_2\ (s) + H_2\ (g)$

Na: $2\ Na\ (s) + 2\ H_2O\ (l) \rightarrow 2\ NaOH\ (aq) + H_2\ (g)$

The above can displace H_2 from water, steam, or acids

Mg: $Mg\ (s) + 2\ H_2O\ (g) \rightarrow Mg(OH)_2\ (s) + H_2\ (g)$

Al: $2\ Al\ (s) + 6\ H_2O\ (g) \rightarrow 2\ Al(OH)_3\ (s) + 3\ H_2\ (g)$

Mn: $Mn\ (s) + 2\ H_2O\ (g) \rightarrow Mn(OH)_2\ (s) + H_2\ (g)$

Zn: $Zn\ (s) + 2\ H_2O\ (g) \rightarrow Zn(OH)_2\ (s) + H_2\ (g)$

Fe: $Fe\ (s) + 2\ H_2O\ (g) \rightarrow Fe(OH)_2\ (s) + H_2\ (g)$

The above can displace H_2 from steam or acids

Ni: $Ni\ (s) + 2\ H^+\ (aq) \rightarrow Ni^{2+}\ (aq) + H_2\ (g)$

Sn: $Sn\ (s) + 2\ H^+\ (aq) \rightarrow Sn^{2+}\ (aq) + H_2\ (g)$

Pb: $Pb\ (s) + 2\ H^+\ (aq) \rightarrow Pb^{2+}\ (aq) + H_2\ (g)$

The above can displace H_2 from acids only

$$H_2 > Cu > Ag > Pt > Au$$

The above cannot displace H_2

70. E is correct.

The Arrhenius acid–base theory states that acids produce H^+ ions (protons) in H_2O solution, and bases produce OH^- ions (hydroxide) in H_2O solution. $Al(OH)_3\ (s)$ is insoluble in water; therefore, it cannot function as an Arrhenius base.

71. A is correct.

By the Brønsted-Lowry acid–base theory:

An acid (reactant) dissociates a proton to become the conjugate base (product).

A base (reactant) gains a proton to become the conjugate acid (product).

The definition is expressed in terms of an equilibrium expression:

acid + base $\leftrightarrow$ conjugate base + conjugate acid.

72. D is correct.

Henderson-Hasselbach equation:

$$pH = pK_a + \log(A^- / HA)$$

A buffer is an aqueous solution that consists of a weak acid and its conjugate base, or vice versa.

Buffered solutions resist changes in pH and are often used to keep the pH at a nearly constant value in many chemical applications. It does this by readily absorbing or releasing protons (H^+) and ^-OH.

When an acid is added to the solution, the buffer releases ^-OH and accepts H^+ ions from the acid.

73. C is correct.

The weakest acid has the smallest K_a (or largest pK_a).

The weakest acid has the strongest (i.e., least stable) conjugate base.

74. E is correct.

The balanced reaction:

$$2\ H_3PO_4 + 3\ Ba(OH)_2 \rightarrow Ba_3(PO_4)_2 + 6\ H_2O$$

There are 2 moles of H_3PO_4 in a balanced reaction.

However, acids are categorized by the number of H^+ per mole of acid.

For example:

HCl is monoprotic (has one H^+ to dissociate).

H_2SO_4 is diprotic (has two H^+ to dissociate).

H_3PO_4 is a triprotic acid (has three H^+ to dissociate).

75. E is correct.

Because Na^+ forms a strong base (NaOH) and S forms a weak acid (H_2S), it undergoes hydrolysis in water:

$$Na_2S\ (aq) + 2\ H_2O\ (l) \rightarrow 2\ NaOH\ (aq) + H_2S$$

Ionic equation for individual ions:

$$2\ Na^+ + S^{2-} + 2\ H_2O \rightarrow 2\ Na^+ + 2\ OH^- + H_2S$$

Removing Na^+ ions from both sides of the reaction:

$$S^{2-} + 2\ H_2O \rightarrow 2\ OH^- + H_2S$$

Removing H_2O from both sides of the reaction:

$$S^{2-} + H_2O \rightarrow OH^- + HS^-$$

76. A is correct.

HNO_3 is a strong acid, which means that $[HNO_3] = [H_3O^+] = 0.0765$.

$$pH = -\log[H_3O^+]$$

$$pH = -\log(0.0765)$$

$$pH = 1.1$$

77. C is correct.

Strong acids dissociate a proton to produce the weakest conjugate base (i.e., most stable anion).

Weak acids dissociate a proton to produce the strongest conjugate base (i.e., least stable anion).

Acetic acid (CH_3COOH) has a pK_a of about 4.8 and is considered a weak acid because it does not dissociate completely. Therefore, it does not readily dissociate into H^+ and CH_3COO^-.

Each of the listed strong acids forms an anion stabilized by resonance.

HBr (hydrobromic acid) has a pK_a of –9.

HNO_3 (nitrous acid) has a pK_a of –1.4.

H_2SO_4 (sulfuric acid) is a diprotic acid and has a pK_a of –3 and 1.99.

HCl (hydrochloric acid) has a pK_a of –6.3.

78. A is correct.

It is important to recognize the chromate ions:

Dichromate: $Cr_2O_7^{2-}$

Chromium (II): Cr^{2+}

Chromic/Chromate: CrO_4^-

Trichromic acid: does not exist.

79. B is correct.

The Brønsted-Lowry acid–base theory focuses on the ability to accept and donate protons (H^+).

A Brønsted-Lowry acid is a term for a substance that donates a proton in an acid–base reaction, while a Brønsted-Lowry base is a substance that accepts a proton.

80. C is correct.

In the Arrhenius theory, acids dissociate in an aqueous solution to produce H^+ (hydrogen ions).

In the Arrhenius theory, bases dissociate in an aqueous solution to produce ^-OH (hydroxide ions).

==

Practice Set 5: Questions 81–100

==

81. B is correct.

The acid requires two equivalents of base to be fully titrated and therefore is a diprotic acid (see diagram).

Using the fully protonated sulfuric acid (H_2SO_4) as an example:

At point A, the acid is 50% fully protonated, 50% singly deprotonated: (50% H_2SO_4: 50% HSO_4^-).

At point B, the acid exists in the singly deprotonated form only (100% HSO_4^-).

At point C, the acid exists as 50% singly deprotonated HSO_4^- and 50% doubly deprotonated SO_4^{2-}.

At point D, the acid exists as 100% SO_4^{2-}.

Point A is known as pK_{a1}, and point C is pK_{a2} (i.e., strongest buffering regions).

Point B and D are known as equivalence points (i.e., weakest buffering region)

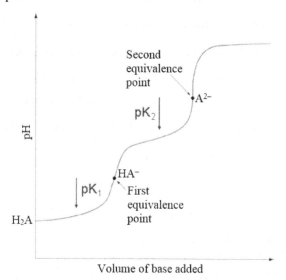

Titration curve: addition of a strong base to diprotic acid (H_2A)

82. C is correct.

Point C on the graph for the question is pK_{a2}. At this point, the acid exists as 50% singly deprotonated (e.g., HSO_4^-) and 50% doubly deprotonated (e.g., SO_4^{2-}).

The Henderson-Hasselbalch equation:

$$pH = pK_{a2} + \log[\text{salt} / \text{acid}]$$

$$pH = pK_{a2} + \log[50\% / 50\%]$$

$$pH = pK_{a2} + \log[1]$$

$$pH = pK_{a2} + 0$$

$$pH = pK_{a2}$$

83. A is correct.

Point A is pK_{a1}. At this point, the acid exists as 50% fully protonated (e.g., H_2SO_4) and 50% singly deprotonated (e.g., HSO_4^-).

The Henderson-Hasselbalch equation:

$$pH = pK_{a1} + \log[\text{salt} / \text{acid}]$$

$$pH = pK_{a1} + \log[50\% / 50\%]$$

$$pH = pK_{a1} + \log[1]$$

$$pH = pK_{a1} + 0$$

$$pH = pK_{a1}$$

84. B is correct.

At point B on the graph (i.e., equivalence point), the acid exists in the singly deprotonated form only (100% HSO_4^- for the example of H_2SO_4).

85. B is correct.

The buffer region is the flattest region on the curve (the region that resists pH increases with added base). This diprotic acid has two buffering regions: pK_{a1} (around point A) and pK_{a2} (around point C).

A buffer is an aqueous solution that consists of a weak acid and its conjugate base, or vice versa.

Buffered solutions resist changes in pH and are often used to keep the pH at a nearly constant value in many chemical applications. It does this by readily absorbing or releasing protons (H^+) and ^-OH.

When an acid is added to the solution, the buffer releases ^-OH and accepts H^+ ions from the acid.

When a base is added, the buffer accepts ^-OH ions from the base and releases protons (H^+).

86. D is correct.

Sodium hydroxide turns into soap (i.e., saponification) from the reaction with the fatty acid esters and oils on the fingertips (i.e., skin). Fatty acid esters react with NaOH by releasing free fatty acids, which act as soap surrounding grease with their nonpolar (i.e., hydrophobic) ends, while their polar (i.e., hydrophilic) ends orient towards water molecules. This decreases friction and accounts for the slippery feel of NaOH interacting with the skin.

87. E is correct.

I: water is produced in a neutralization reaction.

II: this reaction is a common example of neutralization, but it is not always the case. For example, weak bases react with strong acids: $AH + B \rightleftharpoons A^- + BH^+$

Or strong bases react with weak acids: $AH + H_2O \rightleftharpoons H_3O^+ + A^-$

III: neutralization occurs when acid donates a proton to the base.

88. E is correct.

Strong acids dissociate entirely, or almost completely, when in water. They dissociate into a positively-charged hydrogen ion (H^+) and another ion that is negatively charged. An example is hydrochloric acid (HCl), which dissociates into H^+ and Cl^- ions.

Polarity refers to the distribution of electrons in a bond. If a molecule is polar, one side has a partial positive charge, and the other side has a partial negative charge. The more polar the bond, the easier it is for a molecule to dissociate into ions, and therefore the acid is more strongly acidic.

89. D is correct.

The Brønsted-Lowry acid–base theory focuses on the ability to accept and donate protons (H^+).

A Brønsted-Lowry acid is a term for a substance that donates a proton in an acid–base reaction, while a Brønsted-Lowry base is a substance that accepts a proton.

90. E is correct.

Formula to calculate pH of an acidic buffer:

$$[H^+] = ([acid] / [salt]) \times K_a$$

Calculate H^+ from pH:

$$[H^+] = 10^{-pH}$$

$$[H^+] = 10^{-4}$$

Calculate Ka from pK_a:

$$K_a = 10^{-pKa}$$

$$K_a = 10^{-3}$$

Substitute values into the buffer equation:

$$[H^+] = ([acid] / [salt]) \times K_a$$

$$10^{-4} = ([acid] / [salt]) \times 10^{-3}$$

$$10^{-1} = [acid] / [salt]$$

$$[salt] / [acid] = 10$$

Therefore, the ratio of salt to acid is 10:1.

91. D is correct.

An electrolyte is a substance that dissociates into cations (i.e., positive ions) and anions (i.e., negative ions) when placed in solution. An electrolyte produces an electrically conducting solution when dissolved in a polar solvent (e.g., water). The dissolved ions disperse uniformly through the solvent.

If an electrical potential (i.e., voltage) is applied to such a solution, the cations of the solution migrate towards the electrode (i.e., an abundance of electrons). In contrast, the anions migrate towards the electrode (i.e., a deficit of electrons).

If a high proportion of the solute dissociates to form free ions, it is a strong electrolyte.

If most of the solute does not dissociate, it is a weak electrolyte.

The more free ions present, the better the solution conducts electricity.

92. E is correct.

Litmus (i.e., dyes extracted from lichens) tests whether a solution is acidic or basic.

Wet litmus paper can be used to test for water-soluble acidic or alkaline gases that dissolve in the water and produce color changes in the litmus paper.

Neutral litmus paper is purple, with color changes occurring over the pH range of 4.5 to 8.3.

Red litmus paper turns blue when exposed to alkaline ammonia gas (basic conditions).

Blue litmus paper turns red under acidic conditions.

93. C is correct.

The more acidic molecule has a lower pK_a.

94. E is correct.

Acetic acid ($CH_3COOH = HC_2H_3O_2$) is commonly known as vinegar. It has a pK_a of about 4.8 and is considered a weak acid because it does not dissociate completely.

HI (hydroiodic acid) has a pK_a of -10.

$HClO_4$ (perchloric acid) has a pK_a of -10.

HCl (hydrochloric acid) has a pK_a of -6.3.

HNO_3 (nitrous acid) has a pK_a of -1.4.

95. C is correct.

An acid anhydride is a compound that has two acyl groups bonded to the same oxygen atom.

$$H_3CCO_3CCH_3$$

Anhydride means *without water* and is formed via dehydration (i.e., removal of H_2O).

96. E is correct.

Acids and bases react to form H_2O and salts: $NaOH + HCl \rightleftharpoons Na^+Cl^-$ (a salt) $+ H_2O$ (water)

The Brønsted-Lowry acid–base theory focuses on the ability to accept and donate protons (H^+). A Brønsted-Lowry acid is a substance that donates a proton in an acid–base reaction, while a Brønsted-Lowry base is a substance that accepts a proton.

The Arrhenius acid–base theory states that acids produce H^+ ions in H_2O solution and bases produce ^-OH ions in H_2O solution.

Lewis acids are defined as electron-pair acceptors, whereas Lewis bases are electron-pair donors.

97. D is correct.

The Bronsted-Lowry acid–base theory focuses on the ability to accept and donate protons.

The definition is expressed in terms of an equilibrium expression:

acid + base ↔ conjugate base + conjugate acid.

With an acid, HA, the equation can be written symbolically as:

$HA + B \leftrightarrow A^- + HB^+$

98. C is correct.

By the Brønsted-Lowry acid–base theory:

An acid (reactant) dissociates a proton to become the conjugate base (product).

A base (reactant) gains a proton to become the conjugate acid (product).

The definition is expressed in terms of an equilibrium expression:

acid + base ↔ conjugate base + conjugate acid.

99. D is correct.

The addition of hydroxide (i.e., NaOH) decreases the solubility of magnesium hydroxide due to the common ion effect. Therefore, the amount of undissociated $Mg(OH)_2$ increases.

100. A is correct.

The weakest acid has the highest pK_a (and lowest K_a).

===

Practice Set 6: Questions 101–120

===

101. C is correct.

Strong acids dissociate completely (or almost completely) because they form a stable anion.

Hydrogen cyanide (HCN) has a pK_a of 9.3, and hydrogen sulfide (H_2S) has a pK_a of 7.0.

In this example, the strong acids include HI, HBr, HCl, H_3PO_4 and H_2SO_4.

102. D is correct.

Bromothymol blue is a pH indicator often used for solutions with neutral pH near 7 (e.g., managing the pH of pools and fish tanks). Bromothymol blue acts as a weak acid in a solution that can be protonated or deprotonated. It appears yellow when protonated (lower pH), blue when deprotonated (higher pH), and bluish-green in neutral solution.

Methyl red has a pK_a of 5.1 and is a pH indicator dye that changes color in acidic solutions: it turns red in pH under 4.4, orange in pH between 4.4 and 6.2, and yellow in pH over 6.2.

Phenolphthalein is used as an indicator for acid–base titrations. It is a weak acid, which can dissociate protons (H^+ ions) in solution. The phenolphthalein molecule is colorless, and the phenolphthalein ion is pink. It turns colorless in acidic solutions and pink in basic solutions. With basic conditions, the phenolphthalein (neutral) $\rightleftharpoons$ ions (pink) equilibrium shifts to the right, leading to more ionization as H^+ ions are removed.

103. A is correct.

Anhydride means *without water* and is formed via dehydration (i.e., removal of H_2O).

Therefore, a basic anhydride is a base without water.

104. C is correct.

The problem is asking for K_a of an acid, which indicates that the acid in question is a weak acid.

In aqueous solutions, a hypothetical weak acid HX partly dissociates and create this equilibrium:

$$HX \leftrightarrow H^+ + X^-$$

with an acid equilibrium constant, or

$$K_a = [H^+] \cdot [X^-] / [HX]$$

To calculate K_a, the concentration of all species is needed.

From the given pH, the concentration of H^+ ions can be calculated:

$$[H^+] = 10^{-pH}$$

$$[H^+] = 10^{-7} \text{ M}$$

The number of H^+ and X^- ions are equal; concentration of X^- ions is also 10^{-7} M.

Those ions came from the dissociated acid molecules. According to the problem, only 24% of the acid is dissociated. Therefore, the rest of acid molecules (100% – 24% = 76%) did not dissociate. Use the simple proportion of percentages to calculate the concentration of HX:

$$[HX] = (76\% / 24\%) \times 1 \times 10^{-7} \, M$$

$$[HX] = 3.17 \times 10^{-7} \, M$$

Use the concentration values to calculate K_a:

$$K_a = [H^+] \cdot [X^-] / [HX]$$

$$K_a = [(1 \times 10^{-7}) \times (1 \times 10^{-7})] / (3.17 \times 10^{-7})$$

$$K_a = 3.16 \times 10^{-8}$$

Calculate pK_a:

$$pK_a = -\log K_a$$

$$pK_a = -\log (3.16 \times 10^{-8})$$

$$pK_a = 7.5$$

105. E is correct.

With a neutralization reaction, the cations and anions on reactants switch pairs, resulting in salt and water.

106. D is correct.

The strongest acid has the largest K_a value (or the lowest pK_a).

107. E is correct.

If the $[H_3O^+]$ is greater than 1×10^{-7}, it is an acidic solution.

If the $[H_3O^+]$ is less than 1×10^{-7}, it is a basic solution.

If the $[H_3O^+]$ equals 1×10^{-7}, it has a pH of 7, and the solution is neutral.

108. A is correct.

An electrolyte is a substance that dissociates into cations (i.e., positive ions) and anions (i.e., negative ions) when placed in solution. An electrolyte produces an electrically conducting solution when dissolved in a polar solvent (e.g., water). The dissolved ions disperse uniformly through the solvent.

If an electrical potential (i.e., the voltage generated by the battery) is applied to such a solution, the cations of the solution migrate towards the electrode (i.e., an abundance of electrons).

The anions migrate towards the electrode (i.e., a deficit of electrons).

109. D is correct.

$$H_3PO_4 \rightarrow H_2PO_4^- \rightarrow HPO_4^{2-}$$

In the Arrhenius theory, acids are defined as substances that dissociate in an aqueous solution to produce H^+ (hydrogen ions).

In the Arrhenius theory, bases are defined as substances that dissociate in an aqueous solution to produce ^-OH (hydroxide ions).

In the Brønsted–Lowry theory, acids and bases are defined by the way they react.

The definition is expressed in terms of an equilibrium expression:

acid + base $\leftrightarrow$ conjugate base + conjugate acid.

With an acid, HA, the equation can be written symbolically as:

$$HA + B \leftrightarrow A^- + HB^+$$

110. E is correct.

The Arrhenius acid–base theory states that acids produce H^+ ions in H_2O solution and bases produce OH^- ions in H_2O solution.

The Arrhenius acid–base theory states that neutralization happens when acid–base reactions produce water and salt and that these reactions must occur in an aqueous solution.

There are additional ways to classify acids and bases.

The Brønsted-Lowry acid–base theory focuses on the ability to accept and donate protons.

The Lewis acid–base theory focuses on the ability to accept and donate electrons.

111. A is correct.

The Brønsted-Lowry acid–base theory focuses on the ability to accept and donate protons (H^+).

A Brønsted-Lowry acid is a term for a substance that donates a proton in an acid–base reaction, while a Brønsted-Lowry base is a substance that accepts a proton.

The Arrhenius acid–base theory focuses on the ability to produce H^+ and ^-OH ions. The Arrhenius acid–base theory states that neutralization happens when acid–base reactions produce water and salt and that these reactions must occur in an aqueous solution.

The Lewis acid–base theory focuses on the ability to accept and donate electrons.

112. B is correct. By the Brønsted-Lowry acid–base theory:

An acid (reactant) dissociates a proton to become the conjugate base (product).

A base (reactant) gains a proton to become the conjugate acid (product).

The definition is expressed in terms of an equilibrium expression:

acid + base $\leftrightarrow$ conjugate base + conjugate acid.

113. C is correct.

Water is neutral and has an equal concentration of hydroxide ($^-$OH) and hydronium (H_3O^+) ions.

114. D is correct.

The pI (isoelectric point) for an amino acid is defined as the pH for which an ionizable molecule has a net charge of zero. In general, the net charge on the molecule is affected by pH, as it can become more positively or negatively charged due to the gain or loss of protons (H^+), respectively.

An amphoteric compound can react as an acid (i.e., donates protons) as well as a base (i.e., accepts protons).

A zwitterion is a molecule with a positive (cation) and negative (anion) region within the same molecule. A zwitterion is amphoteric. Amino acids (amino and carboxyl end) are an example of amphoteric molecules.

115. B is correct.

Spectator ions exist as a reactant and product in a chemical equation. A net ionic equation ignores the spectator ions that were part of the original equation.

A zwitterion is a molecule with a positive (cation) and negative (anion) region within the same molecule. A zwitterion is amphoteric. Amino acids (amino and carboxyl end) are an example of amphoteric molecules.

116. A is correct.

Salts that result from the reaction of strong acids and strong bases are neutral.

For example:

$$\text{HCl} + \text{NaOH} \leftrightarrow \text{NaCl} + H_2O$$
strong acid strong base neutral

117. D is correct.

Polyprotic means two H^+ that can dissociate.

118. E is correct.

An acid dissociates a proton to form the conjugate base, while a conjugate base accepts a proton to form the acid.

119. C is correct.

The products of a neutralization reaction (e.g., salt and water) are not corrosive.

NaOH and HCl are very corrosive.

However, after a neutralization reaction, they form NaCl, or common table salt, which is not corrosive.

120. C is correct.

Start by calculating the moles of $CaCO_3$:

Moles of $CaCO_3$ = mass of $CaCO_3$ / molecular mass of $CaCO_3$

Moles of $CaCO_3$ = 0.5 g / 100.09 g/mol

Moles of $CaCO_3$ = 0.005 mol

Use the coefficients in the reaction equation to find moles of HNO_3:

Moles of HNO_3 = (coefficient of HNO_3 / coefficient of $CaCO_3$) × moles of $CaCO_3$

Moles of HNO_3 = (2/1) × 0.005 mol

Moles of HNO_3 = 0.01 mol

Divide moles by volume to calculate molarity:

Molarity of HNO_3 = moles of HNO_3 / volume of HNO_3

Molarity of HNO_3 = 0.01 mol / (25 mL × 0.001 L/mL)

Molarity of HNO_3 = 0.4 M

===

Practice Set 7: Questions 121–140

===

121. D is correct.

Strong acids dissociate a proton to produce the weakest conjugate base (i.e., most stable anion).

Weak acids dissociate a proton to produce the strongest conjugate base (i.e., least stable anion).

Hydrofluoric acid (HF) has a pK_a of about 3.8 and is considered a weak acid because it does not dissociate completely. The F^- anion is the least stable halogen anion (due to its small valence shell).
^-OH is the conjugate base of H_2O.

$NaNH_2$ (sodium amide) is a strong base and has a pK_a of 38.

HNO_3 (nitrous acid) has a pK_a of –1.4.

HI (hydroiodic acid) has a pK_a of –10.

122. C is correct.

Methyl red has a pK_a of 5.1 and is a pH indicator dye that changes color in acidic solutions: it turns red in pH under 4.4, orange in pH 4.4-6.2, and yellow in pH over 6.2.

Phenolphthalein is used as an indicator for acid–base titrations. It is a weak acid, which can dissociate protons (H^+ ions) in solution. The phenolphthalein molecule is colorless, and the phenolphthalein ion is pink. It turns colorless in acidic solutions and pink in basic solutions. With basic conditions, the phenolphthalein (neutral) ⇌ ions (pink) equilibrium shifts to the right, leading to more ionization as H^+ ions are removed.

Bromothymol blue is a pH indicator often used for solutions with neutral pH near 7 (e.g., managing the pH of pools and fish tanks). Bromothymol blue acts as a weak acid in a solution that can be protonated or deprotonated. It appears yellow when protonated (lower pH), blue when deprotonated (higher pH) and bluish-green in neutral solution.

123. C is correct.

Balanced reaction:

$$H_2CO_3 + 2\ KOH\ (aq) \rightarrow K_2CO_3\ (aq) + 2\ H_2O\ (l)$$

With a neutralization reaction, the cations and anions on reactants switch pairs, resulting in salt and water.

124. D is correct.

Strong acids completely dissociate protons into the aqueous solution. The resulting anion is stable, which accounts for the ~100% ionization of the acid.

125. B is correct.

A solution is acidic if the $[H_3O^+]$ is greater than 1×10^{-7}.

If the $[H_3O^+]$ is less than 1×10^{-7}, it is a basic solution.

If the $[H_3O^+]$ equals 1×10^{-7}, it has a pH of 7, and the solution is neutral.

126. E is correct.

Electrolytes dissociate into ions when dissolved in water.

CH_4, as with most hydrocarbon compounds, is not an electrolyte and will not dissociate.

127. B is correct.

Weak acids form unstable conjugate bases, and therefore the equilibrium lies on the side of the acid.

128. B is correct.

The Arrhenius acid–base theory states that acids produce H^+ ions (protons) in H_2O solution and bases produce ^-OH ions (hydroxide) in H_2O solution.

The Brønsted-Lowry acid–base theory focuses on the ability to accept and donate protons (H^+).

A Brønsted-Lowry acid is a substance that donates a proton in an acid–base reaction, while a Brønsted-Lowry base is a substance that accepts a proton.

129. B is correct.

The Arrhenius acid–base theory states that acids produce H^+ ions (protons) in H_2O solution and bases produce ^-OH ions (hydroxide) in H_2O solution.

130. D is correct.

The Brønsted-Lowry acid–base theory focuses on the ability to accept and donate protons (H^+).

A Brønsted-Lowry acid is a substance that donates a proton in an acid–base reaction, while a Brønsted-Lowry base is a substance that accepts a proton.

131. E is correct.

By the Brønsted-Lowry acid–base theory:

An acid (reactant) dissociates a proton to become the conjugate base (product).

A base (reactant) gains a proton to become the conjugate acid (product).

The definition is expressed in terms of an equilibrium expression:

acid + base ↔ conjugate base + conjugate acid.

132. A is correct.

Because HNO_3 is a strong acid, $[HNO_3] = [H_3O^+] = 0.045$ M.

$$pH = -\log[H_3O^+]$$

$$pH = -\log(0.045)$$

$$pH = 1.35$$

133. D is correct.

The pI (isoelectric point) for an amino acid is defined as the pH for which an ionizable molecule has a net charge of zero. In general, the net charge on the molecule is affected by pH, as it can become more positively or negatively charged due to the gain or loss of protons (H^+), respectively.

134. C is correct.

The Brønsted-Lowry acid–base theory focuses on the ability to accept and donate protons (H^+).

A Brønsted-Lowry acid is a substance that donates a proton in an acid–base reaction, while a Brønsted-Lowry base is a substance that accepts a proton.

The Arrhenius acid–base theory states that acids produce H^+ ions in H_2O solution and bases produce ^-OH ions in H_2O solution.

135. E is correct.

All reactions either increase the production of H^+ or ^-OH or decrease H^+ or ^-OH production and resist changes in pH (a feature of a buffer system).

A buffer is an aqueous solution that consists of a weak acid and its conjugate base, or vice versa.

Buffered solutions resist changes in pH and are often used to keep the pH at a nearly constant value in many chemical applications. It does this by readily absorbing or releasing protons (H^+) and ^-OH.

When an acid is added to the solution, the buffer releases ^-OH and accepts H^+ ions from the acid.

When a base is added, the buffer accepts ^-OH ions from the base and releases protons (H^+).

136. A is correct. A standard solution contains a precisely known concentration of an element or substance, usually determined to 3-4 significant digits.

Standard solutions are often used to determine the concentration of other substances, such as solutions in titrations.

137. B is correct.

Amphoteric means that the compound can act as an acid or base.

138. E is correct.

The inflection point is halfway between the beginning of the curve (before any titrant has been added) and the equivalence point (where the titrant has neutralized the starting material). At the inflection point, the concentrations of the two species (the acid and conjugate base) are equal.

Henderson-Hasselbalch equation:

$$pH = pK_a + log[conjugate\ base] / [acid]$$

A buffer is an aqueous solution that consists of a weak acid and its conjugate base, or vice versa.

Buffered solutions resist changes in pH and are often used to keep the pH at a nearly constant value in many chemical applications. It does this by readily absorbing or releasing protons (H^+) and ^-OH.

When an acid is added to the solution, the buffer releases ^-OH and accepts H^+ ions from the acid.

139. A is correct.

Three possible ways to balance this reaction:

$2\ LiOH\ (aq) + H_2SO_4\ (aq) = 2\ H_2O\ (l) + Li_2SO_4\ (aq)$ – salt, so this is probably the correct balance

$2\ LiOH\ (aq) + H_2SO_4\ (aq) = 2\ H_2O + Li_2SO_4\ (aq)$

$LiOH\ (aq) + H_2SO_4\ (aq) = H_2O\ (l) + LiHSO_4\ (aq)$ – *hydrogen sulfate (bisulfate)*

140. B is correct.

Protons (H^+) migrate between amino acid and solvent, depending on the pH of solvent and pK_a of functional groups on the amino acid.

Carboxylic acid groups can donate protons, while amine groups can receive protons.

For the carboxylic acid group:

   If the pH of solution $< pK_a$: group is protonated and neutral

   If the pH of solution $> pK_a$: group is deprotonated and negative

For the amine group:

   If the pH of solution $< pK_a$: group is protonated and positive

   If the pH of solution $> pK_a$: group is deprotonated and neutral

Notes for active learning

Electrochemistry – Detailed Explanations

==

Practice Set 1: Questions 1–20

==

1. D is correct.

In electrochemical (i.e., galvanic) cells, oxidation occurs at the anode, and reduction occurs at the cathode.

Br^- is oxidized at the anode, not the cathode.

2. C is correct.

As ionization energy increases, it is more difficult for electrons to be released by a substance. The release of electrons would increase the charge of the substance, which is the definition of oxidation.

Substances with higher ionization energy are more likely to be reduced. If a substance is reduced in a redox reaction, it is considered an oxidizing agent because it facilitates the oxidation of the other reactant.

Therefore, with the increase of ionization energy, a substance will be more likely to be reduced or be a stronger oxidizing agent.

3. E is correct.

Half-reaction:

$$H_2S \rightarrow S_8$$

Balancing half-reaction in acidic conditions:

Step 1: Balance all atoms except for H and O

$$8\,H_2S \rightarrow S_8$$

Step 2: To balance oxygen, add H_2O to the side with fewer oxygen atoms

There is no oxygen at all, so skip this step.

Step 3: To balance hydrogen, add H^+:

$$8\,H_2S \rightarrow S_8 + 16\,H^+$$

Balance charges by adding electrons to the side with higher/more positive total charge.

Total charge on left side: 0

Total charge on right side: $16(+1) = +16$

Add 16 electrons to right side:

$$8\,H_2S \rightarrow S_8 + 16\,H^+ + 16\,e^-$$

4. B is correct.

An electrolytic cell is a nonspontaneous electrochemical cell that requires the supply of electrical energy (e.g., battery) to initiate the reaction.

The anode is positive, and the cathode is the negative electrode.

For electrolytic and galvanic cells, oxidation occurs at the anode, while reduction occurs at the cathode.

Therefore, Co metal is produced at the cathode because it reduces the product (from an oxidation number of +3 on the left to 0 on the right). Co will not be produced at the anode.

5. C is correct.

The anode in galvanic cells attracts anions.

Anions in solution flow toward the anode, while cations flow toward the cathode.

Oxidation (i.e., loss of electrons) occurs at the anode.

Positive ions are formed while negative ions are consumed at the anode.

Therefore, negative ions flow toward the anode to equalize the charge.

6. E is correct.

By convention, the reference standard for potential is hydrogen reduction.

7. C is correct.

Calculate the oxidation numbers of species involved and look for the oxidized species (increase in oxidation number). Cd's oxidation number increases from 0 on the left side of the reaction to +2 on the right.

8. D is correct.

The oxidation number of Fe increases from 0 on the left to +3 on the right.

9. D is correct.

Salt bridge contains cations (positive ions) and anions (negative ions).

Anions flow towards the oxidation half-cell because the oxidation product is positively charged, and the anions are required to balance the charges within the cell. The opposite is true for cations; they flow towards the reduction half-cell.

10. E is correct.

A redox reaction, or oxidation-reduction reaction, involves the transfer of electrons between two reacting substances. An oxidation reaction refers explicitly to a substance that is losing electrons, and a reduction reaction refers explicitly to a substance that is gaining reactions.

The oxidation and reduction reactions alone are half-reactions because they occur together to form a whole reaction.

The key to this question is that it is about the *transfer* of electrons between *two* species, referring to the whole redox reaction in its entirety, not just one half-reaction.

An electrochemical reaction takes place during the passage of electric current and does involve redox reactions. However, it is not the correct answer to the question posed.

11. A is correct.

The cell reaction is:

$$Co\ (s) + Cu^{2+}\ (aq) \rightarrow Co^{2+}\ (aq) + Cu\ (s)$$

Separate it into half reactions:

$$Cu^{2+}\ (aq) \rightarrow Cu\ (s)$$

$$Co\ (s) \rightarrow Co^{2+}\ (aq)$$

The potentials provided are written in this format:

$$Cu^{2+}\ (aq)\ |\ Cu\ (s)$$

$$+0.34\ V$$

It means that for the reduction reaction:

$$Cu^{2+}\ (aq) \rightarrow Cu\ (s), \text{ the potential is } +0.34\ V.$$

The reverse reaction or oxidation reaction is:

$$Cu\ (s) \rightarrow Cu^{2+}\ (aq) \text{ has opposing potential value: } -0.34\ V.$$

Obtain the potential values for both half-reactions.

Reverse sign for potential values of oxidation reactions:

$$Cu^{2+}\ (aq) \rightarrow Cu\ (s) = +0.34\ V \ \ (reduction)$$

$$Co\ (s) \rightarrow Co^{2+}\ (aq) = +0.28\ V \ \ (oxidation)$$

Determine the standard cell potential:

Standard cell potential = sum of half-reaction potential

Standard cell potential = 0.34 V + 0.28 V

Standard cell potential = 0.62 V

12. E is correct.

The other methods listed involve two or more energy conversions between light and electricity.

It is an electrical device that converts light energy directly into electricity by the photovoltaic effect (i.e., chemical and physical processes).

The operation of a photovoltaic (PV) cell has the following requirements:

      1) Light is absorbed, which excites electrons.

      2) Separation of charge carries opposite types.

      3) The separated charges are transferred to an external circuit.

In contrast, solar panels supply heat by absorbing sunlight.

A photoelectrolytic / photoelectrochemical cell refers either to a type of photovoltaic cell or to a device that splits water into hydrogen and oxygen using only solar illumination.

13. A is correct.

Each half-cell contains an electrode; two half-cells are required to complete a reaction.

14. B is correct.

A: electrolysis can be performed on any metal, not only iron.

C: electrolysis will not boil water nor raise the ship.

D: it is probably unlikely that the gases would stay in the compartments.

E: electrolysis will not reduce the weight enough to float the ship.

15. C is correct.

Balanced reaction:

$$\text{Zn } (s) + \text{CuSO}_4 \ (aq) \rightarrow \text{Cu } (s) + \text{ZnSO}_4 \ (aq)$$

Zn is a stronger reducing agent (more likely to be oxidized) than Cu. This can be determined by each element's standard electrode potential ($E°$).

16. E is correct.

The positive cell potential indicates that the reaction is spontaneous, which means it favors the formation of products.

  Cells that generate electricity spontaneously are considered galvanic cells.

17. B is correct.

A joule is a unit of energy.

18. C is correct.

Electrolysis of aqueous sodium chloride yields hydrogen and chlorine, with aqueous sodium hydroxide remaining in the solution.

Sodium hydroxide is a strong base, which means the solution will be basic.

19. C is correct.

Balanced equation:

$$Ag^+ + e^- \rightarrow Ag \,(s)$$

Formula to calculate deposit mass:

mass of deposit = (atomic mass × current × time) / 96,500 C

mass of deposit = [107.86 g × 3.50 A × (12 min × 60 s/min)] / 96,500 C

mass of deposit = 2.82 g

20. E is correct.

The anode is the electrode where oxidation occurs.

A salt bridge provides electrical contact between the half-cells.

The cathode is the electrode where reduction occurs.

A spontaneous electrochemical cell is called a galvanic cell.

===

Practice Set 2: Questions 21–40

===

21. A is correct.

Ionization energy (IE) is the energy required to release one electron from an element.

Higher IE means the element is more stable.

Elements with high IE usually need one or two more electrons to achieve stable configuration (e.g., complete valence shell – 8 electrons, such as the noble gases, or 2 electrons as in H). It is more likely for these elements to gain an electron and reach stability than lose an electron. When an atom gains electrons, its oxidation number goes down, and it is reduced. Therefore, elements with high IE are easily reduced.

In oxidation-reduction reactions, species that undergo reduction are oxidizing agents because their presence allows the other reactant to be oxidized. Because elements with high IE are easily reduced, they are strong oxidizing agents.

Reducing agents are species that undergo oxidation (i.e., lose electrons).

Elements with high IE do not undergo oxidation readily and therefore are weak reducing agents.

22. D is correct.

Balancing a half-reaction in basic conditions:

The first few steps are identical to balancing reactions in acidic conditions.

Step 1: Balance all atoms except for H and O

$\qquad$ $C_8H_{10} \rightarrow C_8H_4O_4{}^{2-}$ $\qquad$ C is already balanced

Step 2: To balance oxygen, add H_2O to the side with fewer oxygen atoms

$\qquad$ $C_8H_{10} + 4\ H_2O \rightarrow C_8H_4O_4{}^{2-}$

Step 3: To balance hydrogen, add H^+ to the opposing side of H_2O added in the previous step

$\qquad$ $C_8H_{10} + 4\ H_2O \rightarrow C_8H_4O_4{}^{2-} + 14\ H^+$

This next step is the unique additional step for basic conditions.

Step 4: Add equal amounts of OH^- on both sides. The number of OH^- should match the number of H^+ ions. Combine H^+ and OH^- on the same side to form H_2O. If H_2O molecules are on both sides, subtract accordingly to end up with H_2O on one side only.

There are 14 H^+ ions on the right, so add 14 OH^- ions on both sides:

$\qquad$ $C_8H_{10} + 4\ H_2O + 14\ OH^- \rightarrow C_8H_4O_4{}^{2-} + 14\ H^+ + 14\ OH^-$

Combine H^+ and OH^- ions to form H_2O:

$\qquad$ $C_8H_{10} + 4\ H_2O + 14\ OH^- \rightarrow C_8H_4O_4{}^{2-} + 14\ H_2O$

H_2O molecules are on both sides, which cancel, and some H_2O remain on one side:

$$C_8H_{10} + 14 \, OH^- \rightarrow C_8H_4O_4{}^{2-} + 10 \, H_2O$$

Step 5: Balance charges by adding electrons to the side with a higher/more positive total charge

Total charge on the left side: $14(-1) = -14$

Total charge on the right side: -2

Add 12 electrons to the right side:

$$C_8H_{10} + 14 \, OH^- \rightarrow C_8H_4O_4{}^{2-} + 10 \, H_2O + 12 \, e^-$$

23. A is correct.

In electrochemical (i.e., galvanic) cells, oxidation occurs at the anode, and reduction occurs at the cathode.

Therefore, CO_2 is produced at the anode because it is an oxidation product (carbon's oxidation number increases from 0 on the left to +2 on the right).

24. A is correct.

Cu^{2+} is being reduced (i.e., gains electrons), while Sn^{2+} is being oxidized (i.e., loses electrons).

Salt bridge contains both cations (positive ions) and anions (negative ions).

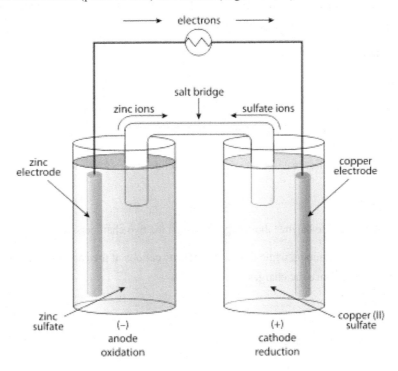

A schematic example of a Zn–Cu galvanic cell

Anions flow towards the oxidation half-cell because the oxidation product is positively charged, and the anions are required to balance the charges within the cell. The opposite is true for cations; they flow towards the reduction half-cell.

25. B is correct.

In cells, reduction occurs at the cathode, while oxidation occurs at the anode.

26. C is correct.

ZnO is being reduced; the oxidation number of Zn decreases from +2 on the left to 0 on the right.

27. E is correct.

Consider the reactions:

$$Ni\ (s) + Ag^+\ (aq) \rightarrow Ag\ (s) + Ni^{2+}\ (aq)$$

Because it is spontaneous, Ni is more likely to be oxidized than Ag.

Ni is oxidized, while Ag is reduced in this reaction.

Arrange the reactions, so the metal that was oxidized in a reaction is reduced in the following reaction:

$$Ni\ (s) + Ag^+\ (aq) \rightarrow Ag\ (s) + Ni^{2+}\ (aq)$$

$$Cd\ (s) + Ni^{2+}\ (aq) \rightarrow Ni\ (s) + Cd^{2+}\ (aq)$$

$$Al\ (s) + Cd^{2+}\ (aq) \rightarrow Cd\ (s) + Al^{3+}\ (aq)$$

Lastly, Ag (s) + H$^+$ (aq) → no reaction should be first in the order because Ag was not oxidized in this reaction, so it should be before a reaction where Ag is reduced.

$$Ag\ (s) + H^+\ (aq) \rightarrow no\ reaction$$

$$Ni\ (s) + Ag^+\ (aq) \rightarrow Ag\ (s) + Ni^{2+}\ (aq)$$

$$Cd\ (s) + Ni^{2+}\ (aq) \rightarrow Ni\ (s) + Cd^{2+}\ (aq)$$

$$Al\ (s) + Cd^{2+}\ (aq) \rightarrow Cd\ (s) + Al^{3+}\ (aq)$$

The metal being oxidized in the last reaction has the highest tendency to be oxidized.

28. A is correct.

The purpose of the salt bridge is to balance the charges between the two chambers/half-cells.

Oxidation creates cations at the anode, while reduction reduces cations at the cathode. The ions in the bridge travel to those chambers to balance the charges.

29. B is correct.

Since $G° = -nFE$, when E° is positive, G is negative.

30. C is correct.

The electrons travel through the wires that connect the cells instead of the salt bridge. The function of the salt bridge is to provide ions to balance charges at the cathode and anode.

31. A is correct.

Electrochemistry is the branch of physical chemistry that studies chemical reactions at the interface of an ionic conductor (i.e., the electrolyte) and an electrode.

Electric charges move between the electrolyte and the electrode through a series of redox reactions, and chemical energy is converted to electrical energy.

32. B is correct.

Light energy from the sun causes the electron to move towards the silicon wafer. This starts the process of electric generation.

33. B is correct.

Oxidation number (or *oxidation state*) indicates the degree of oxidation (i.e., loss of electrons) in an atom.

If an atom is electron-poor, this means that it has lost electrons and would, therefore, have a positive oxidation number.

If an atom is electron-rich, this means that it has gained electrons and would, therefore, have a negative oxidation number.

34. B is correct.

Reduction potential measures a substance's ability to acquire electrons (i.e., undergo reduction). The more positive the reduction potential, the more likely it is that the substance will be reduced. Reduction potential is generally measured in volts.

35. D is correct.

Batteries run down and need to be recharged, while fuel cells do not run down because they can be refueled.

36. A is correct.

Disproportionation reaction is a reaction where a species undergoes oxidation and reduction in the same reaction.

The first step in balancing this reaction is to write the substance undergoing disproportionation twice on the reactant side.

37. A is correct.

The terms spontaneous electrochemical, galvanic, and voltaic are synonymous.

An example of a voltaic cell is an alkaline battery – it generates electricity spontaneously.

Electrons flow from the anode (oxidation half-cell) to the cathode (reduction half-cell).

38. B is correct.

Reaction at the anode:

$$2 \, H_2O \rightarrow O_2 + 4 \, H^+ + 4 \, e^-$$

Oxygen gas is released, and H^+ ions are added to the solution, which causes the solution to become acidic (i.e., lowers the pH).

39. D is correct.

Nonspontaneous electrochemical cells are electrolysis cells.

In electrolysis, the metal being reduced is produced at the cathode.

40. B is correct.

Electrolysis is nonspontaneous because it needs an electric current from an external source to occur.

===

Practice Set 3: Questions 41–60

===

41. C is correct.

Electronegativity indicates the tendency of an atom to attract electrons.

An atom that attracts electrons strongly would be more likely to pull electrons from another atom. As a result, this atom would be a strong oxidizer because it would cause other atoms to be oxidized when the electrons are pulled towards the atom with high electronegativity.

A strong oxidizing agent is a weak reducing agent – they are opposing attributes.

42. E is correct.

MnO_2 is being reduced into Mn_2O_3.

The oxidation number increases from +4 on the left to +3 on the right.

43. C is correct.

Electrolytic cell: needs electrical energy input

Battery: spontaneously produces electrical energy. A dry cell is a type of battery.

Half-cell: does not generate energy by itself.

44. B is correct.

The anode in galvanic cells attracts anions.

Anions in solution flow toward the anode, while cations flow toward the cathode. Oxidation (i.e., loss of electrons) occurs at the anode.

Positive ions are formed, while negative ions are consumed at the anode.

Therefore, negative ions flow toward the anode to equalize the charge.

45. B is correct.

In electrochemical (i.e., galvanic) cells, oxidation occurs at the anode, and reduction occurs at the cathode.

Therefore, Al metal is produced at the cathode because it is a reduction product (from an oxidation number of +3 on the left to 0 on the right).

46. B is correct.

Anions in solution flow toward the anode, while cations flow toward the cathode.

Oxidation is the loss of electrons. Sodium is a group I element and therefore has a single valence electron. During oxidation, the Na becomes Na^+ with a complete octet.

---

47. A is correct.

A salt bridge contains both cations (positive ions) and anions (negative ions).

Anions flow towards the oxidation half-cell because the oxidation product is positively charged, and the anions are required to balance the charges within the cell.

The opposite is true for cations; they flow towards the reduction half-cell.

Anions in the salt bridge should flow from Cd to Zn half-cell.

48. B is correct.

This can be determined using the electrochemical series:

Equilibrium	E°
$Li^+ (aq) + e^- \leftrightarrow Li (s)$	–3.03 volts
$K^+ (aq) + e^- \leftrightarrow K (s)$	–2.92
*$Ca^{2+} (aq) + 2 e^- \leftrightarrow Ca (s)$	–2.87
*$Na^+ (aq) + e^- \leftrightarrow Na (s)$	–2.71
$Mg^{2+} (aq) + 2 e^- \leftrightarrow Mg (s)$	–2.37
$Al^{3+} (aq) + 3 e^- \leftrightarrow Al (s)$	–1.66
$Zn^{2+} (aq) + 2 e^- \leftrightarrow Zn (s)$	–0.76
$Fe^{2+} (aq) + 2 e^- \leftrightarrow Fe (s)$	–0.44
$Pb^{2+} (aq) + 2 e^- \leftrightarrow Pb (s)$	–0.13
$2 H^+ (aq) + 2 e^- \leftrightarrow H_2 (g)$	0.0
$Cu^{2+} (aq) + 2 e^- \leftrightarrow Cu (s)$	+0.34
$Ag^+ (aq) + e^- \leftrightarrow Ag (s)$	+0.80
$Au^{3+} (aq) + 3 e^- \leftrightarrow Au (s)$	+1.50

For a substance to act as an oxidizing agent for another substance, the oxidizing agent must be located below the substance being oxidized in the series.

Cu is being oxidized, which means only Ag^+ or Au^{3+} are capable of oxidizing Cu.

49. E is correct.

Anions in solution flow toward the anode, while cations flow toward the cathode.

Oxidation (i.e., loss of electrons) occurs at the anode.

Positive ions are formed while negative ions are consumed at the anode.

Therefore, negative ions flow toward the anode to equalize the charge.

50. B is correct.

Electrochemistry is the branch of physical chemistry that studies chemical reactions at the interface of an ionic conductor (i.e., the electrolyte) and an electrode.

Electric charges move between the electrolyte and the electrode through a series of redox reactions, and chemical energy is converted to electrical energy.

51. E is correct.

E° tends to be negative and G positive, because electrolytic cells are nonspontaneous.

Electrons must be forced into the system for the reaction to proceed.

52. C is correct.

To create chlorine gas (Cl_2) from chloride ion (Cl^-), the ion needs to be oxidized.

In electrolysis, oxidation occurs at the anode, which is the positive electrode.

The positively-charged electrode would absorb electrons from the ion and transfer them to the cathode for reduction.

53. B is correct.

In electrolytic cells:

The anode is a positively charged electrode where oxidation occurs.

The cathode is a negatively charged electrode where reduction occurs.

54. C is correct.

Nonspontaneous electrochemical cells are electrolysis cells.

In electrolysis, the metal that is being oxidized dissolves at the anode.

55. D is correct.

Dry-cell batteries are the common disposable household batteries. They are based on zinc and manganese dioxide cells.

56. C is correct.

Disproportionation is a type of redox reaction in which a species is simultaneously reduced and oxidized to form two different products.

Unbalanced reaction: $HNO_2 \rightarrow NO + HNO_3$

Balanced reaction: $3\ HNO_2 \rightarrow 2\ NO + HNO_3 + H_2O$

   $2\ H_2O \rightarrow 2\ H_2 + O_2$: decomposition

   $H_2SO_3 \rightarrow H_2O + SO_2$: decomposition

   $Mg + H_2SO_4 \rightarrow MgSO_4 + H_2$: single replacement

57. D is correct.

Electrolysis is the same process in reverse for the chemical process inside a battery.

Electrolysis is often used to separate elements.

58. A is correct.

Fuel cell automobiles are fueled by hydrogen, and the only emission is water (the statement above is reverse of the true statement).

59. A is correct.

Impure copper is oxidized so that it would happen at the anode. Then, pure copper would plate out (i.e., be reduced) on the cathode.

60. B is correct.

Electrolysis is a chemical reaction that results when electrical energy is passed through a liquid electrolyte.

Electrolysis utilizes **direct electric current (DC)** to drive an otherwise non-spontaneous chemical reaction. The voltage needed for electrolysis is called the **decomposition potential**

==

Practice Set 4: Questions 61–80

==

61. D is correct.

Calculate the oxidation numbers of all species involved and identify the oxidized species (increase in oxidation number).

Zn's oxidation number increases from 0 on the left side of reaction I to +2 on the right.

62. C is correct.

The lack of an aqueous solution makes it a dry cell. An example of a dry cell is an alkaline battery.

63. B is correct.

Half-reaction: $C_2H_6O \rightarrow HC_2H_3O_2$

Balancing half-reaction in acidic conditions:

Step 1: Balance all atoms except for H and O

$\qquad C_2H_6O \rightarrow HC_2H_3O_2$ (C is already balanced)

Step 2: To balance oxygen, add H_2O to the side with fewer oxygen atoms

$\qquad C_2H_6O + H_2O \rightarrow HC_2H_3O_2$

Step 3: To balance hydrogen, add H^+ to the opposing side of H_2O added in the previous step

$\qquad C_2H_6O + H_2O \rightarrow HC_2H_3O_2 + 4\,H^+$

Step 4: Balance charges by adding electrons to the side with a higher/more positive total charge

Total charge on the left side: 0

Total charge on the right side: $4(+1) = +4$

Add 4 electrons to the right side:

$\qquad C_2H_6O + H_2O \rightarrow HC_2H_3O_2 + 4\,H^+ + 4\,e^-$

64. B is correct.

In electrochemical (i.e., galvanic) cells, oxidation occurs at the anode, and reduction occurs at the cathode.

Therefore, CCl_4 will be produced at the anode because it is an oxidation product (carbon's oxidation number increases from 0 on the left to +4 on the right).

65. C is correct.

Battery: spontaneously produces electrical energy. A dry cell is a type of battery.

Half-cell: does not generate energy by itself

Electrolytic cell: needs electrical energy input

66. D is correct.

Anions in solution flow toward the anode, while cations flow toward the cathode. Oxidation (i.e., loss of electrons) occurs at the anode.

Positive ions are formed, while negative ions are consumed at the anode.

Therefore, negative ions flow toward the anode to equalize the charge.

67. A is correct.

A reducing agent is a reactant that is oxidized (i.e., loses electrons).

Therefore, the best reducing agent is most easily oxidized.

Reversing each of the half-reactions shows that the oxidation of Cr (*s*) has a potential of +0.75 V, which is greater than the potential (+0.13 V) for the oxidation of Sn^{2+} (*aq*).

Thus, Cr (*s*) is a stronger reducing agent.

68. B is correct.

A salt bridge contains both cations (positive ions) and anions (negative ions).

Anions flow towards the oxidation half-cell because the oxidation product is positively charged, and the anions are required to balance the charges within the cell.

The opposite is true for cations; they flow toward the reduction half-cell.

Anions in the salt bridge should flow from Cd to Zn half-cell.

69. B is correct.

When zinc is added to HCl, the reaction is:

$$Zn + HCl \rightarrow ZnCl_2 + H_2O,$$

which means that Zn is oxidized into Zn^{2+}.

The number provided in the problem is reduction potential, so to obtain the oxidation potential, flip the reaction:

$$Zn \ (s) \rightarrow Zn^{2+} + 2 \ e^-$$

The $E°$ is inverted and becomes +0.76 V.

Because the $E°$ is higher than hydrogen's value (which is set to 0), the reaction occurs.

70. E is correct.

Calculate the mass of metal deposited in cathode:

Step 1: Calculate total charge using current and time

$$Q = \text{current} \times \text{time}$$

$$Q = 1 \text{ A} \times (10 \text{ minutes} \times 60 \text{ s/minute})$$

$$Q = 600 \text{ A·s} = 600 \text{ C}$$

Step 2: Calculate moles of electron that has the same amount of charge

$$\text{moles e}^- = Q / 96{,}500 \text{ C/mol}$$

$$\text{moles e}^- = 600 \text{ C} / 96{,}500 \text{ C/mol}$$

$$\text{moles e}^- = 6.22 \times 10^{-3} \text{ mol}$$

Step 3: Calculate moles of metal deposit

Half-reaction of zinc ion reduction:

$$Zn^{2+} (aq) + 2 \text{ e}^- \rightarrow Zn (s)$$

$$\text{moles of Zn} = (\text{coefficient Zn} / \text{coefficient e}^-) \times \text{moles e}^-$$

$$\text{moles of Zn} = (\tfrac{1}{2}) \times 6.22 \times 10^{-3} \text{ mol}$$

$$\text{moles of Zn} = 3.11 \times 10^{-3} \text{ mol}$$

Step 4: Calculate mass of metal deposit

$$\text{mass Zn} = \text{moles Zn} \times \text{molecular mass of Zn}$$

$$\text{mass Zn} = 3.11 \times 10^{-3} \text{ mol} \times (65 \text{ g/mol})$$

$$\text{mass Zn} = 0.20 \text{ g}$$

71. A is correct.

The oxidation number of Mg increases from 0 to +2, which means that Mg loses electrons (i.e., is oxidized).

Conversely, the oxidation number of Cu decreases, which means that it gains electrons (i.e., is reduced).

72. D is correct.

Electrolysis is a method of using a direct electric current to provide electricity to a nonspontaneous redox process to drive the reaction. The direct electric current must be passed through an ionic substance or solution that contains electrolytes.

Electrolysis is often used to separate elements.

73. D is correct.

Oxidation is the loss of electrons, while reduction is the gain of electrons.

74. D is correct.

E° for the cell is always positive.

75. B is correct.

Galvanic cells are spontaneous and generate electrical energy.

76. D is correct.

In electrolytic cells:

The anode is a positively charged electrode where oxidation occurs.

The cathode is a negatively charged electrode where reduction occurs.

77. B is correct.

Balanced equation:

$$Cu^{2+} + 2\ e^- \rightarrow Cu\ (s)$$

Formula to calculate deposit mass:

mass of deposit = atomic mass × moles of electron

Since Cu has 2 electrons per atom, multiply the moles by 2:

mass of deposit = atomic mass × (2 × moles of electron)

mass of deposit = atomic mass × (2 × current × time) / 96,500 C

4.00 g = 63.55 g × (2 × 2.50 A × time) / 96,500 C

time = 4,880 s

time = (4,880 s × 1 min/60s × 1 hr/60 min) = 1.36 hr

78. C is correct.

Alkaline and dry-cell are common disposable batteries and are non-rechargeable.

Fuel cells require fuel that is going to be consumed and are non-rechargeable.

79. E is correct.

Electrolytic cells do not occur spontaneously; the reaction only occurs with the addition of external electrical energy.

80. D is correct.

A redox reaction, or oxidation-reduction reaction, involves the transfer of electrons between two reacting substances. An oxidation reaction specifically refers to the substance that is losing electrons, and a reduction reaction specifically refers to the substance that is gaining electrons.

The oxidation and reduction reactions alone are half-reactions because they occur together to form a whole reaction.

Therefore, half-reaction can represent either a separate oxidation process or a separate reduction process.

Notes for active learning

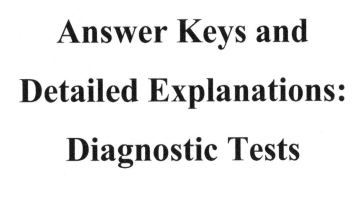

Answer Keys and
Detailed Explanations:
Diagnostic Tests

Diagnostic Test 1 – Detailed Explanations and Answer Key

1	E	Electronic Structure & Periodic Table	31	A	Thermochemistry
2	E	Bonding	32	A	Kinetics Equilibrium
3	A	Phases & Phase Equilibria	33	E	Solution Chemistry
4	B	Stoichiometry	34	D	Acids & Bases
5	E	Thermochemistry	35	D	Electronic Structure & Periodic Table
6	B	Kinetics Equilibrium	36	D	Bonding
7	C	Solution Chemistry	37	E	Phases & Phase Equilibria
8	E	Acids & Bases	38	A	Stoichiometry
9	C	Electrochemistry	39	E	Thermochemistry
10	B	Electronic Structure & Periodic Table	40	A	Kinetics Equilibrium
11	B	Bonding	41	D	Solution Chemistry
12	C	Phases & Phase Equilibria	42	C	Acids & Bases
13	B	Stoichiometry	43	E	Electrochemistry
14	B	Thermochemistry	44	D	Electronic Structure & Periodic Table
15	B	Kinetics Equilibrium	45	D	Bonding
16	A	Solution Chemistry	46	E	Phases & Phase Equilibria
17	A	Acids & Bases	47	A	Stoichiometry
18	E	Electronic Structure & Periodic Table	48	D	Thermochemistry
19	C	Bonding	49	C	Kinetics Equilibrium
20	A	Phases & Phase Equilibria	50	C	Solution Chemistry
21	D	Stoichiometry	51	E	Acids & Bases
22	A	Thermochemistry	52	C	Electronic Structure & Periodic Table
23	E	Kinetics Equilibrium	53	E	Bonding
24	D	Solution Chemistry	54	D	Phases & Phase Equilibria
25	B	Acids & Bases	55	E	Stoichiometry
26	A	Electrochemistry	56	A	Thermochemistry
27	A	Electronic Structure & Periodic Table	57	E	Kinetics Equilibrium
28	E	Bonding	58	E	Solution Chemistry
29	B	Phases & Phase Equilibria	59	C	Acids & Bases
30	E	Stoichiometry	60	E	Electrochemistry

1. E is correct.

Metals are excellent conductors of heat and electricity because the molecules in metal are very closely packed (i.e., have a high density).

Metals are generally solid at room temperature.

Metals are very malleable, meaning that they can be pressed or hammered into different shapes without breaking.

2. E is correct.

The number of valence electrons in a molecule = sum of valence electrons in each atom.

Check the periodic table to determine valence electrons of each element:

   S = 6 electrons; O = 6 electrons; F = 7 electrons

   SOF_2: 6 e$^-$ + 6 e$^-$ + (2 × 7 e$^-$) = 26 electrons

3. A is correct.

Deposition is the thermodynamic process where gas transitions directly into a solid without first becoming a liquid.

An example of deposition occurs in sub-freezing air when water vapor changes directly into ice without becoming a liquid.

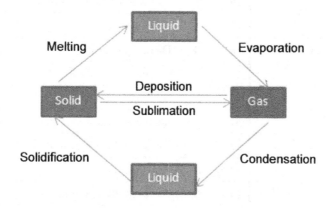

Interconversion of states of matter

4. B is correct.

A spectator ion is an ion that exists as both a reactant and a product in a chemical equation.

5. E is correct.

Heat = mass × specific heat × change in temperature.

$$q = m \times c \times \Delta T$$

Rearrange the equation to solve for mass:

$$m = q / (c \times \Delta T)$$

$$m = 488 \text{ J} / [0.130 \text{ J/g·°C} \times (34.4 \text{ °C} - 21.8 \text{ °C})]$$

$$m = 488 \text{ J} / (0.130 \text{ J/g·°C} \times 12.6 \text{ °C})$$

$$m = 298 \text{ g}$$

6. B is correct.

This is a propagation reaction because the free radical (i.e., unpaired electron) has been transferred.

Initiation: a neutral molecule undergoes homolytic cleavage to generate radicals.

Propagation: a radical species exists on both sides of the reaction.

Termination: radicals combine to generate a neutral molecule.

7. C is correct.

Most ionic compounds, or compounds with a metal cation and nonmetal anion, produce ions when dissolved in water.

Here are some types of ionic compounds that dissolve easily in water:

    - most group IA compounds (e.g., sodium, potassium, etc.)

    - nitrate (NO_3) compounds

Organic compounds (mainly consisting of C, H, O) generally do not produce ions when dissolved in water.

Organic compound formaldehyde (CH_2O) is an aldehyde and does not dissociate into ions.

The resulting anion would have a high pK_a because it would be too unstable.

8. E is correct.

Balanced reaction:

$$H_2SO_4 \ (aq) + 2 \ NaOH \ (aq) \rightarrow Na_2SO_4 \ (aq) + 2 \ H_2O \ (l)$$

In a neutralization reaction, cations and anions from acid and base exchange partners, forming salt and water.

9. C is correct.

In standard notation for electrolysis reactions, the oxidation reaction is written first and reduction reaction second.

Zinc is undergoing oxidation, so it is on the left side. Therefore, Ag is on the right side.

It is essential to note the format of the reactions provided. Both reactions provided are in the form of oxidation reactions.

Therefore, for the reduction of (Ag), reverse the reaction from Ag (s)|Ag$^+$ (aq) to Ag$^+$ (aq)|Ag (s).

10. B is correct.

The n = 3 shell contains subshells 3s, 3p and 3d.

An s shell can contain 2 electrons, a p shell can contain up to 6 electrons, and a d shell can contain up to 10 electrons. 2 + 6 + 10 = 18, so the maximum number of electrons in the n = 3 shell is 18.

11. B is correct.

Carbon is the central atom bonded to three substituents (i.e., two chlorine atoms and a double bond to oxygen).

The hybridization of carbon is sp^2 with a bond angle of 120°. Atoms that are sp^2 hybridized have a trigonal planar geometry.

12. C is correct.

Gases do *not* have definite volumes and shapes; rather, they take up the volume and shape of the container that they are in. This is different from liquids (which have a definite volume but take up the shape of the container that they are in) and solids (which have both a definite volume and a definite shape).

13. B is correct.

An electrolyte is a substance that produces an electrically conducting solution when dissolved in a polar solvent (e.g., water).

The dissolved electrolyte separates into positively-charged cations and negatively-charged anions.

Strong electrolytes dissociate completely (or nearly completely) because the resulting ions are stable in the solution.

14. B is correct.

Enthalpy change is defined by the following equation: $\Delta H = H_f - H_i$

If the standard enthalpy of the products is less than the standard enthalpy of the reactants, the standard enthalpy of reaction is negative, and the reaction is exothermic (i.e., releases heat).

If the standard enthalpy of the products is more than the standard enthalpy of the reactants, the standard enthalpy of reaction is positive, and the reaction is endothermic (i.e., absorbs heat).

In contrast to exothermic reactions, where heat is a product, endothermic reactions absorb heat (as a reactant) from the surroundings.

15. B is correct.

Increasing the temperature of a reaction increases the reaction rate. An increase in the rate constant equals a faster reaction rate.

16. A is correct.

Find concentrations of ions of $Mg(NO_3)_2$ and K_2SO_4 upon complete dissociation:

$$Mg(NO_3)_2 \rightarrow Mg^{2+} + 2\ NO_3^-$$
$$0.04\ \text{mole} \qquad 0.04\ \text{mole} \quad 0.08\ \text{mole}$$

1 liter $Mg(NO_3)_2 \times$ (0.04 mole/liter) = 0.04 moles $Mg(NO_3)_2$

1 liter $K_2SO_4 \times$ (0.08 mole/liter) = 0.24 moles K_2SO_4

Calculate the concentration of Mg^{2+} in the mixture:

$$[Mg^{2+}] = (0.04\ \text{mole} / 4\ \text{liters total volume}) = 0.01\ M$$

Calculate the concentration SO_4^{2+}:

$$[SO_4^{2-}] = (0.24\ \text{mole} / 4\ \text{liters total volume}) = 0.06\ M$$

The ion product is Q:

$$Q = [Mg^{2+}]\cdot[SO_4^{2-}]$$

$$Q = [0.01\ M]\cdot[0.06\ M]$$

$$Q = [1 \times 10^{-2}\ M]\cdot[6 \times 10^{-2}\ M]$$

$$Q = 6 \times 10^{-4}\ M$$

Because $Q > K_{sp}$ ($6 \times 10^{-4} > 4 \times 10^{-5}$), more solid forms to decrease the ion product to the K_{sp} value. Therefore, precipitation occurs.

17. A is correct.

Strong bases are unstable anions that become more stable by abstracting a proton (Arrhenius or Brønsted-Lowry definitions) or sharing electrons (Lewis definition).

The hydroxides of Group I and II metals are considered strong bases. Examples include LiOH (lithium hydroxide), NaOH (sodium hydroxide), KOH (potassium hydroxide), RbOH (rubidium hydroxide), CsOH (cesium hydroxide), $Ca(OH)_2$, calcium hydroxide, $Sr(OH)_2$ strontium hydroxide, and $Ba(OH)_2$ barium hydroxide.

18. E is correct.

Isotopes are variants of a particular element, which differ in the number of neutrons. Isotopes of the element have the same number of protons and occupy the same position on the periodic table.

The number of protons within the atom's nucleus is the atomic number (Z) and is equal to the number of electrons in the neutral (non-ionized) atom. Each atomic number identifies a specific element, but not the isotope; an atom of a given element may have a wide range in its number of neutrons. The number of protons and neutrons (i.e., nucleons) in the nucleus is the atom's mass number (A), and each isotope of an element has a different mass number.

19. C is correct.

HCN is polar due to the electronegativity of the nitrogen atom.

Nitrogen is partial negative, while hydrogen is partial positive.

20. A is correct.

Gay-Lussac's Law:

$$(P_1 / T_1) = (P_2 / T_2)$$

or

$$(P_1 T_2) = (P_2 T_1)$$

According to Gay-Lussac Law:

$$P / T = \text{constant}$$

If the pressure of a gas sample is doubled, the temperature is doubled as well.

21. D is correct.

Balanced equation (synthesis):

$$4\ P\ (s) + 5\ O_2\ (g) \rightarrow 2\ P_2O_5\ (s)$$

22. A is correct.

The enthalpy (i.e., internal energy) of a system cannot be measured directly; the *enthalpy change* is measured.

Enthalpy change is defined by the following equation: $\Delta H = H_f - H_i$

If the standard enthalpy of the products is less than the standard enthalpy of the reactants, the standard enthalpy of reaction is negative, and the reaction is exothermic (i.e., releases heat).

If the standard enthalpy of the products is greater than the standard enthalpy of the reactants, the standard enthalpy of reaction is positive, and the reaction is endothermic (i.e., absorbs heat).

23. E is correct.

If the temperature increases, the reaction shifts toward the endothermic side. Because the reaction is endothermic, the reaction shifts towards the product.

For exothermic reactions, when the temperature is increased, it shifts towards the reactants.

24. D is correct.

Calculate the concentration of each option.

Because solutions have the same solute and solvent, concentration expressed in g/mL can be used to compare concentration.

A: 2.8 g / 2 mL = 1.4 g/mL

B: 2.8 g / 5 mL = 0.56 g/mL

C: 25 g / 50 mL = 0.5 g/mL

D: 40 g / 160 mL = 0.25 g/mL

E: 50 g / 180 mL = 0.28 g/mL

25. B is correct.

$pH = -\log[H^+]$

$pH = -\log(1.4 \times 10^{-3})$

$pH = 2.86$

26. A is correct.

Half-reaction:

$NO_3^- \rightarrow NH_4^+$

Balancing half-reaction in acidic conditions:

Step 1: Balance atoms except for H and O

$NO_3^- \rightarrow NH_4^+$ (N is already balanced)

Step 2: To balance oxygen, add H_2O to the side with fewer oxygen atoms

$NO_3^- \rightarrow NH_4^+ + 3\ H_2O$

Step 3: To balance hydrogen, add H^+ to the opposing side of H_2O added in the previous step

$NO_3^- + 10\ H^+ \rightarrow NH_4^+ + 3\ H_2O$

Step 4: Balance charges by adding electrons to the side with a higher/more positive total charge

Total charge on the left side: $10(+1) - 1 = +9$

Total charge on the right side: $+1$

Add 8 electrons to the left side:

$NO_3^- + 10\ H^+ + 8\ e^- \rightarrow NH_4^+ + 3\ H_2O$

27. A is correct.

Metalloids are semimetallic elements (i.e., between metals and nonmetals). The metalloids are boron (B), silicon (Si), germanium (Ge), arsenic (As), antimony (Sb) and tellurium (Te). Some literature reports polonium (Po) and astatine (At) as metalloids.

They have properties between metals and nonmetals. They typically have a metallic appearance but are only fair conductors of electricity (as opposed to metals which are excellent conductors), which makes them useable in the semiconductor industry. Metalloids tend to be brittle, and chemically they behave more like nonmetals.

28. E is correct.

Hydrogen bonds (H is bonded directly to fluorine, oxygen or a nitrogen atom) are the strongest intermolecular forces (between molecules), followed by dipole-dipole and then van der Waals forces (i.e., London dispersion).

Hydrogens, bonded directly to N, F, and O, participate in hydrogen bonds. The hydrogen is partially positive (i.e., delta plus) due to the bond to these (F, O, N) electronegative atoms. The lone pair of electrons on the F, O or N interacts with the partially positive hydrogen to form a hydrogen bond.

Polar molecules have stronger dipole-dipole force between molecules because the partially positive end of one molecule bonds with the partially negative end of another molecule.

Nonpolar molecules have induced dipole-induced dipole intermolecular force, which is weak compared to dipole-dipole force. The strength of induced dipole-induced dipole interaction correlates with the number of electrons and protons of an atom.

29. B is correct.

Vapor pressure is inversely proportional to the strength of the intermolecular force.

With stronger intermolecular forces, the molecules are more likely to stick together in the liquid form, and less of them participate in the liquid-vapor equilibrium.

30. E is correct.

Assume that H and O are in their most common oxidation states, +1 and –2, respectively.

Because the sum of oxidation numbers in a neutral molecule is zero:

0 = oxidation state of H + oxidation state of Cl + (4 × oxidation state of O)

0 = 1 + oxidation state of Cl + 4(–2)

oxidation state of Cl = +7

31. A is correct.

The overall reaction is the sum of the two individual reactions.

Enthalpy change is defined by the following equation:

$\Delta H = H_f - H_i$

ΔH_{rxn}: (–806 kJ/mlol) + (–86 kJ/mol) = –892 kJ/mol

32. A is correct.

Most nitrate $(NO_3)_2$ ions are soluble in aqueous solutions.

The addition of $Pb(NO_3)_2$ increases the concentration of Pb^{2+} (product), shifting equilibrium towards reactants.

33. E is correct.

Molarity = moles of solute / liter of solution

To obtain molarity: start with the chemical formula. .

Using the formula, the molecular weight of the molecule can be determined using the periodic table.

Once the molecular weight is determined, the mass of the solute is required to calculate the number of moles.

Finally, divide moles by the volume of solution to obtain molarity.

34. D is correct.

Acids are known to have a sour taste (e.g., lemon juice) because the sour taste receptors on the tongue detect the dissolved hydrogen (H^+) ions.

A pH greater than 7 and feels slippery are qualities of bases, not acids. An acid is a chemical substance with a pH of less than 7, producing H^+ ions in water. An acid can be neutralized by a base (i.e., a substance with a pH above 7) to form a salt.

Litmus paper is red under acidic conditions and blue under basic conditions. However, acids are not known to have a slippery feel; this is a characteristic of bases. Bases feel slippery because they dissolve the fatty acids and oils from the skin and reduce the friction between the skin cells.

35. D is correct.

The atomic number (Z) refers to the number of protons and defines the element.

36. D is correct.

With little or no difference in electronegativity between the atoms (i.e., Pauling units < 0.4), the bond is a nonpolar covalent bond; the electrons are shared equally between the two bonding atoms.

Polar covalent bonded compounds are covalently bonded atoms that participate in unequal sharing of electrons due to large electronegativity differences between the atoms (i.e., Pauling units of 0.4 to 1.7).

When the difference in electronegativity is greater than 1.7 Pauling units, the compounds form ionic bonds. Ionic bonds involve transferring an electron from the electropositive element (along the left-hand column/group) to the electronegative element (along the right-hand column/groups) on the periodic table.

In covalent and polar covalent bonds, the electrons are shared so that each atom acquires a noble gas configuration.

Ionic bonds (in dry conditions) are stronger than covalent bonds. In aqueous conditions, ionic bonds are weak, and the compound spontaneously dissociates into ions (e.g., table salt in a glass of water).

The electropositive atom loses an electron(s) to become a cation (with a complete octet), and the electronegative atom gains an electron(s) to become an anion (with a complete octet).

37. E is correct.

The kinetic theory assumes random motion of molecules, elastic collisions, no significant volume occupied by molecules, and little attraction between molecules.

At the same temperature, the average kinetic energy of molecules of different gases is the same, regardless of differences in mass.

38. A is correct.

A single-replacement reaction is a chemical reaction in which one element is substituted for another element in a compound, making a new compound and an element.

A double-replacement reaction occurs when parts of two ionic compounds exchange to make two new compounds.

$BaCl_2 + H_2SO_4 \rightarrow BaSO_4 + 2\ HCl$ is a double-replacement reaction with two products ($BaSO_4$ and 2 HCl).

$F_2 + 2\ NaCl \rightarrow Cl_2 + 2\ NaF$ and $Fe + CuSO_4 \rightarrow Cu + FeSO_4$ are correctly classified as single-replacement reactions because they each produce a compound (2NaF and $FeSO_4$, respectively) and an element (Cl_2 and Cu).

$2\ NO_2 + H_2O_2 \rightarrow 2\ HNO_3$ is correctly classified as a synthesis reaction because two species are combining to form a more complex chemical compound as the product.

39. E is correct.

Lattice energy is the amount of energy released when two gaseous ions combine to create a molecule.

The combined energy of both ions is lower than the sum of the ions' initial energies. The resulting molecule has lower potential energy and is more stable.

40. A is correct.

Reagents that do not affect the rate are 0th order.

41. D is correct.

The *like dissolves like* rule applies when a solvent is miscible with a solute that has similar properties.

A polar solute is miscible with a polar solvent.

An ionic compound (separates into polar electrolytes) is miscible in a polar solvent.

42. C is correct.

If a solution is a good conductor, it has a high concentration of ions. Therefore, it is highly ionized.

Conductivity does not have any correlation with reactivity.

43. E is correct.

The starting materials are NaCl and H_2O. There are 4 ions in the solution: Na^+, H^+, Cl^- and ^-OH.

Analyze the cations and anions separately.

Comparing Na^+ and H^+: according to the electrochemical series, H^+ has a higher reduction potential than Na^+. H is reduced (i.e., gains electrons) to form H_2 while Na^+ remains unchanged.

Comparing Cl^- and ^-OH: according to the electrochemical series, Cl^- is more electronegative than OH^-, so Cl^- is oxidized (i.e., loss of electrons) to form Cl_2, while ^-OH remains unchanged.

44. D is correct.

Electron configuration: $1s^2 2s^2 2p^6 3s^2$

The fastest way to identify an element using the periodic table is by using the element's atomic number.

The atomic number equals the number of protons (and equals the same number for electrons in a neutral atom).

In this problem, the number of electrons can be determined by adding the electrons in the provided electron configuration: $2 + 2 + 6 + 2 = 12$.

Locate element #12 in the periodic table – Mg.

45. D is correct.

The valence shell is the outermost shell of electrons around an atom.

Atoms with a complete valence shell (i.e., containing the maximum number of electrons), such as noble gases, are the most non-reactive elements.

Atoms with only one electron in their valence shells (alkali metals) or those just missing one electron from a complete valence shell (halogens) are the most reactive elements.

46. E is correct.

Start by applying the ideal gas equation:

$$PV = nRT$$

Next, convert the given temperature units from Celsius to Kelvin:

$$T_1 = 24 \ °C + 273 = 297 \ K$$

$$T_2 = 64 \ °C + 273 = 337 \ K$$

Then, convert the given volume units from mL to L:

$$V_1 = 350 \ mL = 0.350 \ L$$

$$V_2 = 300 \ mL = 0.300 \ L$$

Set the initial and final P, V and T conditions equal:

$$(P_1V_1 / T_1) = (P_2V_2 / T_2)$$

Solve for the final pressure:

$$P_2 = (P_1V_1T_2) / (V_2T_1)$$

$$P_2 = [(1.2 \text{ atm}) \times (0.350 \text{ L}) \times (337 \text{ K})] / [(0.300 \text{ L}) \times (297 \text{ K})]$$

Notice that the question does not ask to solve for the final pressure.

47. A is correct.

Balance the overall reaction:

$$6 \text{ HCl } (aq) + 2 \text{ Fe } (s) \rightarrow 2 \text{ FeCl}_3 (aq) + 3 \text{ H}_2 (g)$$

Identify the component that undergoes oxidation (increasing oxidation number).

For this problem, it would be Fe because it went from 0 in elemental iron to Fe^{3+} in $FeCl_3$.

Write the reactions as a half-reaction with its coefficients:

$$2 \text{ Fe} \rightarrow 2 \text{ Fe}^{3+}$$

Add electrons to balance the charges:

$$2 \text{ Fe} \rightarrow 2 \text{ Fe}^{3+} + 6 \text{ e}^-$$

48. D is correct.

Thermal radiation is the emission of electromagnetic waves (i.e., the energy carried via photons) from matter with a temperature (i.e., kinetic energy) greater than absolute zero.

A mirror-like surface reflects radiation, keeping heat inside the thermos.

49. C is correct.

Temperature is the only change that would affect the constant.

Changing pressure, volume or concentration shift the equilibriums, but the equilibrium constant would not change.

A catalyst increases the reaction rate (k) but has no effect on the stability of the products and therefore does not change the equilibrium constant (K_{eq}).

50. C is correct.

For dilution problems:

The concentration (C) × volume (V) of the solutions before mixing = concentration × volume of final solution:

$(CV)_{\text{Soln 1}} + (CV)_{\text{Soln 2}} = (CV)_{\text{Final soln}}$

$(0.03 \text{ M}) \cdot (20 \text{ mL}) + (0.06 \text{ M}) \cdot (15 \text{ mL}) = C(35 \text{ mL})$

$[(0.03 \text{ M}) \cdot (20 \text{ mL}) + (0.06 \text{ M}) \cdot (15 \text{ mL})] / 35 \text{ mL} = C$

$(0.6 \text{ M} \cdot \text{mL}) + (0.9 \text{ M} \cdot \text{mL}) / 35 \text{ mL} = C$

$C = 0.043 \text{ M}$

51. E is correct.

For K_a of an acid: a higher value indicates a stronger acid (i.e., more stable anion).

HF (hydrofluoric acid) has the highest K_a value and is, therefore, the strongest acid.

The order of acid strength (from strongest to weakest) is HF (hydrofluoric acid), HNO_2 (nitrous acid), H_2CO_3 (carbonic acid), HClO (hypochlorous acid), and then HCN (hydrocyanic acid).

52. C is correct.

Charged plates are a pair of plates with a positive charge on one plate and a negative on the other. Particles are shot at high speed between the plates. If a particle is deflected, it travels in a curve towards one of the plates.

A particle is deflected if it has a positive or negative charge and travels towards a plate with the opposite charge.

A hydrogen atom has an equal number of protons and electrons, which means that the overall charge on the atom is zero. Thus, it is not affected by positive or negative charges on the plates.

Alpha particles, protons, and cathode rays are charged particles and are deflected by charged plates.

53. E is correct.

Not all elements can form double or triple covalent bonds because there must be at least two vacancies in an atom's valence electron shell for the double bond formation and at least three vacancies for triple bond formation.

The elements of Group VIIA have seven valence electrons with one vacancy. Therefore, covalent bonds formed by these elements are single covalent bonds.

54. D is correct.

The molecules of an ideal gas exert no attractive forces. Therefore, a real gas behaves most nearly like an ideal gas at high temperature and low pressure. Under these conditions, the molecules are far apart from each other and exert little or no attractive forces on each other.

55. E is correct.

The mass number is an approximation of the atomic weight of the element as amu (or g/mole)

> 14 grams = 1 mole
>
> 28 grams = 2 moles

56. A is correct.

In this problem, heat flows from Au to Ag.

> q released by Au = q captured by Ag

Let Au be sample 1 and Ag be sample 2.

> $(m_1 \times c_1 \times \Delta T_1) = (m_2 \times c_2 \times \Delta T_2)$

Solving for the unknown (mass of Ag or m_2):

> $m_2 = (m_1 \times c_1 \times \Delta T_1) / (c_2 \times \Delta T_2)$
>
> $m_2 = [26 \text{ g} \times 0.130 \text{ J/g·°C} \times (97.2 \text{ °C} - 29.4 \text{ °C})] / [0.240 \text{ J/g·°C} \times (34 \text{ °C} - 21.4 \text{ °C})]$
>
> $m_2 = (26 \text{ g} \times 0.130 \text{ J/g·°C} \times 67.8 \text{ °C}) / (0.240 \text{ J/g·°C} \times 12.6 \text{ °C})$
>
> $m_2 = 75.8 \text{ g}$

57. E is correct.

All aspects of collision affect the rate of chemical reactions.

Another important factor in reaction rates is temperature.

58. E is correct.

> %v/v of alcohol = (volume of alcohol / volume of wine) × 100%
>
> %v/v of alcohol = (25 mL / 225 mL) × 100%
>
> %v/v of alcohol = 11.1%

59. C is correct.

When given a choice of multiple explanations, choose the most descriptive one.

However, check the statement because sometimes the most descriptive option is not accurate.

In this case, the longest option is accurate, so that would be the most specific description.

60. E is correct.

Oxidation of Mg (s):	$Mg\ (s) \rightleftarrows Mg^{2+} + 2\ e^-$	$E° = 2.35$ V
Reduction of Pb (s):	$Pb^{2+} + 2\ e^- \rightleftarrows Pb\ (s)$	$E° = -0.13$ V
Net Reaction:	$Mg\ (s) + Pb^{2+} \rightleftarrows Pb\ (s) + Mg^{2+}$	$E° = 2.22$ V

Note: when a reaction reverses, the sign of E° changes.

Notes for active learning

Notes for active learning

Diagnostic Test 2 – Detailed Explanations and Answer Key

1	A	Electronic Structure & Periodic Table	31	D	Thermochemistry
2	A	Bonding	32	C	Kinetics Equilibrium
3	E	Phases & Phase Equilibria	33	A	Solution Chemistry
4	E	Stoichiometry	34	D	Acids & Bases
5	E	Thermochemistry	35	E	Electronic Structure & Periodic Table
6	B	Kinetics Equilibrium	36	E	Bonding
7	E	Solution Chemistry	37	C	Phases & Phase Equilibria
8	A	Acids & Bases	38	B	Stoichiometry
9	E	Electrochemistry	39	E	Thermochemistry
10	C	Electronic Structure & Periodic Table	40	D	Kinetics Equilibrium
11	D	Bonding	41	B	Solution Chemistry
12	D	Phases & Phase Equilibria	42	B	Acids & Bases
13	D	Stoichiometry	43	B	Electrochemistry
14	B	Thermochemistry	44	A	Electronic Structure & Periodic Table
15	A	Kinetics Equilibrium	45	B	Bonding
16	E	Solution Chemistry	46	D	Phases & Phase Equilibria
17	A	Acids & Bases	47	C	Stoichiometry
18	D	Electronic Structure & Periodic Table	48	C	Thermochemistry
19	E	Bonding	49	C	Kinetics Equilibrium
20	E	Phases & Phase Equilibria	50	E	Solution Chemistry
21	E	Stoichiometry	51	C	Acids & Bases
22	E	Thermochemistry	52	D	Electronic Structure & Periodic Table
23	E	Kinetics Equilibrium	53	B	Bonding
24	A	Solution Chemistry	54	B	Phases & Phase Equilibria
25	E	Acids & Bases	55	D	Stoichiometry
26	B	Electrochemistry	56	B	Thermochemistry
27	A	Electronic Structure & Periodic Table	57	C	Kinetics Equilibrium
28	A	Bonding	58	B	Solution Chemistry
29	C	Phases & Phase Equilibria	59	A	Acids & Bases
30	D	Stoichiometry	60	D	Electrochemistry

1. A is correct.

Neutrons are neutral particles located inside the nucleus of an atom.

1 amu is 1/12 the mass of ^{12}C atoms.

1 amu is approximately the mass of 1 proton.

Neutrons have approximately the same mass as a proton.

2. A is correct.

Covalent bonds are chemical bonds that involve the sharing of electron pairs between atoms. These electrons can be shared equally (for atoms with the same electronegativity) or unequally (atoms with different electronegativity).

With little or no difference in electronegativity between the atoms (Pauling units < 0.4), it is a nonpolar covalent bond, whereby the electrons are shared equally between the two bonding atoms.

Polar covalent bonded atoms are covalently bonded compounds that involve unequal sharing of electrons due to large electronegativity differences between the atoms (Pauling units of 0.4 to 1.7).

However, the correct answer must be "covalent" because the other choices do not involve the sharing of electrons.

Ionic bonds involve the transfer of electrons within the molecule.

Dipole, London, and van der Waals are weak intermolecular (i.e., between molecules) forces.

3. E is correct.

Vapor pressure is proportional to temperature.

Vapor pressure is the pressure exerted by a vapor in thermodynamic equilibrium with its condensed phases (i.e., solid or liquid) at a given temperature in a closed system.

The equilibrium vapor pressure is an indicator of a liquid's evaporation rate.

4. E is correct.

Definition:

mole = weight / molecular mass

Usually, the moles are calculated for each answer choice.

Since all choices have the same weight, compare the molecular masses instead.

A molecule with the largest molecular mass has the lowest number of moles.

From the periodic table:

CH_4 = 16.04 g/mol

Si = 28.09 g/mol

CO = 28.01 g/mol

N_2 = 28.01 g/mol

AlH_3 = 30.01 g/mol

5. E is correct.

6. B is correct.

Endergonic reactions are nonspontaneous, with the energy of the products higher than the energy of the reactants. ΔG is greater than 0.

Exergonic reactions are spontaneous, with the energy of the reactants higher than the energy of the products. ΔG is less than 0.

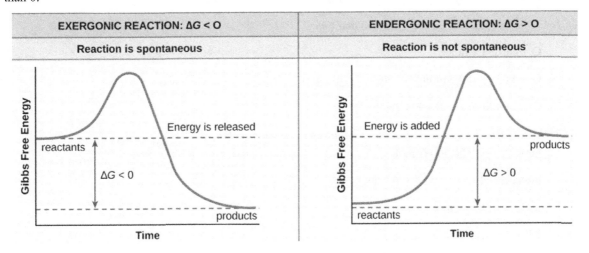

7. E is correct.

A colloid is a solution with particles larger than those of the solution but not large enough to precipitate due to gravity.

Colloidal particles are too small to be extracted by simple filtration.

8. A is correct.

Phenolphthalein is used as an indicator for acid-base titrations. It is a weak acid, which can dissociate protons (H^+ ions) in solution. The phenolphthalein molecule is colorless, and the phenolphthalein ion is pink. It turns colorless in acidic solutions and pink in basic solutions.

With basic conditions, the phenolphthalein (neutral) ⇌ ions (pink) equilibrium shifts to the right, leading to more ionization as H^+ ions are removed.

Bromothymol blue is a pH indicator that is often used for solutions with neutral pH near 7 (e.g., managing the pH of pools and fish tanks). Bromothymol blue acts as a weak acid in a solution that can be protonated or deprotonated. It appears yellow when protonated (lower pH), blue when deprotonated (higher pH), and bluish-green in neutral solution.

Methyl red has a pK_a of 5.1 and is a pH indicator dye that changes color in acidic solutions: it turns red in pH under 4.4, orange in pH 4.4-6.2 and yellow in pH over 6.2.

9. E is correct.

Calculatie the mass of metal deposited in the cathode.

Step 1: Calculate total charge using current and time

$$Q = \text{current} \times \text{time}$$

$$Q = 4.8 \text{ A} \times (50.0 \text{ minutes}) \times (60 \text{ s} / 1 \text{ minute})$$

$$Q = 14,400 \text{ A·s} = 14,400 \text{ C}$$

Step 2: Calculate moles of electrons that have that same amount of charge

$$\text{moles } e^- = Q / 96,500 \text{ C/mol}$$

$$\text{moles } e^- = 14,400 \text{ C} / 96,500 \text{ C/mol}$$

$$\text{moles } e^- = 0.149 \text{ mol}$$

Step 3: Calculate moles of metal deposit

The solution contains chromium (III) sulfate. Half-reaction of chromium (III) ion reduction:

$$Cr^{3+} (aq) + 3e^- \rightarrow Cr (s)$$

Use the moles of electron from the previous calculation to calculate moles of Cr:

$$\text{moles Cr} = (\text{coefficient Cr} / \text{coefficient electron}) \times \text{moles electron}$$

$$\text{moles Cr} = (1 / 3) \times 0.149 \text{ mol}$$

$$\text{moles Cr} = 0.0497 \text{ mol}$$

Step 4: Calculate mass of metal deposit

$$\text{mass Cr} = \text{moles Cr} \times \text{molecular mass of Cr}$$

$$\text{mass Cr} = 0.0497 \text{ mol} \times 52.00 \text{ g/mol}$$

$$\text{mass Cr} = 2.58 \text{ g}$$

10. C is correct.

Atoms in excited states have electrons in lower energy orbitals promoted to higher energy orbitals.

$1s^2 2s^2 2p^6 3s^2 3p^6 4s^1$ is the electron configuration of potassium in the ground state.

$1s^2 2s^2 2p^6 3s^2 3p^6$ has only 18 electrons, not the 19 electrons of potassium.

$1s^2 2s^2 2p^6 3s^2 3p^6 4s^2$ has 20 electrons, not the 19 electrons of potassium.

$1s^2 2s^2 2p^6 3s^2 3p^7$ has 19 electrons, but $3p^7$ exceeds the limit of 6 electrons for the p orbital.

$1s^2 2s^2 2p^6 3s^2 3p^2 4s^6$ has 20 electrons the $4s^6$ exceeds the limit of 2 electrons for the s orbital.

11. D is correct.

The charge on Ca is +2, while the charge on SO_4 is −2.

The proper formula is $CaSO_4$.

12. D is correct.

Boyle's Law:

$$(P_1 V_1) = (P_2 V_2)$$

or

$$P \times V = \text{constant: pressure and volume are inversely proportional.}$$

If the volume of a gas increases, its pressure decreases proportionally.

13. D is correct.

An electrolyte is a substance that produces an electrically conducting solution when dissolved in a polar solvent (e.g., water).

The dissolved electrolyte separates into positively-charged cations and negatively-charged anions.

Strong electrolytes dissociate entirely (or nearly completely) because the resulting ions are stable in the solution.

14. B is correct.

From the equation for Gibbs free energy:

$$\Delta G = \Delta H - T\Delta S$$

With no change in entropy, $T\Delta S = 0$.

Therefore, $\Delta G = \Delta H$

The reaction is spontaneous if ΔG is negative.

The reaction is exothermic (releases heat as a product) if ΔH is negative.

Endothermic reactions are nonspontaneous with heat as a reactant. The products are less stable than the reactants, and ΔG is positive.

Exothermic reactions are spontaneous with heat as a product. The products are more stable than the reactants, and ΔG is negative.

Endergonic refers to a positive ΔG, while exergonic refers to a negative ΔG.

	ΔH < 0	ΔH > 0
ΔS > 0	Spontaneous at all T (ΔG < 0)	Spontaneous at high T (TΔS is large)
ΔS < 0	Spontaneous at low T (TΔS is small)	Nonspontaneous at all T (ΔG > 0)

15. A is correct.

Increasing the concentration of reactants increases the frequency of collision and, therefore, the relative rate of the reaction.

16. E is correct.

Spectator ions appear on both sides of the ionic equation.

Ionic equation of the reaction:

$2 K^+ (aq) + SO_4^{2-} (aq) + Ba^{2+} (aq) + 2 NO_3^- (aq) \rightarrow BaSO_4 (s) + 2 K^+ (aq) + 2 NO_3^- (aq)$

17. A is correct.

Neutralization reactions form water and salt (e.g., $Cr_2(SO_4)_3$).

18. D is correct.

Nonmetals tend to be highly volatile (i.e., easily vaporized), have low density, and are good insulators of heat and electricity. They tend to have high ionization energy and electronegativity and share (or gain) an electron when bonding with other elements.

Seventeen elements are generally classified as nonmetals. Most are gases (hydrogen, helium, nitrogen, oxygen, fluorine, neon, chlorine, argon, krypton, xenon, and radon). One nonmetal is a liquid (bromine), and a few are solids (carbon, phosphorus, sulfur, selenium, and iodine).

19. E is correct.

Electronegativity is a chemical property to describe an atom's tendency to attract electrons to itself. The most common use of electronegativity pertains to polarity along the sigma (single) bond.

The greater the difference in electronegativity between two atoms, the more polar the bonds these atoms form. The atom with the higher electronegativity is the partial (delta) negative end of the dipole.

In the periodic table, electronegativity increases from left to right.

Metals are located towards the left side and nonmetals towards the right.

Therefore, nonmetals should have higher electronegativity values than metals.

20. E is correct.

The sodium is a cation (positive ion), while water has a dipole due to the differences in electronegativity between the electronegative oxygen and the bonded hydrogens.

$$\delta^+ H \overset{\delta^-}{\underset{}{O}} \delta^+ H$$

The dipole in a water molecule

The delta minus on the electronegative oxygen is attracted to the positive sodium ion.

21. E is correct.

The mass of an electron is less than $1/10^{th}$ of one percent of a proton (or neutron) mass.

Neutrons and protons have approximately the same mass.

The element with the greatest sum for protons and neutrons has the greatest mass:

Mass number: 35 protons + 35 neutrons = 70

22. E is correct.

Heat = mass × specific heat × change in temperature:

$$q = m \times c \times \Delta T$$

Rearrange to solve for ΔT:

$$\Delta T = q / (m \times c)$$

$$\Delta T = 340 \text{ J} / (30 \text{ g} \times 4.184 \text{ J/g·°C})$$

$$\Delta T = 2.7 \text{ °C}$$

The problem indicates that heat is removed from H_2O, so temperature decreases:

$$\text{Final T} = \text{initial T} - \Delta T$$

$$\text{Final T} = 19.8 \text{ °C} - 2.7 \text{ °C}$$

$$\text{Final T} = 17.1 \text{ °C}$$

23. E is correct.

Catalysts provide an alternative pathway for the reaction to proceed to product formation. It lowers the energy of activation (i.e., relative energy between reactants and transition state) and, therefore, speeds the reaction rate. Catalysts do not affect the Gibbs free energy (ΔG: stability of products vs. reactants) or the enthalpy (ΔH: bond breaking in reactants or bond making in products).

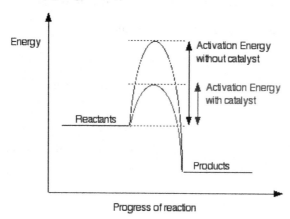

24. A is correct.

Mass volume percent:

(mass of the solute / volume of solution) × 100%

(14.6 g / 260 mL) × 100% = 5.62%

25. E is correct.

In the Arrhenius theory, acids are defined as substances that dissociate in aqueous solutions to donate H^+ (hydrogen ions).

In the Arrhenius theory, bases are defined as substances dissociating in aqueous solutions to donate ^-OH (hydroxide ions).

A strong acid dissociates completely (or nearly so) because the resulting anion is stable (i.e., low pK_a for the acid).

In contrast, a weak acid does not dissociate entirely because the resulting anion is unstable (i.e., high pK_a for the acid).

26. B is correct.

A galvanic (i.e., voltaic or electrochemical) cell involves a spontaneous redox reaction ($\Delta G < 0$) that uses the energy released to generate electricity.

An electrolytic cell utilizes electrical energy from an external source to drive a nonspontaneous redox reaction ($\Delta G > 0$).

Oxidation occurs at the anode, while reduction occurs at the cathode.

In reaction II, each I^- anion loses 1 e^- to produce neutral I_2 during oxidation.

27. A is correct.

The three coordinates that come from Schrodinger's wave equations are the principal (n), angular (l), and magnetic (m) quantum numbers. These quantum numbers describe the size, shape, and orientation in the space of the orbitals on an atom, respectively.

The principal quantum number (n) describes the size of the orbital and the energy of an electron, and the most probable distance of the electron from the nucleus.

The angular momentum quantum number (l) describes the shape of the orbital (i.e., subshells).

The magnetic quantum number (m) determines the number of orbitals and their orientation within a subshell. Consequently, its value depends on the orbital angular momentum quantum number (l). Given a certain l, m is an interval ranging from $-l$ to $+l$ (i.e., it can be zero, a negative integer, or a positive integer).

The s is the spin quantum number (e.g., $+\frac{1}{2}$ or $-\frac{1}{2}$).

28. A is correct.

The valence shell is the outermost shell (i.e., highest principal quantum number n) of an atom.

Valence electrons are those of the outermost electron shell that can participate in a chemical bond.

The number of valence electrons for an element can be determined by its group (i.e., vertical column) on the periodic table. Except for the transition metals (i.e., groups 3-12), the group number identifies how many valence electrons are associated with a particular element: elements of the same group have the same number of valence electrons.

Electron dot structure (or *Lewis dot structure*) is a visual representation of the valence electron configuration.

It is drawn by placing dots that represent valence electrons around a chemical symbol.

29. C is correct.

The ideal gas law:

$$PV = nRT$$

where P is pressure, V is volume, n is the number of molecules, R is the ideal gas constant, and T is the temperature of the gas.

The volume and temperature are directly proportional; if one increases, the other increases as well.

30. D is correct.

$ClCH_2$ total mass:

$$[Cl = 35.5 \text{ g/mol} + C = 12 \text{ g/mol} + (4 \times H = 4 \text{ g/mol})] \approx 50 \text{ g/mol}$$

Two hydrogens have a % by mass of about 2 / 50, which is 4%.

ClC_2H_5 mass:

$$(35.5 \text{ g/mol}) + (2 \times 12 \text{ g/mol}) + (5 \times 1 \text{ g/mol}) = 65 \text{ g/mol}$$

Five hydrogens have a % by mass of about 5 / 65, which is greater than 4%.

Cl_2CH_2 mass:

$(2 \times 35.5 \text{ g/mol}) + 12 \text{ g/mol} + (2 \times 1 \text{ g/mol}) = 85 \text{ g/mol}$

Two hydrogens have a % by mass of 2 / 85 for the H, which is less than 4%.

$ClCH_3$ mass:

$(35.5 \text{ g/mol}) + 12 \text{ g/mol} + 3 \times 1 \text{ g/mol}) = 50 \text{ g/mol}$

Three hydrogens have a % mass of 3 / 50 for the H, which is about 6%.

CCl_4: zero hydrogens have a % mass of 0%.

31. D is correct.

The enthalpy (i.e., internal energy) of a system cannot be measured directly; the *enthalpy change* is measured instead.

Enthalpy change is defined by the following equation:

$\Delta H = H_f - H_i$

If the standard enthalpy of the products is less than the standard enthalpy of the reactants, the standard enthalpy of reaction is negative, and the reaction is exothermic (i.e., releases heat).

If the standard enthalpy of the products is more than the standard enthalpy of the reactants, the standard enthalpy of reaction is positive, and the reaction is endothermic (i.e., absorbs heat).

If the bonds formed are stronger than bonds broken, that would be an exothermic reaction because some energy is released to the surroundings.

32. C is correct.

Chemical equilibrium refers to a dynamic process whereby the *rate* at which a reactant molecule is transformed into a product is the same as the *rate* for a product molecule to be transformed into a reactant.

33. A is correct.

The addition of dissolved F^- (by adding NaF) decreases the solubility of CaF_2.

The common ion effect is responsible for reducing the solubility of an ionic precipitate when a soluble compound containing one of the precipitate ions is added to the solution in equilibrium with the precipitate.

If the concentration of any of the ions is increased, some of the ions in excess combine with the oppositely charged ions and are effectively removed from the solution (consistent with Le Châtelier's principle). Then, some of the salt is precipitated until the ion product is equal to the solubility product.

34. D is correct.

The production of H⁺ ions in water is characteristic of acid, not a base.

A base is a chemical substance with a pH greater than 7 and produces hydroxide (OH⁻) ions in water.

A base is neutralized by an acid (i.e., a substance with a pH below 7) to form a salt.

Bases are known to have a slippery or soapy feel because they dissolve the fatty acids and oils from the skin, reducing the amount of friction.

They have a bitter taste; for example, coffee is bitter because it contains caffeine, a base.

35. E is correct.

The alkaline earth metals (group IIA) include beryllium (Be), magnesium (Mg), calcium (Ca), strontium (Sr), barium (Ba), and radium (Ra).

Representative elements from the periodic table are groups IA & IIA (on the left) and groups IIIA – VIIIA (on the right).

The transition elements have characteristics that are not found in other elements, which results from the partially filled d shell. These include the formation of compounds whose color is due to d–d electronic transitions and compounds in many oxidation states due to the relatively low reactivity of unpaired d electrons.

The alkali metals (group IA) include lithium (Li), potassium (K), sodium (Na), rubidium (Rb), cesium (Cs), and francium (Fr).

Alkali metals lose one electron to become +1 cations, and the resulting ion has a complete octet of valence electrons. The alkali metals have low electronegativity and react violently with water (e.g., the violent reaction of sodium metal with water).

On the periodic table, the lanthanides and the actinides are shown as two additional rows below the main body of the table.

36. E is correct.

The Pauli exclusion principle states that two identical electrons cannot have the same quantum numbers. It is not possible for two electrons to have the same values of the four quantum numbers: the principal quantum number (n), the angular momentum quantum number (ℓ), the magnetic quantum number (m_ℓ), and the spin quantum number (m_s).

For two electrons residing in the same orbital, n, ℓ, and m_ℓ are the same, m_s must be different, and the electrons have opposite half-integer spins, $+\frac{1}{2}$ and $-\frac{1}{2}$.

37. C is correct.

Hydrogens, bonded directly to F, O or N, participate in hydrogen bonds. The hydrogen is partially positive (i.e., $\partial+$) due to the bond to these electronegative atoms. The lone pair of electrons on the F, O or N interacts with the partial positive ($\partial+$) hydrogen to form a hydrogen bond.

38. B is correct.

Balanced equation (double replacement):

Spark

$$2\ C_2H_6\ (g) + 7\ O_2\ (g) \rightarrow 4\ CO_2\ (g) + 6\ H_2O\ (g)$$

39. E is correct.

If V, R, and T are constants, then the ideal gas law,

$PV = nRT$ implies that P is proportional to n.

Therefore, increasing n causes an increase in P.

40. D is correct.

The reaction is exothermic, which means that heat is a product. Lowering the temperature causes more products to form.

Summary of temperature effects:

Increasing the temperature of a system in dynamic equilibrium favors the

endothermic reaction because the system counteracts the change by

absorbing the extra heat.

Decreasing the temperature of a system in dynamic equilibrium favors

the exothermic reaction, because the system counteracts the change by

producing more heat.

41. B is correct.

% (m/m) means that solute and solvent are measured by their mass.

If the solution contains 14% (m/m), it means that the rest of it is water, or:

100% − 14% = 86%

Use this percentage to determine the mass:

mass of water = 86% × 160 grams

mass of water = 138 g

42. B is correct.

First, determine the reaction product by writing down the complete equation.

When an acid reacts with a base, it forms salt and water.

Exchange the cations and anions from acid and base to form the new salt:

$$NaOH + H_2SO_4 \rightarrow Na_2SO_4 + H_2O$$

Check that a reaction is balanced before calculating. That equation was not balanced, so balance it first:

$$2\ NaOH + H_2SO_4 \rightarrow Na_2SO_4 + 2\ H_2O$$

The residue mentioned in the problem must be Na_2SO_4 residue. Calculate moles of Na_2SO_4:

moles Na_2SO_4 = mass Na_2SO_4 / molecular mass Na_2SO_4

moles Na_2SO_4 = (840 mg × 0.001 g/mg) / [(2 × 22.99 g/mol) + 32.06 g/mol + (4 × 16 g/mol)]

moles Na_2SO_4 = (840 mg × 0.001 g/mg) / (45.98 g/mol) + 32.06 g/mol + (64 g/mol)]

moles Na_2SO_4 = (840 mg × 0.001 g/mg) / (142.04 g/mol)

moles Na_2SO_4 = 5.9×10^{-3} mole

Because coefficient Na_2SO_4 = coefficient H_2SO_4, moles Na_2SO_4 = moles H_2SO_4 = 6.06×10^{-3} mole.

Use this information to determine the molarity:

Molarity of H_2SO_4 = moles of H_2SO_4 / volume of H_2SO_4

Molarity of H_2SO_4 = 5.9×10^{-3} mole / (40 mL × 0.001 L / mL)

Molarity of H_2SO_4 = 0.148 M

43. B is correct.

Half-reaction: $C_2H_6O \rightarrow HC_2H_3O_2$

Balancing half-reaction in acidic conditions:

Step 1: Balance all atoms except for H and O

$C_2H_6O \rightarrow HC_2H_3O_2$ (C is already balanced)

Step 2: To balance oxygen, add H_2O to the side with fewer oxygen atoms

$C_2H_6O + H_2O \rightarrow HC_2H_3O_2$

Step 3: To balance hydrogen, add H^+ to the opposing side of H_2O added in the previous step

$C_2H_6O + H_2O \rightarrow HC_2H_3O_2 + 4\ H^+$

Step 4: Balance charges by adding electrons to the side with a higher/more positive total charge

Total charge on the left side: 0

Total charge on the right side: 4(+1) = +4

Add 4 electrons to the right side:

$C_2H_6O + H_2O \rightarrow HC_2H_3O_2 + 4\ H^+ + 4\ e^-$

44. A is correct.

Isotopes are variants of a particular element, which differ in the number of neutrons. Isotopes of the element have the same number of protons and occupy the same position on the periodic table.

The number of protons within the atom's nucleus is the atomic number (Z) and is equal to the number of electrons in the neutral (non-ionized) atom. Each atomic number identifies a specific element, but not the isotope; an atom of a given element may have a wide range in its number of neutrons.

The number of protons and neutrons (i.e., nucleons) in the nucleus is the atom's mass number (A), and each isotope of an element has a different mass number.

The ^1H is the most common hydrogen isotope with an abundance of more than 99.98%. The ^1H is protium, and its nucleus consists of a single proton.

The ^2H isotope is deuterium, the ^3H tritium.

Protium	Deuterium	Tritium
1 proton	1 proton	1 proton
	1 neutron	2 neutrons

Three isotopes of hydrogen

45. B is correct.

Resonance spreads the double-bond character equally among three S–O bonds.

Resonance structures of the sulfite ion

46. D is correct.

The boiling point depends on the intermolecular bonding of the molecules. The options are nonpolar halogen molecules. The dominant intermolecular force is dispersion force (or induced dipole).

As the number of electron shells increases, the valence electrons are further from the nucleus, and these molecules would be more prone to the temporary induced dipole. Therefore, larger molecules have stronger bonds and higher boiling points.

47. C is correct.

The oxidizing agent is the substance that causes the other reactant(s) to be oxidized.

The oxidizing agent also undergoes reduction.

In this reaction, HCl is the oxidizing agent because it is reduced to H_2 and oxidizes Co.

48. C is correct.

Convection is defined as the movement of heat through liquids and gases.

The lid prevents the fluid (e.g., air) from escaping the vessel.

49. C is correct.

General formula for the equilibrium constant of a reaction:

$$aA + bB \leftrightarrow cC + dD$$

$$K_{eq} = ([C]^c \times [D]^d) / ([A]^a \times [B]^b)$$

For equilibrium constant calculation, only include species in aqueous or gas phases (in this case, is all of them):

$$K_{eq} = [NO_2]^2 / [N_2O_4]$$

Substitute in the values for each concentration:

$$K_{eq} = (0.400)^2 / 0.800$$

$$K_{eq} = 0.200$$

50. E is correct.

Most hydroxide salts are only slightly soluble.

In general, hydroxide salts of Group I elements are soluble.

Hydroxide salts of Group II elements (Ca, Sr and Ba) are slightly soluble.

Hydroxide salts of transition metals and Al^{3+} are insoluble (e.g., $Fe(OH)_3$, $Al(OH)_3$, $Co(OH)_2$).

Salts containing Cl^-, Br^-, I^- are generally soluble. Exceptions to this rule are halide salts of Ag^+, Pb^{2+}, and $(Hg_2)^{2+}$. For example, $AgCl$, $PbBr_2$, and Hg_2Cl_2 are insoluble.

51. C is correct.

Use this formula for pH:

$$pH = -\log[H^+]$$

$$pH = -\log[0.000001]$$

$$pH = 6$$

52. D is correct.

Electrons cannot be precisely located at any point in time, and orbitals describe probability regions for finding the electrons.

The one s subshell is spherically symmetrical, the three p orbitals have dumbbell shapes, and the five d orbitals have four lobes.

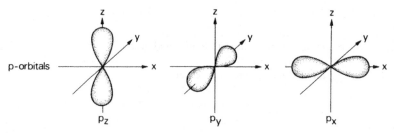

Representation of the three p orbitals

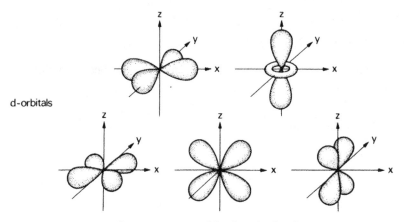

Representation of the five d orbitals

53. B is correct.

The trigonal pyramidal molecular geometry results from three bonds (i.e., three substituents) and one lone pair on the central atom in the molecule.

Examples of trigonal pyramidal molecules (shown below) include ammonia (NH_3), xenon trioxide (XeO_3), the chlorate ion (ClO_3^-), and the sulfite ion (SO_3^{2-}).

ammonia   xenon trioxide   chlorate ion

Three resonance structures for the sulfite ion

In ammonia, the trigonal pyramidal molecule undergoes rapid inversion of the lone pair of electrons on the central nitrogen atom.

Molecules with tetrahedral electron pair geometries (and four substituents) have sp^3 hybridization at the central atom and do not undergo inversion.

54. B is correct.

The mixture is 40% CO_2. The partial pressure of CO_2:

    0.4×700 torrs $= 280$ torrs

55. D is correct.

Start by calculating the molecular mass (MW) of glucose:

(6 × atomic mass of C) + (12 × atomic mass of H) + (6 × atomic mass of O)

(6 × 12.01 g/mole) + (12 × 1.01 g/mole) + (6 × 16.00 g/mole)

MW glucose = 180.18 g/mole

Then multiply the number of moles by the molecular mass:

moles of glucose × MW of glucose

8.50 moles × 180.18 g/mole

Mass of glucose = 1,532 g

56. B is correct.

A spontaneous process releases free energy (usually as heat) and moves to a lower, more thermodynamically stable energy state.

A release of free energy from the system corresponds to a negative change in free energy but a positive change for the surroundings.

For a reaction at STP, ΔG is the Gibbs free energy:

$\Delta G = \Delta H - T\Delta S$

The sign of ΔG depends on the signs of the changes in enthalpy (ΔH) and entropy (ΔS), and the absolute temperature (T, in Kelvin).

ΔG changes from positive to negative (or vice versa) when $T = \Delta H / \Delta S$.

For heterogeneous systems, where the reaction species are in different phases and can be mechanically separated, the following is true.

When ΔG is negative, a process or chemical proceeds spontaneously in the forward direction.

When ΔG is positive, the process proceeds spontaneously in reverse.

When ΔG is zero, the process is already in equilibrium, with no net change.

Consider the signs of ΔH and ΔS.

When ΔS is positive, and ΔH is negative, the process is spontaneous.

When ΔS is positive and ΔH is positive, the relative magnitudes of ΔS and ΔH determine if the reaction is spontaneous. High temperatures make the reaction more favorable.

When ΔS is negative and ΔH is negative, the relative magnitudes of ΔS and ΔH determine if the reaction is spontaneous. Low temperatures make the reaction more favorable.

When ΔS is negative and ΔH is positive, a process is not spontaneous at any temperature, but the reverse process is spontaneous.

57. C is correct.

Increasing the concentration of reactants (or increasing pressure) increases the rate of a reaction because there is an increased probability that any two reactant molecules collide with sufficient energy to overcome the energy of activation and to form products.

58. B is correct.

First, determine the moles of NaCl:

Moles of NaCl = mass of NaCl / molecular mass of NaCl

Moles of NaCl = 30 g / (58.45 g/mol)

Moles of NaCl = 0.513 mole

Then, use this information to calculate the molarity:

Molarity = moles/volume

Molarity = 0.513 mole / 0.675 L

Molarity 0.76 M

59. A is correct.

Salt is a combination of an acid and a base.

The pH is determined by the acid and base that created the salt.

Combination of:

weak acid and strong base: salt is basic

strong acid and weak base: salt is acidic

strong acid and strong base: salt is neutral

weak acid and weak base: could be anything (acidic, basic, or neutral)

In this problem, NaF is a salt created from a strong base (NaOH) and a weak acid (HF), which means NaF is basic.

Examples of strong acids and bases that commonly appear in chemistry problems:

Strong acids:

HCl (hydrochloric acid) HNO_3 (nitric acid)

H_2SO_4 (sulfuric acid)   HBr (hydrobromic acid)

HI (hydroiodic acid) $HClO_3$ (chloric acid)

$HClO_4$ (perchloric acid)

Strong bases:

LiOH (lithium hydroxide)   NaOH (sodium hydroxide)

KOH (potassium hydroxide) $Ca(OH)_2$ (calcium hydroxide)

RbOH (rubidium hydroxide) $Sr(OH)_2$, (strontium hydroxide)

CsOH (cesium hydroxide) $Ba(OH)_2$ (barium hydroxide)

60. D is correct.

CCl_4 is an oxidation product because the oxidation number of C increases from 0 to +4.

Oxidations occur at the anode.

Notes for active learning

Diagnostic Test 3 – Detailed Explanations and Answer Key

1	D	Electronic Structure & Periodic Table	31	D	Thermochemistry
2	C	Bonding	32	E	Kinetics Equilibrium
3	A	Phases & Phase Equilibria	33	E	Solution Chemistry
4	B	Stoichiometry	34	B	Acids & Bases
5	A	Thermochemistry	35	B	Electronic Structure & Periodic Table
6	B	Kinetics Equilibrium	36	A	Bonding
7	B	Solution Chemistry	37	A	Phases & Phase Equilibria
8	C	Acids & Bases	38	B	Stoichiometry
9	A	Electrochemistry	39	B	Thermochemistry
10	A	Electronic Structure & Periodic Table	40	B	Kinetics Equilibrium
11	A	Bonding	41	A	Solution Chemistry
12	D	Phases & Phase Equilibria	42	D	Acids & Bases
13	A	Stoichiometry	43	A	Electrochemistry
14	A	Thermochemistry	44	E	Electronic Structure & Periodic Table
15	B	Kinetics Equilibrium	45	B	Bonding
16	E	Solution Chemistry	46	D	Phases & Phase Equilibria
17	C	Acids & Bases	47	B	Stoichiometry
18	A	Electronic Structure & Periodic Table	48	D	Thermochemistry
19	E	Bonding	49	B	Kinetics Equilibrium
20	E	Phases & Phase Equilibria	50	E	Solution Chemistry
21	D	Stoichiometry	51	D	Acids & Bases
22	C	Thermochemistry	52	A	Electronic Structure & Periodic Table
23	E	Kinetics Equilibrium	53	A	Bonding
24	B	Solution Chemistry	54	D	Phases & Phase Equilibria
25	E	Acids & Bases	55	B	Stoichiometry
26	D	Electrochemistry	56	D	Thermochemistry
27	E	Electronic Structure & Periodic Table	57	D	Kinetics Equilibrium
28	A	Bonding	58	A	Solution Chemistry
29	D	Phases & Phase Equilibria	59	E	Acids & Bases
30	D	Stoichiometry	60	A	Electrochemistry

1. D is correct.

A proton has a charge of +1 and a mass of 1 atomic mass unit (amu).

A neutron has no charge and a mass of 1 atomic mass unit (amu).

An electron has a charge of –1 and effectively no mass.

2. C is correct.

Sulfur has 6 valence electrons. Similar to oxygen (6 valence electrons), neutral sulfur makes 2 covalent bonds.

Like for oxygen, a single bond to S results in a formal charge on sulfur of –1, while three bonds to sulfur produce a +1 charge.

3. A is correct.

Pressure is directly proportional to the number of moles.

Determine the number of moles for O_2 and H_2

(15 g O_2) × (1 mole / 32 grams) = 0.47 moles O_2

(15 g H_2) × (1 mole / 2 grams) = 7.5 moles H_2

Fifteen grams of O_2 has substantially fewer moles than fifteen grams of H_2 because the molecular weight of O_2 is much greater.

Therefore, the pressure of the H_2 is greater than the pressure for the O_2.

4. B is correct.

One mole of H_2SO_4 contains 32.07 g of S

One mole of H_2SO_4 contains (4 × 16 g/mol) = 64 g of O

Therefore, the % mass of O is twice the % mass of S.

Calculate the molecular mass of H_2SO_4:

(2 × atomic mass of H) + atomic mass of S + (4 × atomic mass of O)

(2 × 1.01 g/mole) + 32.07 g/mole + (4 x 16.00 g/mole)

Molecular mass of H_2SO_4 = 98.09 g/mole

To obtain the percent mass composition of each element, calculate the mass of each element, divide it by the molecular mass and multiply by 100%:

Mass % of hydrogen = [(2 × 1.01 g/mole) / 98.09 g/mole] × 100%

Mass % of hydrogen = 2.1%

Mass % of sulfur = (32.07 g/mole) / (98.09 g/mole) × 100%

Mass % of sulfur = 32.7%

Mass % of oxygen = (4 × 16.00 g/mole) / (98.09 g/mole) × 100%

Mass % of oxygen = 65.3%

5. A is correct.

Gibbs free energy:

$$\Delta G = \Delta H - T\Delta S$$

T is always positive.

When ΔH is negative and ΔS is positive, then ΔG is negative.

6. B is correct.

From the rate law given:

[X] is raised to the first power, while [Y] is squared.

Tripling [Y] has a greater effect on the rate.

7. B is correct.

Calculate moles of KOH required:

moles of KOH = molarity × volume

moles of KOH = 0.65 M × (26.0 mL × 0.001 L/mL)

moles of KOH = 0.0169 moles

Then, use the number of moles to calculate mass:

mass of KOH = moles of KOH × molar mass of KOH

mass of KOH = 0.0169 × (39.1 g/mol + 16 g/mol + 1.01 g/mol)

mass of KOH = 0.0169 × (53.11 g/mol)

mass of KOH = 0.898 g

8. C is correct.

Use the equation:

$$C_1V_1 = C_2V_2$$

where C_1 = original concentration of the solution (before dilution)

C_2 = final concentration of the solution, after dilution

V_1 = volume about to be diluted

V_2 = final volume after dilution

Phosphoric acid is triprotic, but 0.1 N indicates that there are 0.1 moles of acidic protons.

$$V_1 = C_2V_2 / C_1$$

$$(V_1) = [(0.1 \text{ M}) \cdot (40 \text{ ml})] / (0.045 \text{ M})$$

$$V_1 = 88.9 \text{ ml}$$

When working with acids and bases, normality is sometimes used instead of molarity.

Normality indicates the concentration of H^+ ions in acids and ^-OH ions in bases.

Some acids and bases have multiple H^+ or ^-OH ions per molecule.

For example, H_2SO_4 has 2 hydrogen ions, which means that a 1 M solution of H_2SO_4 has 2 M of hydrogen ions. When expressed in normality, the concentration of this solution is 2 N.

The problem states that there are 40 ml of 0.1 NH_3PO_4, which means $[H^+]$ = 0.1 M.

Moles of H^+ = volume of acid × normality

Moles of H^+ = (40 ml × 0.001 L / mL) × 0.1 N

Moles of H^+ = 0.004 mole

A titration is complete when H^+ ions completely reacted with ^-OH ions.

Therefore, the amount of ^-OH required to titrate the acid is also 0.004 mole.

Because NaOH is a strong base,

$[NaOH] = [^-OH]$ = 0.045 M

Calculate volume of NaOH required:

volume of NaOH = moles of NaOH / molarity of NaOH

volume of NaOH = 0.004 mole / 0.045 M

volume of NaOH = 0.089 L

Convert to milliliters:

0.089 L × 1,000 mL / L = 88.9 mL

9. A is correct.

Galvanic cells are spontaneous electrochemical cells.

The two half-cells occur in different chambers and are connected through a salt bridge.

A salt bridge is not just a pathway for ions to travel, but it contains ions that would travel to either half-cell deficient in ions.

10. A is correct.

If n = 4, the subshell must be either a 4*s*, 4*p*, 4*d* or 4*f*

If l = 2, the l values (a nonnegative integer) has a range from 0 to 2 (i.e., 0, 1, 2).

The third subshell of the 4th shell is the 4*d* subshell.

If l = 1, the second subshell is 4*p*

If l = 0, the first subshell is 4*s*

11. A is correct.

A polyatomic (i.e., molecular) ion is a charged chemical species composed of two or more atoms covalently bonded (or of a metal complex) acting as a single unit.

An example is the hydroxide ion; one oxygen and one hydrogen atom; hydroxide has a charge of -1.

Oxidation state	-1	$+1$	$+3$	$+5$	$+7$
Anion name	chloride	hypochlorite	chlorite	chlorate	perchlorate
Formula	Cl^-	ClO^-	ClO_2^-	ClO_3^-	ClO_4^-

A polyatomic ion does bond with other ions.

A polyatomic ion has various charges.

A polyatomic ion might contain only metals or nonmetals.

A polyatomic ion is not neutral.

12. D is correct.

Hydrogen bonds are the strongest intermolecular forces (i.e., between molecules), followed by dipole-dipole, dipole-induced dipole, and van der Waals forces (i.e., London dispersion).

Hydrogens, bonded directly to N, F and O, participate in hydrogen bonds. The hydrogen is partial positive (i.e., delta plus) due to the bond to these (F, O, N) electronegative atoms. The lone pair of electrons on the F, O or N interacts with the partially positive hydrogen to form a hydrogen bond.

In CH_3CH_2OH and H_2O, both molecules exhibit hydrogen bonding.

Other bonds affecting the molecules:

H_2S : dipole-dipole

CH_4 : dispersion (i.e., London dispersion or Van der Waals)

NH_3 : hydrogen bonding

CH_3CH_3 : dispersion

CCl_4 : dispersion

CH_3OH : hydrogen bonding

13. A is correct.

Break down the molecules into their constituent ions:

$$Ca^{2+} + 2\ ^-OH + 2\ H^+ + 2\ Cl^- \rightarrow Ca^{2+} + 2\ Cl^- + 2\ H_2O$$

Remove species that appear on both sides of the reaction: $2\ ^-OH + 2\ H^+ \rightarrow 2\ H_2O$

Divide by 2 to simplify the equation:

$$^-OH + H^+ \rightarrow H_2O$$

14. A is correct.

Enthalpy is designated by the letter "H."

It consists of the internal energy of the system (U) plus the product of pressure (P) and volume (V) of the system:

$$H = U + PV$$

Since U, P, and V are functions of the state of the system, enthalpy is a thermodynamic state function (i.e., it does not depend on the path by which the system arrived at its present state).

The symbol for a change of entropy (i.e., disorder) is ΔS.

15. B is correct.

The equilibrium constant of a chemical reaction is the value of the reaction quotient when the reaction has reached equilibrium.

Pure solids and pure liquids are never part of the equilibrium and are not included in the equilibrium constant expression.

16. E is correct.

Ionic product constant of water:

$$K_w = [H_3O^+] \cdot [^-OH]$$

Rearrange equation to solve for [OH^-]:

$$[^-OH] = K_w / [H_3O^+]$$

$$[^-OH] = [1 \times 10^{-14}] / [1 \times 10^{-4}]$$

$$[^-OH] = 1 \times 10^{-10}$$

17. C is correct.

Calculate moles of $NaHCO_3$:

moles of $NaHCO_3$ = mass of $NaHCO_3$ / molecular mass of $NaHCO_3$

moles of $NaHCO_3$ = 0.90 g / 84.01 g/mol

moles of $NaHCO_3$ = 0.011 mol

Using the coefficients in the reaction equation, calculate moles of H_2SO_4:

moles H_2SO_4 = (coefficient H_2SO_4 / coefficient $NaHCO_3$) × moles $NaHCO_3$

moles H_2SO_4 = (½) × 0.011 mol

moles H_2SO_4 = 0.0055 mol

Divide moles by volume to determine molarity:

Molarity of H_2SO_4 = moles of H_2SO_4 / volume of H_2SO_4

Molarity of H_2SO_4 = 0.0055 mol / (40 mL × 0.001 L/mL)

Molarity of H_2SO_4 = 0.138 M

18. A is correct.

The *s* subshell has 1 orbital and can accommodate 2 electrons.

The *p* subshell has 3 orbitals and can accommodate 6 electrons.

The *d* subshell has 5 orbitals and can accommodate 10 electrons.

The *f* subshell has 7 orbitals and can accommodate 14 electrons.

19. E is correct.

Carbon has 4 valence electrons. The molecular geometry of carbon with 4 substituents (e.g., CH_4) is tetrahedral. The removal of 1 of the 4 groups surrounding the central atom leaves the central carbon with three groups and an unhybridized *p* orbital.

This unhybridized *p* orbital is vacant for a carbocation, contains a single electron for a radical (·CH) and a pair of electrons for carbanions ($^-CH_3$).

20. E is correct.

Balance the chemical equation: $N_2 + 3 H_2 \rightarrow 2 NH_3$

Using the balanced coefficients from the written equation, apply dimensional analysis to solve for the volume of H_2 needed to react to completion with 14.5 L N_2:

V_{H2} = V_{N2} × (mole H_2 / mole N_2)

V_{H2} = (14.5 L N_2) × (3 moles H_2 / 1 mole N_2)

V_{H2} = 43.5 L H_2

21. D is correct.

The strongest oxidizing agent is the substance that causes the other reactant(s) to be oxidized.

The oxidizing agent undergoes reduction (i.e., gains electrons).

In this reaction, Sn^{2+} is the oxidizing agent because it is reduced to Sn and oxidizes Mg.

22. C is correct.

Endothermic reactions are nonspontaneous with heat as a reactant. The products are less stable than the reactants, and ΔG is positive.

Exothermic reactions are spontaneous with heat as a product. The products are more stable than the reactants, and ΔG is negative.

23. E is correct.

Compare reactions 1 and 3 to find the rate for H_2 (NO is constant):

   Doubling the $[H_2]$ increases the rate by a factor of 4.

   Since $2^2 = 4$, the reaction is second order for H_2.

Compare reactions 1 and 2 to find the rate for NO (H_2 is constant):

   Doubling the $[NO]$ doubles the rate.

   Since $2^1 = 2$, the reaction is first order for NO.

Therefore, the rate = $k[H_2]^2 \times [NO]$.

24. B is correct.

Electrolytes dissociate into ions in an aqueous solution and conduct electricity when current is applied to the solution.

Sucrose is a disaccharide with the molecular formula of $C_{12}H_{22}O_{11}$. It does not dissociate into ions; only individual sucrose molecules separate.

25. E is correct.

The hydronium ion is H_3O^+.

   $pH = -\log[H_3O^+]$

   $pH = -\log 3.82 \times 10^{-9}$

   $pH = 8.42$

26. D is correct.

The redox reaction for a spontaneous electrochemical cell (galvanic) has the two half-cells occurring in different chambers connected through a salt bridge.

A spontaneous electrochemical (galvanic) cell produces electrical energy.

$Cu^{2+} + 2 e^- \rightarrow Cu$ (reduction)

$Sn \rightarrow Sn^{2+} + 2 e^-$ (oxidation)

Oxidation occurs at the anode, while reduction occurs at the cathode.

Sn loses electrons (i.e., oxidation) and Cu^{2+} gains (i.e., reduction) electrons.

27. E is correct.

Neutrons, like protons, are nucleons. They are particles that make up the nucleus of an atom.

Neutrons are closely associated with protons, held together by the nuclear force.

Neutrons and protons have a mass of about 1 amu and are much more massive than electrons.

Neutrons are neutral and thus have a charge of 0. This makes them more difficult to detect than protons (which have a +1 charge) and electrons (which have a –1 charge).

28. A is correct.

Atomic nuclei are positively charged, while electrons are negatively charged.

A chemical bond is an attraction between two atoms, and it is caused by the electrostatic force of attraction between the oppositely charged nuclei and electrons.

29. D is correct.

Charles' Law:

$(V_1 / T_1) = (V_2 / T_2)$

In Charles' Law and gas-related laws, T is expressed in K.

Convert the initial temperature to K:

46 °C + 273 = 319 K

Solve for the final temperature:

$T_2 = (3.8 \text{ L}) \times [(319 \text{ K}) / (4.8 \text{ L})]$

$T_2 = 273 \text{ K}$

Convert the temperature to Celsius:

252.5 K – 273 = –20.5 °C

---

30. D is correct.

Calculate moles of cobalt:

Moles of cobalt = mass of cobalt sample / atomic mass of cobalt

Moles of cobalt = 58.9 g / 58.9 g/mol

Moles of cobalt = 1.00 mole

Then, use this value and Avogadro's number to calculate the number of atoms:

Number of atoms = number of moles × Avogadro's number

Number of atoms = (1.00 mole) × (6.02 × 10^{23} atoms/mole)

Number of atoms = 6.02 × 10^{23} atoms

31. D is correct.

The change in enthalpy (ΔH) is defined as the quantity of heat absorbed (i.e., energy change) when a reaction occurs at constant pressure.

ΔE refers to the energy change independent of pressure.

ΔG refers to the change in free energy.

ΔS refers to the change in entropy or randomness of a system.

ΔP refers to the change in pressure.

32. E is correct.

The final amount of NOBr is known (0.54 mol), which means the amount of NO and Br_2 used can be calculated by comparing coefficients:

NO = (2 / 2) × 0.54 mol = 0.54 mol NO

Br_2 = (1 / 2) × 0.54 mol = 0.27 mol Br_2

Calculate final concentration of the reactants:

NO = 0.68 mol – 0.54 mol = 0.14 mol NO

Br_2 = 0.42 mol – 0.27 mol = 0.15 mol Br_2

Final composition: 0.14 mol NO and 0.15 mol Br_2

33. E is correct.

Solubility is proportional to pressure. Use simple proportions to calculate solubility at a different pressure:

($Solubility_2$ / $Solubility_1$) = ($Pressure_2$ / $Pressure_1$)

$Solubility_2$ = ($Pressure_2$ / $Pressure_1$) × $Solubility_1$

$Solubility_2$ = (10.0 atm / 1.0 atm) × 1.60 g/L = 16.0 g/L

34. B is correct.

A buffer is an aqueous solution that consists of a weak acid and its conjugate base, or vice versa.

Buffered solutions resist changes in pH and are often used to keep the pH at a nearly constant value in many chemical applications. They do this by readily absorbing or releasing protons (H^+) and ^-OH.

When an acid is added to the solution, the buffer releases ^-OH and accepts H^+ ions from the acid.

When a base is added, the buffer accepts ^-OH ions from the base and releases protons (H^+).

35. B is correct.

Determine the atomic number of Ar from the periodic table.

Ar is located in Group VIII/18 and its atomic number is 18.

The atomic mass of $^{40}Ar = 40$ g/mol.

To determine the number of neutrons, subtract the atomic number from the atomic mass:

$40 - 18 = 22$ neutrons

36. A is correct.

Covalent bonds involve the sharing of electron pairs between atoms.

Covalent single bonds involve the sharing of one electron pair, while covalent double bonds involve the sharing of two electron pairs. Double bonds are stronger and shorter than single bonds.

37. A is correct.

Gay-Lussac's Law:

$(P_1 / T_1) = (P_2 / T_2)$

or

$(P_1 T_2) = (P_2 T_1)$

According to Gay-Lussac Law:

$P / T = $ constant

If the temperature of a gas sample is increased, the pressure is increased.

38. B is correct.

Balanced equation (double replacement):

$2\ C_8H_{18}\ (g) + 25\ O_2\ (g) \rightarrow 16\ CO_2\ (g) + 18\ H_2O\ (g)$

39. B is correct.

A closed system can exchange energy (as heat or work) but not matter or temperature with its surroundings.

An isolated system cannot exchange any heat, work, matter, or temperature with the surroundings.

An open system can exchange energy, matter, and temperature with the surroundings.

It is not an isolated system because work is being done on the system, but it is a closed system because no matter moves in or out.

40. B is correct.

Increased pressure or increased concentration of reactants increases the probability that the reactants collide with sufficient energy and orientation to overcome the energy of activation barrier and proceed toward products.

This increases the reaction rate, but no change in the equilibrium, K_{eq} of the reaction as the reactants and products have the same relative energies.

41. A is correct.

First, calculate the number of moles:

Moles of sucrose = mass of sucrose / molar mass of sucrose

Moles of sucrose = 12.0 g / 342.0 g/mol

Moles of sucrose = 0.0351 mol

Then, divide moles by volume to calculate molarity:

Molarity of sucrose = moles of sucrose / volume of sucrose

Molarity = 0.0351 mol / (100.0 mL × 0.001 L/mL)

Molarity = 0.351 M

42. D is correct.

Calculate the moles of KOH (convert the volume to liters):

moles KOH = molarity × volume

moles KOH = 0.600 M × 30.0 mL × 0.001 L/mL

moles KOH = 0.018 mol

The coefficients of KOH and HNO_3 are equal:

moles KOH = moles HNO_3.

Calculate the volume of HNO_3 required:

volume HNO_3 = moles HNO_3 / molarity HNO_3

volume HNO_3 = 0.018 mol / 0.350 M

volume HNO_3 = 0.05 L

Convert to liters: 0.05 L × 1,000 mL/L = 50 mL

43. A is correct.

Reduction is the gain of electrons.

Oxidation is the loss of electrons.

In electrochemical (galvanic) cells, like in an electrolytic cell, reduction occurs on the cathode side, while oxidation occurs at the anode.

44. E is correct.

Alkali metals (group IA) include lithium (Li), potassium (K), sodium (Na), rubidium (Rb), cesium (Cs) and francium (Fr).

Alkali metals lose one electron to become +1 cations, and the resulting ion has a complete octet of valence electrons.

The alkali metals have low electronegativity and react violently with water (e.g., highly exothermic reaction of sodium metal with water).

45. B is correct.

There are six alkaline earth metals in group IIA of the periodic table: beryllium (Be), magnesium (Mg), calcium (Ca), strontium (Sr), barium (Ba), and radium (Ra).

They have similar properties of being shiny, silvery-white, and relatively reactive metals at standard temperature and pressure.

In group IIA, they contain 2 electrons in their *s* shell. They form cations with a charge of +2 and have a complete octet (like the noble gases). Their oxidation state (oxidation number) is +2.

46. D is correct.

Condensation refers to a change of state from a gas to a liquid.

For example, condensation occurs when atmospheric water vapor in clouds condenses to form droplets of liquid water that become heavy enough to fall due to gravity. This results in rain, an essential part of the Earth's water cycle.

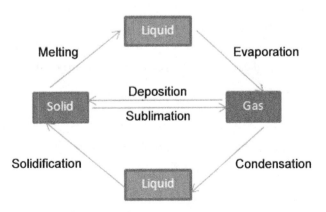

Interconversion of states of matter

47. B is correct.

> Moles of Mg = (mass of Mg) / (atomic mass of Mg)
>
> Moles of Mg = (4.80 g) / (24.31 g/mole)
>
> Moles of Mg = 0.20 mole

48. D is correct.

Specific heat is the heat required to increase the temperature of *1 gram* of sample by 1 °C.

Heat capacity is the amount of heat required to increase the temperature of *the whole sample* by 1 °C.

Each phase of water has a different specific heat, so those values are required for liquid and steam.

Because water changes phase from liquid to vapor at 100 °C, the heat of vaporization is required to calculate the total heat energy.

49. B is correct.

The rate depends on the specific reaction and is not related to the order of the reaction.

50. E is correct.

As a general rule, most NH_4, Cl, OH compounds are soluble in water, while sulfide (S) compounds are less likely to be soluble.

51. D is correct.

The molecule that produces the most stable anion dissociates completely.

Anions are stabilized by resonance (delocalization of negative charge among several atoms).

Resonance forms of the NO_3^- anion after ionization in H_2O

52. A is correct.

Atomic radius generally decreases to the right across a row in the periodic table because of the increased effective nuclear charge for atoms of the same principal quantum number (i.e., shell size).

Therefore, comparing the 4th-period elements of calcium and gallium, gallium has a smaller radius.

53. A is correct.

A higher boiling point indicates stronger intermolecular forces.

Molecules with a higher dipole moment have stronger intermolecular bonds compared to non-polar molecules. X has a higher boiling point than Y, which means X is most likely more polar than Y.
Next, consider the fact that X and Y are not miscible. This most likely means that one of them is polar and the other one is nonpolar.

This fact supports the hypothesis that X is more polar than Y: if X has strong polar characteristics, it will be the more polar component, and it is not miscible with the nonpolar Y.

54. D is correct.

Boiling point (BP) elevation:

$$\Delta BP = iKm$$

where i = number of particles produced when molecule dissociates, K = constant (depends on substance), m = molality (moles / kg solvent)

molality = [(80 g) / (180 g/mole)] / 0.5 kg

molality = 0.89 m

BP elev = (1)·(0.52)·(0.89 m)

BP elev = 0.46 K

The BP of H_2O = 373 K

With solute:

H_2O = 373 K + 0.46 K = 373.46 K

55. B is correct.

Since atoms cannot be created or destroyed in chemical reactions, there must be an equal number of atoms on each side of the reaction arrow.

The number of atoms present in a reaction can never vary, even if conditions change in the reaction.

However, the number of products and reactants does not have to be the same. For example, two reactants (on the left side of a reaction arrow) can combine to form one product (on the right side of a reaction arrow).

Similarly, the number of molecules on each side of the reaction arrow does not have to be equal.

56. D is correct.

Both reactions are identical, except for the direction of the reaction.
Therefore, the enthalpy of both reactions has equal value with opposing signs.
Because the reaction with the given enthalpy value has a negative sign (–1,274 kJ/mol), the enthalpy of the reverse reaction is +1,274 kJ/mol.

57. D is correct.

The general formula for the equilibrium constant of a reaction:

$aA + bB \rightarrow cC + dD$

$K_{eq} = ([C]^c \times [D]^d) / ([A]^a \times [B]^b)$

$K_{sp} = [Fe^{3+}]^2 \times [CrO_4^{2-}]^3$

The powers in the given solubility product expression indicate the number of ions present (2 for Fe, 3 for CrO_4). The problem indicates ion charges.

58. A is correct.

Calculate the number of moles:

Moles of Na_2CO_3 = mass Na_2CO_3 / moles of Na_2CO_3

Moles of Na_2CO_3 = 0.130 g / 106.0 g/mol

Moles of Na_2CO_3 = 1.23×10^{-3} mol

Check the reaction to obtain coefficients of HCl and Na_2CO_3:

Moles of HCl required = (coefficient of HCl / coefficient of Na_2CO_3) × moles Na_2CO_3

Moles of HCl required = (2/1) × 1.23×10^{-3} mol

Moles of HCl required = 2.46×10^{-3} mol

Then, divide moles by molarity to find the volume:

volume of HCl = moles of HCl / molarity of HCl

volume of HCl = 2.46×10^{-3} mol / 0.125 M

volume of HCl = 0.0197 L

Convert to milliliters:

0.0197 L × (1,000 mL/L) = 19.7 mL

59. E is correct.

With a neutralization reaction, the cations and anions in reactants switch pairs, resulting in salt and water.

60. A is correct.

Br_2 is gaining electrons and undergoes reduction:

$Br_2 \rightarrow 2\ Br^-$

Balance the charges by adding electrons:

$Br_2 + 2\ e^- \rightarrow 2\ Br^-$

According to the balanced reaction, Br_2 gained 2 electrons (i.e., reduction) to form Br^- ions.

Notes for active learning

Notes for active learning

Appendix

Periodic Table of the Elements

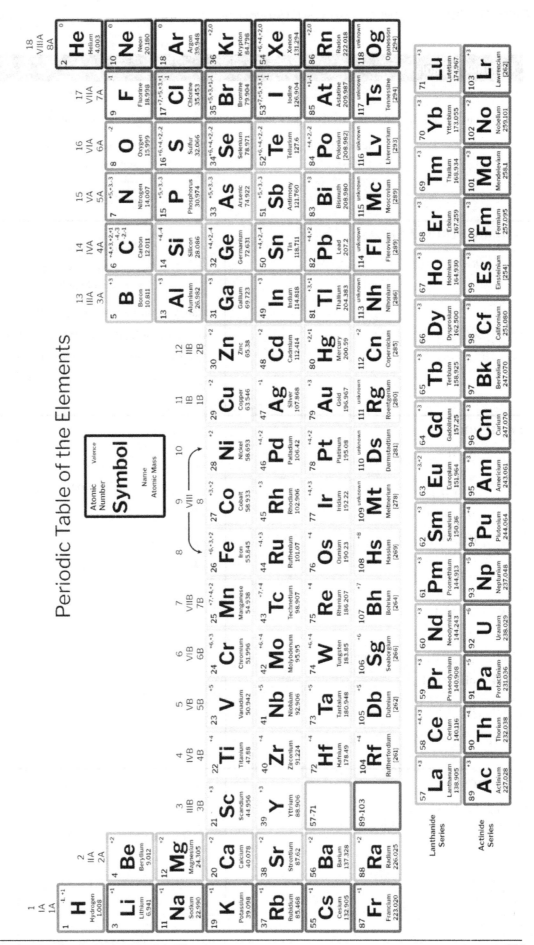

Common Chemistry Equations

Throughout the test the following symbols have the definitions specified unless otherwise noted.

L, mL	=	liter(s), milliliter(s)	mm Hg	=	millimeters of mercury
g	=	gram(s)	J, kJ	=	joule(s), kilojoule(s)
nm	=	nanometer(s)	V	=	volt(s)
atm	=	atmosphere(s)	mol	=	mole(s)

ATOMIC STRUCTURE

$E = h\nu$

$c = \lambda\nu$

E = energy

ν = frequency

λ = wavelength

Planck's constant, $h = 6.626 \times 10^{-34}$ J s

Speed of light, $c = 2.998 \times 10^{8}$ m s^{-1}

Avogadro's number $= 6.022 \times 10^{23}$ mol^{-1}

Electron charge, $e = -1.602 \times 10^{-19}$ coulomb

EQUILIBRIUM

$K_c = \dfrac{[C]^c[D]^d}{[A]^a[B]^b}$, where $a\,A + b\,B \rightleftarrows c\,C + d\,D$

$K_p = \dfrac{(P_C)^c(P_D)^d}{(P_A)^a(P_B)^b}$

$K_a = \dfrac{[H^+][A^-]}{[HA]}$

$K_b = \dfrac{[OH^-][HB^+]}{[B]}$

$K_w = [H^+][OH^-] = 1.0 \times 10^{-14}$ at 25°C

$\quad = K_a \times K_b$

$pH = -\log[H^+]$, $pOH = -\log[OH^-]$

$14 = pH + pOH$

$pH = pK_a + \log\dfrac{[A^-]}{[HA]}$

$pK_a = -\log K_a$, $pK_b = -\log K_b$

Equilibrium Constants

K_c (molar concentrations)

K_p (gas pressures)

K_a (weak acid)

K_b (weak base)

K_w (water)

KINETICS

$\ln[A]_t - \ln[A]_0 = -kt$

$\dfrac{1}{[A]_t} - \dfrac{1}{[A]_0} = kt$

$t_{1/2} = \dfrac{0.693}{k}$

k = rate constant

t = time

$t_{1/2}$ = half-life

GASES, LIQUIDS, AND SOLUTIONS

$$PV = nRT$$

$$P_A = P_{total} \times X_A, \text{ where } X_A = \frac{\text{moles A}}{\text{total moles}}$$

$$P_{total} = P_A + P_B + P_C + \dots$$

$$n = \frac{m}{M}$$

$$K = {}^{\circ}C + 273$$

$$D = \frac{m}{V}$$

$$KE \text{ per molecule} = \frac{1}{2}mv^2$$

Molarity, M = moles of solute per liter of solution

$$A = abc$$

P = pressure
V = volume
T = temperature
n = number of moles
m = mass
M = molar mass
D = density
KE = kinetic energy
v = velocity
A = absorbance
a = molar absorptivity
b = path length
c = concentration

Gas constant, R = 8.314 J mol^{-1} K^{-1}

= 0.08206 L atm mol^{-1} K^{-1}

= 62.36 L torr mol^{-1} K^{-1}

1 atm = 760 mm Hg

= 760 torr

STP = 0.00°C and 10^5 Pa

THERMOCHEMISTRY/ ELECTROCHEMISTRY

$$q = mc\Delta T$$

$$\Delta S^{\circ} = \sum S^{\circ} \text{ products} - \sum S^{\circ} \text{ reactants}$$

$$\Delta H^{\circ} = \sum \Delta H_f^{\circ} \text{ products} - \sum \Delta H_f^{\circ} \text{ reactants}$$

$$\Delta G^{\circ} = \sum \Delta G_f^{\circ} \text{ products} - \sum \Delta G_f^{\circ} \text{ reactants}$$

$$\Delta G^{\circ} = \Delta H^{\circ} - T\Delta S^{\circ}$$

$$= -RT \ln K$$

$$= -nFE^{\circ}$$

$$I = \frac{q}{t}$$

q = heat
m = mass
c = specific heat capacity
T = temperature
S° = standard entropy
H° = standard enthalpy
G° = standard free energy
n = number of moles
E° = standard reduction potential
I = current (amperes)
q = charge (coulombs)
t = time (seconds)

Faraday's constant, F = 96,485 coulombs per mole of electrons

$$1 \text{ volt} = \frac{1 \text{ joule}}{1 \text{ coulomb}}$$

Notes for active learning

Solubility Rules for Ionic Compounds

Use the following rules to *predict the solubility of ionic compounds* in water. These rules generally relate to the compound's lattice and hydration energies of the individual ions.

One of the factors determining the *lattice energy* is the magnitudes of charges on the ions (i.e., the greater the magnitude of the charges, the greater the lattice energy). Simplified, the greater the lattice energy, the lower the solubility.

The rules focus on the magnitudes of the charges and ignore other influences on the lattice energies, such as hydration energy. In addition, the rules disregard the effect of temperature.

The solubility rules are listed in order of *decreasing importance*. For example, if the first rule applies, disregard the subsequent rules. Similarly, if the second rule is applicable, disregard the third rule. Thus, only use the third rule when rules one and two are not applicable.

The following ions are nearly always soluble: $C_2H_3O_2^-$, ClO_3^-, ClO_4^-. K^+, Na^+, NH_4^+, and NO_3^-.

General principle: Hydrides (H^-) decompose in water to yield H_2 and ^-OH. Many metal oxides react with water to produce hydroxides (^-OH).

Rule 1: compounds containing a +1 or –1 ion are typically soluble

Examples include: $AlCl_3$, $FeCl_2$, K_2SO_4, $NaBr$, Rb_3PO_4

Exceptions (Insoluble): OH^- (other than with cations that produce strong bases)

Hg_2^{2+} (other than with NO_3^-, $C_2H_3O_2^-$, ClO_3^-, and ClO_4^-)

Pb^{2+}, Hg^{2+} with most –1 ions

IB metals (column 11)

Rule 2: compounds containing a +3 or –3 or higher ion are typically NOT soluble

Examples include: $AlPO_4$, $Ca_3(PO_4)_2$, $TiO2$

Exceptions (Soluble): cations combined with sulfate or dichromate

Rule 3: compounds containing –2 ion are typically NOT soluble

Examples include:

$CaCO_3$, FeS, ZnC_2O_4

Exceptions (Soluble):

Dichromates and sulfates (insoluble if combined with Ba, Ca, Hg, Pb, and Sr)

There are exceptions to the solubility rules besides those listed. However, the ions normally encountered follow these rules.

Solubility examples for ionic compounds

AgBr	Insoluble	Rule 1	Ag^+
AgCl	Insoluble	Rule 1	$Ag^+ + Cl^-$
$AgIO_3$	Insoluble	Rule 1	$Ag^+ + IO_3^-$
AgOH	Insoluble	Rule 1	weak base
Ag_2S	Insoluble	Rule 1	Ag^+
Hg_2Cl_2	Insoluble	Rule 1	Hg_2^{2+}
$PbCl_2$	Insoluble	Rule 1	Pb^{2+}
$AgNO_3$	Soluble	Rule 1	$NO_3^- + Ag^+$
$Ba(NO_3)_2$	Soluble	Rule 1	NO_3^-
$Ba(OH)_2$	Soluble	Rule 1	strong base
NaCl	Soluble	Rule 1	$Na^+ + Cl^-$
$NaClO_3$	Soluble	Rule 1	$Na^+ + ClO_3^-$
NaOH	Soluble	Rule 1	strong base
Na_3PO_4	Soluble	Rule 1	Na^+
$(NH_4)_2CO_3$	Soluble	Rule 1	NH_4^+
NH_4IO_3	Soluble	Rule 1	$NH_4^+ + IO_3^-$
$FeAsO_4$	Insoluble	Rule 2	$Fe^{3+} + AsO_4^{3-}$
$Fe_2(SO_4)_3$	Soluble	Rule 2	SO_4^{2-}
$BaCO_3$	Insoluble	Rule 3	CO_3^{2-}
$BaSO_4$	Insoluble	Rule 3	$Ba^{2-} + SO_4^{2-}$
$MgSO_4$	Insoluble	Rule 3	SO_4^{2-}
$PbCrO_4$	Insoluble	Rule 3	CrO_4^{2-}
ZnS	Insoluble	Rule 3	S^{2-}
$Ba(IO_3)_2$	Insoluble	exception	

Glossary of Chemistry Terms

A

Absolute entropy (of a substance) – the increase in the entropy of a substance as it goes from a perfectly ordered crystalline form at 0 K (where its entropy is zero) to the temperature in question.

Absolute zero – the zero point on the absolute temperature scale; –273.15 °C or 0 K; theoretically, the temperature at which molecular motion ceases (i.e., the system does not emit or absorb energy, atoms at rest).

Absorption spectrum – spectrum associated with absorption of electromagnetic radiation by atoms (or other species), resulting from transitions from lower to higher energy states.

Accuracy – how close a value is to the actual value; see *precision*.

Acid – a substance that produces H^+ (*aq*) ions in an aqueous solution and gives a pH of less than 7.0; strong acids ionize entirely or almost entirely in dilute aqueous solution; weak acids ionize only slightly. It turns litmus red.

Acid dissociation constant – an equilibrium constant for dissociating a weak acid.

Acid rain – rainwater with a pH of less than 5.7; caused by the gases NO_2 (vehicle exhaust fumes) and SO_2 (from burning fossil fuels) dissolving in the rain. It kills fish, wildlife, and trees and destroys buildings and lakes.

Acidic salt – contains an ionizable hydrogen atom; does not necessarily produce acidic solutions.

Actinides – the fifteen chemical elements that are between actinium (89) and lawrencium (103).

Activated complex – a structure forming because of a collision between molecules while new bonds form.

Activation energy – the amount of energy that reactants must absorb in their ground states to reach the transition state needed for a reaction can occur.

Active metal – a metal with low ionization energy that loses electrons readily to form cations.

Activity (of a component of ideal mixture) – a dimensionless quantity whose magnitude is equal to the molar concentration in an ideal solution; equal to partial pressure in an ideal gas mixture; 1 for pure solids or liquids.

Activity series – a listing of metals (and hydrogen) in order of decreasing activity.

Actual yield – the amount of a specified pure product obtained from a given reaction; see *theoretical yield*.

Addition reaction – a reaction in which two atoms or groups of atoms are added to a molecule, one on each side of a double or triple bond.

Adhesive forces – forces of attraction between a liquid and another surface.

Adsorption – the adhesion of a species onto the surfaces of particles.

Aeration – the mixing of air into a liquid or a solid.

Alcohol – hydrocarbon derivative containing a ~OH group attached to a carbon atom, not in an aromatic ring.

Alkali metals – the elements of Group IA on the periodic table (e.g., Na, K, Rb).

Alkaline battery – a dry cell in which the electrolyte contains KOH.

Alkaline earth metals – group IIA metals on the periodic table; see *earth metals*.

Allomer – a substance that has a different composition than another but the same crystalline structure.

Allotropes – elements with different structures (therefore different forms), such as carbon (e.g., diamonds, graphite, and fullerene).

Allotropic modifications (allotropes) – different forms of the same element in the same physical state.

Alloy – a mixture of metals. For example, bronze is an alloy formed from copper and tin.

Alloying – mixing metal with other substances (usually other metals) to modify its properties.

Alpha (α) particle – a helium nucleus; helium ion with 2+ charge; an assembly of two protons and two neutrons.

Amorphous solid – a non-crystalline solid with no well-defined ordered structure.

Ampere – unit of electrical current; one ampere equals one coulomb per second.

Amphiprotism – the ability of a substance to exhibit amphiprotic by accepting donated protons.

Amphoterism – the ability to react with both acids and bases; to act as either an acid or a base.

Amplitude – the maximum distance that medium particles carrying the wave move from their rest position.

Anion – a negative ion; an atom or group of atoms that has gained one or more electrons.

Anode – in a cathode ray tube, the positive electrode (electrode at which oxidation occurs); the positive side of a dry cell battery or a cell.

Antibonding orbital – a molecular orbital higher in energy than any of the atomic orbitals from which it is derived; lends instability to a molecule or ion when populated with electrons; denoted with star (*) superscript.

Artificial transmutation – an artificially induced nuclear reaction caused by the bombardment of a nucleus with subatomic particles or small nuclei.

Associated ions – short-lived species formed by the collision of dissolved ions of opposite charges.

Atmosphere – a unit of pressure; the pressure supports a column of mercury 760 mm high at 0 °C.

Atom – a chemical element in its smallest form; it comprises neutrons and protons within the nucleus and electrons circling the nucleus; it is the smallest part of an element that can exist.

Atomic mass unit (amu) – one-twelfth of the mass of an atom of the carbon-12 isotope; used for stating atomic and formula weights; known as a dalton.

Atomic number – represents an element corresponding with the number of protons within the nucleus; the number of protons in the nucleus of the atom.

Atomic orbital (*AO*) – a region or volume in space where the probability of finding electrons is highest.

Atomic radius – radius of an atom.

Atomic weight – weighted average of the masses of the constituent isotopes of an element; the relative masses of atoms of different elements.

Aufbau (or *building up*) principle – describes the order in which electrons fill orbitals in atoms.

Autoionization – an ionization reaction between identical molecules.

Avogadro's Law – at the same temperature and pressure, equal volumes of gases contain the same number of molecules.

Avogadro's number (*N*$_A$) – the number (6.022×10^{23}) of atoms, molecules, or particles found in precisely 1 mole of a substance.

B

Background radiation – extraneous to an experiment; usually the low-level natural radiation from cosmic rays and trace radioactive substances present in the environment.

Band – a series of very closely spaced nearly continuous molecular orbitals that belong to the crystal as a whole.

Band of stability – band containing nonradioactive nuclides in a plot of neutrons *vs.* their atomic number.

Band theory of metals – the theory that accounts for the bonding and properties of metallic solids.

Barometer – a device used to measure the pressure in the atmosphere.

Base – a substance that produces ^-OH (*aq*) ions in an aqueous solution; accepts a proton and has a high pH; strongly soluble bases are soluble in water and are entirely dissociated; weak bases ionize only slightly; a typical example of a base is sodium hydroxide (NaOH). It turns litmus blue.

Basic anhydride – the oxide of a metal that reacts with water to form a base.

Basic salt – a salt containing an ionizable OH group.

Beta (β) particle – an electron emitted from the nucleus when a neutron decays to a proton and an electron.

Binary acid – a binary compound in which H is bonded to one or more electronegative nonmetals.

Binary compound – a compound consisting of two elements; it may be ionic or covalent.

Binding energy (nuclear binding energy) – the energy equivalent ($E = mc^2$) of the mass deficiency of an atom (where E is the energy in joules, m is the mass in kilograms, and c is the speed of light in m/s^2).

Boiling – the phase transition of liquid vaporizing.

Boiling point – the temperature at which the vapor pressure of a liquid is equal to the applied pressure; the *condensation point*.

Boiling point elevation – the increase in the boiling point of a solvent caused by the dissolution of a nonvolatile solute.

Bomb calorimeter – a device used to measure the heat transfer between a system and its surroundings at constant volume.

Bond – the attraction and repulsion between atoms and molecules is a cornerstone of chemistry.

Bond energy – the amount of energy necessary to break one mole of bonds in a substance, dissociating the substance in its gaseous state into atoms of its elements in the gaseous state.

Bond order – half the number of electrons in bonding orbitals minus half the electrons in antibonding orbitals.

Bonding orbital – a molecular orbit lower in energy than any of the atomic orbitals from which it is derived; lends stability to a molecule or ion when populated with electrons.

Bonding pair – pair of electrons involved in a covalent bond.

Boron hydrides – binary compounds of boron and hydrogen.

Born-Haber cycle – a series of reactions (and the accompanying enthalpy changes) which, when summed, represents the hypothetical one-step reaction by which elements in their standard states are converted into crystals of ionic compounds (and the accompanying enthalpy changes).

Boyle's Law – at a constant temperature, the volume occupied by a definite mass of a gas is inversely proportional to the applied pressure.

Breeder reactor – a nuclear reactor that produces more fissionable nuclear fuel than it consumes.

Brønsted-Lowrey acid – a chemical species that donates a proton.

Brønsted-Lowrey base – a chemical species that accepts a proton.

Buffer solution – resists change in pH; contains either a weak acid and a soluble ionic salt of the acid or a weak base and a soluble ionic salt of the base.

Buret – a piece of volumetric glassware, usually graduated in 0.1 mL intervals, used to deliver solutions for titrations in a quantitative (drop-like) manner; also spelled *burette*.

C

Calorie – the amount of heat required to raise the temperature of one gram of water from 14.5 °C to 15.5 °C; 1 calorie = 4.184 joules.

Calorimeter – a device used to measure the heat transfer between a system and its surroundings.

Canal ray – a stream of positively charged particles (cations) that moves toward the negative electrode in cathode ray tubes; observed to pass through canals in the negative electrode.

Capillary – a tube having a very small inside diameter.

Capillary action – the drawing of a liquid up the inside of a small-bore tube when adhesive forces exceed cohesive forces; the depression of the surface of the liquid when cohesive forces exceed the adhesive forces.

Catalyst – a chemical compound used to change the rate (to speed or slow it) of a regenerated reaction (i.e., not consumed) at the end of the reaction.

Catenation – the bonding of atoms of the same element into chains or rings (i.e., the ability of an element to bond with itself).

Cathode – the electrode at which reduction occurs; in a cathode ray tube, the negative electrode.

Cathodic protection – protection of a metal (making a cathode) against corrosion by attaching it to a sacrificial anode of more easily oxidized metal.

Cathode ray tube – a closed glass tube containing gas under low pressure, with electrodes near the ends and a luminescent screen near the positive electrode; produces cathode rays when a high voltage is applied.

Cation – a positive ion; an atom or group of atoms that lost one or more electrons.

Cell potential – the potential difference, E_{cell}, between oxidation and reduction half-cells under nonstandard conditions; the force in a galvanic cell pulls electrons through a reducing agent to an oxidizing agent.

Central atom – an atom in a molecule or polyatomic ion bonded to more than one other atom.

Chain reaction – a reaction that, once initiated, sustains itself and expands; a reaction in which reactive species, such as radicals, are produced in more than one step; these reactive species propagate the chain reaction.

Charles' Law – at constant pressure, the volume occupied by a definite mass of gas is directly proportional to its absolute temperature.

Chemical bonds – the attractive forces holding atoms together in elements or compounds.

Chemical change – when one or more new substances are formed.

Chemical equation – description of a chemical reaction by placing the formulas of the reactants on the left of an arrow and the formulas of the products on the right.

Chemical equilibrium – a state of dynamic balance in which the rates of forward and reverse reactions are equal; there is no net change in concentrations of reactants or products while a system is at equilibrium.

Chemical kinetics – the study of rates and mechanisms of chemical reactions and factors they depend on.

Chemical periodicity – the variations in properties of elements with their position in the periodic table.

Chemical reaction – the change of one or more substances into another or multiple substances.

Cloud chamber – a device for observing the paths of speeding particles as vapor molecules condense on them to form fog-like tracks.

Cobalt chloride paper – water test; water changes the color from blue to pink.

Coefficient of expansion – the ratio of the change in the length or the volume of a body to the original length or volume for a unit change in temperature.

Cohesive forces – the forces of attraction among particles of a liquid.

Colligative properties – physical properties of solutions that depend upon the number but not the kind of solute particles present.

Collision theory – theory of reaction rates that states that effective collisions between reactant molecules must occur for the reaction to occur.

Colloid – a heterogeneous mixture in which solute-like particles do not settle out (e.g., many kinds of milk).

Combination reaction – two substances (elements or compounds) combine to form one compound.

Combustible – classification of liquid substances that burn based on flashpoints; any liquid having a flashpoint at or above 37.8 °C (100 °F) but below 93.3 °C (200 °F), except any mixture having components with flashpoints of 93.3 °C (200 °F) or higher, the total of which makes up 99% or more of the volume of the mixture.

Combustion (or *burning*) – an exothermic reaction between an oxidant and fuel with heat and often light.

Common ion effect – suppression of ionization of a weak electrolyte by the presence in the same solution of a strong electrolyte containing one of the same ions as the weak electrolyte.

Complex ions – ions resulting from coordinating covalent bonds between simple ions and other ions or molecules.

Composition stoichiometry – describes the quantitative (mass) relationships among elements in compounds.

Compound – a substance of two or more chemically bonded elements in fixed proportions; can be decomposed into constituent elements.

Compressed gas – a single or mixture of gases having (in a container) an absolute pressure exceeding 40 psi at 21.1 °C (70 °F).

Compression – an area in a longitudinal wave where the particles are closer and pushed in.

Concentration – the amount of solute per unit volume, the mass of solvent or solution.

Condensation – the phase change from gas to liquid.

Condensed phases – the liquid and solid phases; phases in which particles interact strongly.

Condensed states – the solid and liquid states.

Conduction – heat transfer between substances in direct contact with each other (i.e., must be touching); when particles of a hotter substance vibrate, these molecules bump into nearby particles and transfer some energy.

Conduction band – a vacant or partially filled band of energy levels just higher in energy than a filled band; a band within which, or into which, electrons must be promoted to allow electrical conduction to occur in a solid.

Conductor – material that allows electric flow more freely.

Conjugate acid-base pair – in Brønsted-Lowry terms, a reactant and a product that differ by a proton, H⁺.

Conformations – structures of a compound that differ by the extent of their rotation about a single bond.

Continuous spectrum – contains wavelengths in a specified region of the electromagnetic spectrum.

Control rods – rods of materials such as cadmium or boron steel that act as neutron absorbers (not merely moderators), used in nuclear reactors to control neutron fluxes and therefore fission rates.

Conjugated double bonds – double bonds separated from each other by one single bond –C=C–C=C–

Contact process – the industrial process for sulfur trioxide and sulfuric acid production from sulfur dioxide.

Convection – the physical flow of matter when heat flows by energized molecules from one place to another through the movement of fluids. The transfer of heat through a liquid or a gas when molecules of the liquid or gas move and carry the heat.

Coordinate covalent bond – a covalent bond with shared electrons furnished by the same species; a bond between a Lewis acid and a Lewis base.

Coordination compound or complex – a compound containing coordinate covalent bonds.

Coordination number – the number of donor atoms coordinated to metal; the number of nearest neighbors of an atom or ion in describing crystals.

Coordination sphere – the metal ion and its coordinating ligands but no uncoordinated counter-ions.

Corrosion – oxidation of metals (e.g., rusting) in the presence of air and moisture.

Coulomb – the SI unit of electrical charge; unit symbol – C.

Covalent bond – a force of attraction (chemical bond) formed by the sharing of electron pairs between two atoms.

Covalent compounds – compounds made of two or more nonmetal atoms bonded by sharing valence electrons.

Critical mass – the minimum mass of a particular fissionable nuclide in a given volume required to sustain a nuclear chain reaction.

Critical point – the combination of critical temperature and critical pressure of a substance.

Critical pressure – the pressure required to liquefy a gas (vapor) at its *Critical temperature*.

Critical temperature – the temperature above which a gas cannot be liquefied; the temperature above which a substance cannot exhibit distinct gas and liquid phases.

Crystal – a solid that is packed with ions, molecules or atoms in an orderly fashion.

Crystal field stabilization energy – a measure of the net energy of stabilization gained by a metal ion's nonbonding d electrons due to complex formation.

Crystal field theory – theory of bonding in transition metal complexes in which ligands and metal ions are treated as point charges; a purely ionic model; ligand point charges represent the crystal (electrical) field perturbing the metal's d orbitals containing nonbonding electrons.

Crystal lattice – a pattern of arrangement of particles in a crystal.

Crystal lattice energy – the amount of energy that holds a crystal together; the energy change when a mole of solid forms fom its constituent molecules or ions (for ionic compounds) in their gaseous state (always negative).

Crystalline solid – a solid characterized by a regular, ordered arrangement of particles.

Curie (Ci) – the basic unit to describe the intensity of radioactivity in a sample of material; one curie equals 37 billion disintegrations per second or approximately the amount of radioactivity given off by 1 gram of radium.

Current – a flow of charged particles, such as electrons or ions, moving through an electrical conductor or space. It is measured as the net rate of flow of electric charge; the unit is Ampere (A).

Cuvette – glassware used in spectroscopic experiments; usually made of plastic, glass, or quartz and should be as clean and transparent as possible.

Cyclotron – a device for accelerating charged particles along a spiral path.

D

Dalton's Law (or the *law of partial pressures*) – the pressure exerted by a mixture of gases is the sum of the partial pressures of the individual gases.

Daughter nuclide – nuclide produced in nuclear decay.

Debye – the unit used to express dipole moments.

Degenerate – in orbitals, describes orbitals of the same energy.

Deionization – the removal of ions; in the case of water, mineral ions such as sodium, iron and calcium.

Deliquescence – substances that absorb water from the atmosphere to form liquid solutions.

Delocalization – in reference to electrons, bonding electrons distributed among more than two atoms bonded; occurs in species that exhibit resonance.

Density – mass per unit volume; $D = M \times V$.

Deposition – settling particles within a solution; the direct solidification of vapor by cooling; see *sublimation*.

Derivative – a compound that can be imagined arising from a parent compound by replacing one atom with another atom or group of atoms; used extensively in organic chemistry to identify compounds.

Detergent – a soap-like emulsifier with a sulfate, SO_3, or a phosphate group instead of a carboxylate group.

Deuterium – an isotope of hydrogen whose atoms are twice as massive as ordinary hydrogen; deuterium atoms contain a proton and a neutron in the nucleus.

Dextrorotatory – refers to an optically active substance that rotates plane-polarized light clockwise, also known as *dextro*.

Diagonal similarities – chemical similarities in the Periodic Table of Elements of Period 2 to elements of Period 3 one group to the right, especially evident toward the left of the periodic table.

Diamagnetism – weak repulsion by a magnetic field.

Differential Scanning Calorimetry (DSC) – a technique for measuring temperature, direction, and magnitude of thermal transitions in a sample material by heating/cooling and comparing the amount of energy required to maintain its rate of temperature increase or decrease with an inert reference material under similar conditions.

Differential Thermal Analysis (DTA) – a technique for observing the temperature, direction and magnitude of thermally induced transitions in a material by heating/cooling a sample and comparing its temperature with an inert reference material under similar conditions.

Differential thermometer – a thermometer used to measure very small temperature changes accurately.

Dilution – the process of reducing the concentration of a solute in a solution, usually by mixing with more solvent.

Dimer – molecule formed by combining two smaller (identical) molecules.

Dipole – electric or magnetic separation of charge; charge separation between two covalently bonded atoms.

Dipole-dipole interactions – attractive electrostatic forces between polar molecules (i.e., between molecules with permanent dipoles).

Dipole moment – the product of the distance separating opposite charges of an equal magnitude of charge; a measure of the polarity of a bond or molecule; a measured dipole refers to the dipole moment of an entire molecule.

Dispersing medium – the solvent-like phase in a colloid.

Dispersed phase – the solute-like species in a colloid.

Displacement reactions – reactions in which one element displaces another from a compound.

Disproportionation reactions – redox reactions in which the oxidizing agent and the reducing agent are the same species.

Dissociation – in an aqueous solution, the process by which a solid ionic compound separates into its ions.

Dissociation constant – equilibrium constant for dissociating a complex ion into a simple ion and coordinating species (ligands).

Dissolution or solvation – the spread of ions in a monosaccharide.

Distilland – the material in a distillation apparatus that is to be distilled.

Distillate – the material in a distillation apparatus collected in the receiver.

Distillation – separating a liquid mixture into its components based on differences in boiling points; the process in which components of a mixture are separated by boiling away the more volatile liquid; the vaporization of a liquid by heating and then the condensation of the vapor by cooling.

Domain – a cluster of atoms in a ferromagnetic substance, which align in the same direction in the presence of an external magnetic field.

Donor atom – a ligand atom whose electrons are shared with a Lewis acid.

d-**orbitals** – beginning in the third energy level, a set of five degenerate orbitals per energy level, higher in energy than s and p orbitals of the same energy level.

Dosimeter – a small, calibrated electroscope worn by laboratory personnel to measure incident ionizing radiation or chemical exposure.

Double bond – covalent bond resulting from the sharing of four electrons (two pairs) between two atoms.

Double salt – solid consisting of two co-crystallized salts.

Doublet – two peaks or bands of about equal intensity appearing close on a spectrogram.

Downs cell – electrolytic cell for the commercial electrolysis of molten sodium chloride.

DP number – the degree of polymerization; the average number of monomer units per polymer unit.

Dry cells (voltaic cells) – ordinary batteries for appliances (e.g., flashlights, radios).

Dumas method – a method used to determine the molecular weights of volatile liquids.

Dynamic equilibrium – an equilibrium in which the processes occur continuously with no net change.

E

Earth metal – highly reactive elements in group IIA of the periodic table (includes beryllium, magnesium, calcium, strontium, barium, and radium); see *alkaline earth metal*.

Effective collisions – a collision between molecules resulting in a reaction; one in which the molecules collide with proper relative orientations and sufficient energy to react.

Effective molality – the sum of the molalities of solute particles in a solution.

Effective nuclear charge – the nuclear charge experienced by the outermost electrons of an atom; the actual nuclear charge minus the effects of shielding due to inner-shell electrons (e.g., a set of $d_{x^2-y^2}$ and d_{z^2} orbitals); those d orbitals within a set with lobes directed along the x, y and z-axes.

Electrical conductivity – the measure of how easily an electric current can flow through a substance.

Electric charge – a measured property (coulombs) that determines electromagnetic interaction.

Electrochemical cell – using a chemical reaction's current; electromotive force is made.

Electrochemistry – the study of chemical changes produced by electrical current and electricity production by chemical reactions.

Electrodes – surfaces upon which oxidation and reduction half-reactions occur in electrochemical cells; a conductor dips into an electrolyte and allows the electrons to flow to and from the electrolyte.

Electrode potentials – potentials, E, of half-reactions as reductions versus the standard hydrogen electrode.

Electrolysis – 1) occurs in electrolytic cells; chemical decomposition occurs by passing an electric current through a solution containing ions. 2) producing a chemical change using electricity; used to split up water into H and O_2.

Electrolyte – a substance (i.e., anions, cations) which, when dissolved in water, conducts electricity. An ionic solution that conducts a certain amount of current and split categorically as weak and strong.

Electrolytic cells – electrochemical cells in which electrical energy causes nonspontaneous redox reactions to occur (i.e., forced to occur by applying an outside source of electrical energy).

Electrolytic conduction – electrical current passes by ions through a solution or pure liquid.

Electromagnetic radiation – energy propagated using electric and magnetic fields that oscillate in directions perpendicular to the direction of travel of the energy; a type of wave that can go through vacuums as well as material; classified as a "self-propagating wave."

Electromagnetism – fields of an electric charge and electric properties that change how particles move and interact.

Electromotive force – a device that gains energy as electric charges pass through it.

Electromotive series – the relative order of tendencies for elements and their simple ions to act as oxidizing or reducing agents; also known as the "activity series."

Electron – a subatomic particle having a mass of 0.00054858 amu and a charge of 1–.

Electron affinity – the amount of energy absorbed in the process in which an electron is added to a neutral isolated gaseous atom to form a gaseous ion with a 1– charge; it has a negative value if energy is released.

Electron configuration – the specific distribution of electrons in atomic orbitals of atoms or ions.

Electron-deficient compounds –contain at least one atom (other than H) that shares fewer than eight electrons.

Electron shells – an orbital around the atom's nucleus with a fixed number of electrons (usually two or eight).

Electronic transition – the transfer of an electron from one energy level to another.

Electronegativity – a measure of the relative tendency of an atom to attract electrons to itself when chemically combined with another atom.

Electronic geometry – the geometric arrangement of orbitals containing the shared and unshared electron pairs surrounding the central atom or polyatomic ion.

Electrophile – positively charged or electron-deficient.

Electrophoresis – a technique for separating ions by migration rate and direction of migration in an electric field.

Electroplating – a metal is covered with another metal layer using electricity; plating a metal onto a (cathodic) surface by electrolysis.

Element – a substance that cannot be decomposed into simpler substances by chemical means; defined by its *atomic number*. A substance that cannot be split into simpler substances by chemical means.

Eluant (or eluent) – the solvent used in the process of elution, as in liquid chromatography.

Eluate – a solvent (or mobile phase) which passes through a chromatographic column and removes the sample components from the stationary phase.

Emission spectrum – the emission of electromagnetic radiation by atoms (or other species) resulting from electronic transitions from higher to lower energy states.

Empirical formula – gives the simplest whole-number ratio of atoms of each element present in a compound; also known as the simplest formula.

Emulsifying agent – a substance that coats the particles of the dispersed phase and prevents coagulation of colloidal particles; an emulsifier.

Emulsion – colloidal suspension of a liquid in a liquid.

Endothermic – describes processes that absorb heat energy (*H*).

Endothermicity – the absorption of heat by a system as the process occurs.

Endpoint – the point at which an indicator changes color and a titration is stopped.

Energy – a system's ability to do work.

Enthalpy (*H*) – the heat content of a specific amount of substance; *E*= PV.

Entropy *(S)* – a thermodynamic state or property that measures the degree of disorder (i.e., randomness) of a system; the amount of energy not available for work in a closed thermodynamic system (usually denoted by S).

Enzyme – a protein that acts as a catalyst in biological systems.

Equation of state – an expression describes the behavior of matter in a given state; the van der Waals equation describes the behavior of the gaseous state.

Equilibrium or chemical equilibrium – a state of dynamic balance with the rates of forward and reverse reactions equal; the state of a system when neither forward nor reverse reaction is thermodynamically favored.

Equilibrium constant – a quantity that characterizes the equilibrium position for a reversible reaction; its magnitude is equal to the mass action expression at equilibrium; equilibrium, "K," varies with temperature.

Equivalence point – the point when chemically equivalent amounts of reactants have reacted.

Equivalent weight – an oxidizing or reducing agent whose mass gains (oxidizing agents) or loses (reducing agents) 6.022×10^{23} electrons in a redox reaction.

Evaporation – vaporization of a liquid below its boiling point.

Evaporation rate – the rate at which a particular substance will vaporize (evaporate) compared to the rate of a known substance such as ethyl ether, especially useful for health and fire-hazard considerations.

Excited state – any state other than the ground state of an atom or molecule; see *ground state*.

Exothermic – describes processes that release heat energy (*H*).

Exothermicity – the release of heat by a system as a process occurs.

Explosive – a chemical or compound that causes a sudden, almost instantaneous release of pressure, gas, heat, and light when subjected to sudden shock, pressure, high temperature, or applied potential.

Explosive limits – the range of concentrations over which a flammable vapor mixed with the proper ratios of air will ignite or explode if a source of ignition is provided.

Extensive property – a property that depends upon the amount of material in a sample.

Extrapolate – to estimate the value of a result outside the range of a series of known values; a technique used in standard additions calibration procedure.

F

Faraday constant – a unit of electrical charge widely used in electrochemistry and equal to ~ 96,500 coulombs; represents 1 mole of electrons, or the Avogadro number of electrons: 6.022×10^{23} electrons.

Faraday's law of electrolysis – a two-part law that Michael Faraday published about electrolysis. 1. the mass of a substance altered at an electrode during electrolysis is directly proportional to the quantity of electricity transferred at that electrode. 2. the mass of an elemental material altered at an electrode is directly proportional to the element's equivalent weight; one equivalent weight of a substance is produced at each electrode during the passage of 96,487 coulombs of charge through an electrolytic cell.

Fast neutron – a neutron ejected at high kinetic energy in a nuclear reaction.

Ferromagnetism – the ability of a substance to become permanently magnetized by exposure to an external magnetic field.

Flashpoint – the temperature at which a liquid will yield enough flammable vapor to ignite; there are various recognized industrial testing methods; therefore, the method used must be stated.

Fluorescence – absorption of high energy radiation by a substance and subsequent emission of visible light.

First Law of Thermodynamics – the amount of energy in the universe is constant (i.e., energy is neither created nor destroyed in ordinary chemical reactions and physical changes); known as the Law of Conservation of Energy.

Fluids – substances that flow freely; gases and liquids.

Flux – a substance added to react with the charge, or a product of its reduction; in metallurgy, it is usually added to lower a melting point.

Foam – colloidal suspension of a gas in a liquid.

Formal charge – a method of counting electrons in a covalently bonded molecule or ion; it counts bonding electrons as though they were equally shared between the two atoms.

Formula – a combination of symbols that indicates the chemical composition of a substance.

Formula unit – the smallest repeating unit of a substance; the molecule for nonionic substances.

Formula weight – the mass of one formula unit of a substance in atomic mass units.

Fossil fuels – formed from the remains of plants and animals that lived millions of years ago.

Fractional distillation – when a fractioning column is used in a distillation apparatus to separate the components of a liquid mixture with different boiling points.

Fractional precipitation – removal of some ions from a solution by precipitation while leaving other ions with similar properties in the solution.

Free energy change – the indicator of the spontaneity of a process at constant T and P (e.g., if ΔG is negative, the process is spontaneous).

Free radical – a highly reactive chemical species carrying no charge and having a single unpaired electron in an orbital.

Freezing – phase transition from liquid to solid.

Freezing point depression – the decrease in the freezing point of a solvent caused by the presence of a solute.

Frequency – the number of repeating points on a wave that passes a given observation point per unit time; the unit is 1 hertz = 1 cycle per 1 second.

Fuel – any substance that burns in oxygen to produce heat.

Fuel cells – a voltaic cell that converts the chemical energy of a fuel and an oxidizing agent directly into electrical energy continuously.

G

Gamma (γ) ray – a highly penetrating type of nuclear radiation similar to x-ray radiation, except that it comes from within the nucleus of an atom and has higher energy; energy-wise, very similar to cosmic rays except that cosmic rays originate from outer space.

Galvanic cell – battery made up of electrochemical with two different metals connected by a salt bridge.

Galvanizing – placing a thin layer of zinc on a ferrous material to protect the underlying surface from corrosion.

Gangue – sand, rock and other impurities surrounding the mineral of interest in an ore.

Gas – a state of matter in which the particles have no definite shape or volume, though they fill their container.

Gay-Lussac's Law – the expression used for each of the two relationships named after the French chemist Joseph Louis Gay-Lussac concerning the properties of gases; more usually applied to his law of combining volumes.

Geiger counter – a gas-filled tube that discharges electrically when ionizing radiation passes through it.

Gel – colloidal suspension of a solid dispersed in a liquid; a semi-rigid solid.

Gibbs (free) energy – the thermodynamic state function of a system that indicates the amount of energy available for the system to do useful work at constant T and P; a value that indicates the spontaneity of a reaction (usually denoted by G).

Graham's Law – the rates of effusion of gases are inversely proportional to the square roots of their molecular weights or densities.

Ground state – the lowest energy state or most stable state of an atom, molecule, or ion; see *excited state*.

Group – a vertical column in the periodic table; known as a family.

H

Haber process – a process for the catalyzed industrial production of ammonia from N_2 and H_2 at high temperature and pressure.

Half-cell – the compartment in which the oxidation or reduction half-reaction occurs in a voltaic cell.

Half-life – the time required for half of a reactant to be converted into product(s); the time required for half of a given sample to undergo radioactive decay.

Half-reaction – the oxidation or the reduction part of a redox reaction.

Halogens – group VIIA elements: F, Cl, Br, I; halogens are nonmetals.

Hard water – water high in dissolved minerals that is it difficult to form lather with soap.

Heat – a form of energy that flows between two samples of matter because of their temperature differences.

Heat capacity – the amount of heat required to raise the temperature of a mass one degree Celsius.

Heat of condensation – the amount of heat that must be removed from one gram of vapor at its condensation point to condense the vapor with no change in temperature.

Heat of crystallization – the amount of heat that must be removed from one gram of a liquid at its freezing point to freeze it with no change in temperature.

Heat of fusion – the amount of heat required to melt one gram of a solid at its melting point with no change in temperature; usually expressed in J/g; the molar heat of fusion is the amount of heat required to melt one mole of a solid at its melting point with no change in temperature and is usually expressed in kJ/mol.

Heat of solution – the amount of heat absorbed in forming a solution that contains one mole of the solute; the value is positive if heat is absorbed (endothermic) and negative if heat is released (exothermic).

Heat of vaporization – the amount of heat required to vaporize one gram of a liquid at its boiling point with no change in temperature; usually expressed in J/g; the molar heat of vaporization is the amount of heat required to vaporize one mole of liquid at its boiling point with no change in temperature and is usually expressed as ion kJ/mol.

Heisenberg uncertainty principle – states that it is impossible to accurately determine the *momentum* and *position* of an electron simultaneously.

Henry's Law – the gas pressure above a solution is proportional to the concentration of the gas in the solution.

Hess' Law of heat summation – the enthalpy change for a reaction is the same whether it occurs in one step or a series of steps.

Heterogeneous catalyst – exist in a different phase (solid, liquid, or gas) from the reactants; a contact catalyst.

Heterogeneous equilibria – equilibria involving species in more than one phase.

Heterogeneous mixture – a mixture that does not have uniform composition and properties throughout.

Heteronuclear – consisting of different elements.

High spin complex – crystal field designation for an outer orbital complex; t_{2g} and e_g orbitals are singly occupied before pairing occurs.

Homogeneous catalyst – in the same phase (solid, liquid, or gas) as the reactants.

Homogeneous equilibria – when *reagents* and *products* are of the same phase (i.e., gases, liquids, or solids).

Homogeneous mixture – a mixture which has uniform composition and properties throughout.

Homologous series – compounds with each member differing from the next by a specific number and kind of atoms.

Homonuclear – consisting of only one element.

Hund's rule – orbitals of a given sublevel must be occupied by single electrons before pairing begins; see *Aufbau (or building up) principle*.

Hybridization – mixing atomic orbitals to form a new set of atomic orbitals with the same electron capacity and properties and energies intermediate between the original unhybridized orbitals.

Hydrate – a solid compound that contains a definite percentage of bound water.

Hydrate isomers – crystalline complexes that differ in whether water exists inside or outside the coordination sphere.

Hydration – the reaction of a substance with water.

Hydration energy – the energy change accompanying the hydration of a mole of gas and ions.

Hydride – a binary compound of hydrogen.

Hydrocarbons – compounds that contain only carbon and hydrogen.

Hydrogen bond – a relatively strong dipole-dipole interaction (but still considerably weaker than the covalent or ionic bonds) between molecules containing hydrogen directly bonded to a small, highly electronegative atom, such as N, O or F.

Hydrogenation – the reaction in which hydrogen adds across a double or triple bond.

Hydrogen-oxygen fuel cell – hydrogen is the fuel (reducing agent) and oxygen is the oxidizing agent.

Hydrolysis – the reaction of a substance with water or its ions.

Hydrolysis constant – an equilibrium constant for a hydrolysis reaction.

Hydrometer – a device used to measure the densities of liquids and solutions.

Hydrophilic colloids – colloidal particles that repel water molecules.

I

Ideal gas – a hypothetical gas that obeys the postulates of the kinetic-molecular theory.

Ideal gas law – the product of pressure and the volume of an ideal gas is directly proportional to the number of moles of the gas and the absolute temperature. $PV = nRT$

Ideal solution – obeys Raoult's Law strictly.

Immiscible liquids – do not mix to form a solution (e.g., oil and water).

Indicators – for acid-base titrations, organic compounds that exhibit different colors in solutions of different acidities, determine the point at which the reaction between two solutes is complete.

Inert pair effect – characteristic of the post-transition minerals; the tendency of the electrons in the outermost atomic *s* orbital to remain un-ionized or unshared in compounds of post-transition metals.

Inhibitory catalyst – an inhibitor; a catalyst that decreases the rate of reaction.

Inner orbital complex – valence bond designation for a complex in which the metal ion utilizes d orbitals for one shell inside the outermost occupied shell in its hybridization.

Inorganic chemistry – a part of chemistry concerned with inorganic (non carbon-based) compounds.

Insulator – a material that resists the flow of electric current or heat transfer; does not allow heat to flow easily.

Insoluble compound – a substance that will not dissolve in a solvent, even after mixing.

Integrated rate equation – an expression giving the concentration of a reactant remaining after a specified time; has a different mathematical form for different orders of reactants.

Intermolecular forces – forces between individual particles (atoms, molecules, ions) of a substance.

Ion – a molecule that has gained or lost electrons; an atom or a group of atoms carries an electric charge (Na^+).

Ion product for water – equilibrium constant for water ionization; $Kw = [H_3O^+]\cdot[^-OH] = 1.00 \times 10^{-14}$ at 25 °C.

Ionic bond – electrostatic attraction between oppositely charged ions, resulting from a transfer of electrons.

Ionic bonding – chemical bonding resulting from transferring electrons from one atom or group.

Ionic compounds – compounds containing predominantly ionic bonding.

Ionic geometry – arrangement of atoms (not lone pairs of electrons) about the central atom of a polyatomic ion.

Ionization – the breaking up of a compound into separate ions; in an aqueous solution, the process by which a molecular compound reacts with water and forms ions.

Ionization constant – equilibrium constant for the ionization of a weak electrolyte.

Ionization energy – the minimum amount of energy required to remove the most loosely held electron of an isolated gaseous atom or ion.

Ion exchange – a method of removing hardness from water, and it replaces the positive ions that cause the hardness with H^+ ions.

Ionization isomers – result from the interchange of ions inside and outside the coordination sphere.

Isoelectric – having the same electronic configurations.

Isomers – different substances with the same formula.

Isomorphous – refers to crystals having the same atomic arrangement.

Isotopes – two or more forms of atoms of the same element with different masses; atoms containing the same number of protons but different numbers of neutrons.

IUPAC – acronym for "International Union of Pure and Applied Chemistry."

J

Joule (J) – a unit of energy in the SI system; one joule is $1 \text{ kg·m}^2/\text{s}^2$, which is 0.2390 calories.

K

K capture – absorption of a K shell (n = 1) electron by a proton as it is converted to a neutron.

Kelvin – a unit of measure for temperature based upon an absolute scale.

Kinetics – a sub-field of chemistry specializing in reaction rates.

Kinetic energy (*KE*) – energy that matter processes by its motion.

Kinetic-molecular theory – a theory that attempts to explain macroscopic observations on gases in microscopic or molecular terms.

L

Lanthanides – elements 57 (lanthanum) through 71 (lutetium); grouped because of their similar behavior in chemical reactions.

Lanthanide contraction – a decrease in the radii of the elements following the lanthanides compared to what would be expected if there were no *f*-transition metals.

Latent heat – the energy absorbed or released when a substance changes state without changing temperature.

Lattice – unique arrangement of atoms or molecules in a crystalline liquid or solid.

Law of combining volumes (Gay-Lussac's Law) – at constant temperature and pressure, the volumes of reacting gases (and any gaseous products) can be expressed as ratios of small whole numbers.

Law of conservation of energy – energy cannot be created or destroyed; it can only change form.

Law of conservation of matter – there is no detectable change in the quantity of matter during an ordinary chemical reaction.

Law of conservation of matter and energy – the amount of matter and energy in the universe is fixed.

Law of definite proportions (law of constant composition) – different samples of a pure compound contain the same elements in the same proportions by mass.

Law of partial pressures (or *Dalton's Law*) – the pressure exerted by a mixture of gases is the sum of the partial pressures of the individual gases.

Laws of thermodynamics – physical laws which define quantities of thermodynamic systems describe how they behave and (by extension) set certain limitations such as perpetual motion.

Lead storage battery – secondary voltaic cell used in most automobiles.

Leclanche cell – a common type of *dry cell*.

Le Chatelier's principle – states that a system at equilibrium, or striving to attain equilibrium, responds in such a way as to counteract any stress placed upon it; if stress (change of conditions) is applied to a system at equilibrium, the system will shift in the direction that reduces stress.

Leveling effect – acids stronger than the acid characteristic of the solvent reacts with the solvent produce that acid; a similar statement applies to bases. The strongest acid (base) that can exist in a given solvent is the acid (base) characteristic of the solvent.

Levorotatory – an optically active substance rotates plane-polarized light counterclockwise, known as a *levo*.

Lewis acid – any species that can accept a share in an electron pair.

Lewis base – any species that can make available a share in an electron pair.

Lewis dot formula (electron dot formula) – representation of a molecule, ion or formula unit by showing atomic symbols and only outer shell electrons.

Ligand – a Lewis base in a coordination compound.

Light – that portion of the electromagnetic spectrum visible to the naked eye; known as "visible light."

Limiting reactant – a substance that stoichiometrically limits the number of product(s) that can be formed.

Linear accelerator – a device used for accelerating charged particles along a straight line path.

Line spectrum – an atomic emission or absorption spectrum.

Linkage isomers – a particular ligand bonds to a metal ion through different donor atoms.

Liquid – a state of matter which takes the shape of its container.

Liquid aerosol – colloidal suspension of a liquid in gas.

London dispersion forces – very weak and very short-range attractive forces between short-lived temporary (induced) dipoles; known as "dispersion forces."

Lone pair – pair of electrons residing on one atom and not shared by other atoms; unshared pair.

Low spin complex – crystal field designation for an inner orbital complex; contains electrons paired t_{2g} orbitals before e_g orbitals are occupied in octahedral complexes.

Lubricant – a substance capable of reducing friction (i.e., force that opposes the direction of motion).

M

Magnetic field – a space around a magnet where magnetism can be detected.

Magnetic quantum number – quantum mechanical solution to a wave equation designating the orbital within a given set (s, p, d, f) in which an electron resides.

Manometer – a two-armed barometer.

Mass (m) – a measure of the amount of matter in an object; mass is usually measured in grams or kilograms.

Mass action expression – for a reversible reaction, aA + bB cC + dD; the product of the concentrations of the products (species on right), each raised to the power corresponding to its coefficient in the balanced chemical equation, divided by the product of the concentrations of reactants (species on left), each raised to the power corresponding to its coefficient in the balanced equation; at equilibrium the mass action expression is equal to K.

Mass deficiency – the amount of matter converted into energy when an atom forms from constituent particles.

Mass number – the sum of the numbers of protons and neutrons in an atom; an integer.

Mass spectrometer – an instrument that measures the charge-to-mass ratio of charged particles.

Matter – anything that has mass and occupies space.

Mechanism – the sequence of steps by which reactants are converted into products.

Melting point – the temperature at which liquid and solid coexist in equilibrium.

Meniscus – the shape assumed by the surface of a liquid in a cylindrical container.

Melting – the phase change from a solid to a liquid.

Metal – a chemical element that is a good conductor of electricity and heat and forms cations and ionic bonds with nonmetals; elements below and to the left of the stepwise division (metalloids) in the upper right corner of the periodic table; about 80% of known elements are metals.

Metallic bonding – bonding within metals due to the electrical attraction of positively charged metal ions for mobile electrons that belong to the crystal.

Metallic conduction – conduction of electrical current through a metal or along a metallic surface.

Metalloid – a substance with the properties of metals and nonmetals (B, Al, Si, Ge, As, Sb, Te, Po and At).

Metathesis reactions – reactions in which two compounds react to form two new compounds, with no changes in oxidation number; reactions in which the ions of two compounds exchange partners.

Method of initial rates – method of determining the rate-law expression by carrying out a reaction with different initial concentrations and analyzing the resultant changes in initial rates.

Methylene blue – a heterocyclic aromatic chemical compound with the molecular formula $C_{16}H_{18}N_3SCl$.

Miscible liquids – mix to form a solution (e.g., alcohol and water).

Miscibility – the ability of one liquid to mix with (dissolve in) another liquid.

Mixture – two or more different substances mingled together but not chemically combined. A sample of matter composed of two or more substances, each of which retains its identity and properties.

Moderator – a substance (e.g., deuterium, oxygen, paraffin) capable of slowing fast neutrons upon collision.

Molality – a concentration expressed as a number of moles of solute per kilogram of solvent.

Molarity – the number of moles of solute per liter of solution.

Molar solubility – the number of moles of a solute that dissolves to produce a liter of a saturated solution.

Mole – a measurement of an amount of substance; a single mole contains approximately 6.022×10^{23} units or entities; abbreviated mol.

Molecule – a chemically bonded number of electrically neutral atoms.

Molecular equation – a chemical reaction in which formulas are written as if substances existed as molecules; only complete formulas are used.

Molecular formula – indicates the actual number of atoms present in a molecule of a molecular substance.

Molecular geometry – the arrangement of atoms (not lone pairs of electrons) around a central atom of a molecule or polyatomic ion.

Molecular orbital (*MO*) – resulting from the overlap and mixing of atomic orbitals on different atoms (i.e., a region where an electron can be found in a molecule, as opposed to an atom); an MO belongs to the molecule.

Molecular orbital theory – a theory of chemical bonding based upon postulated molecular orbitals.

Molecular weight – the mass of one molecule of a nonionic substance in atomic mass units.

Molecule – the smallest particle of a compound capable of stable, independent existence.

Mole fraction – the number of moles of a component in a mixture divided by the number of moles in the mixture.

Monoprotic acid – can form only one hydronium ion per molecule; may be strong or weak.

Mother nuclide – nuclide that undergoes nuclear decay.

N

Native state – refers to the occurrence of an element in an uncombined or free state in nature.

Natural radioactivity – spontaneous decomposition of an atom.

Neat – conditions with a liquid reagent or gas performed with no added solvent or co-solvent.

Nernst equation – corrects standard electrode potentials for nonstandard conditions.

Net ionic equation – results from canceling spectator ions and eliminating brackets from a total ionic equation.

Neutralization – the reaction of an acid with a base to form a salt and water; usually, the reaction of hydrogen ions with hydrogen ions to form water molecules.

Neutrino – particle that can travel at speeds close to the speed of light; created due to radioactive decay.

Neutron – a neutral unit or subatomic particle that has no net charge and a mass of 1.0087 amu.

Nickel-cadmium cell (NiCd battery) – a dry cell where the anode is Cd, the cathode is NiO_2, and the electrolyte is basic.

Nitrogen cycle – the complex series of reactions by which nitrogen is slowly but continually recycled in the atmosphere, lithosphere, and hydrosphere.

Noble gases – elements of the periodic Group 0; He, Ne, Ar, Kr, Xe, Rn; known as "rare gases;" formerly called "inert gases."

Nodal plane – a region in which the probability of finding an electron is zero.

Nonbonding orbital – a molecular orbital derived only from an atomic orbital of one atom; lends neither stability nor instability to a molecule or ion when populated with electrons.

Nonelectrolyte – a substance whose aqueous solutions do not conduct electricity.

Nonmetal – an element that is not metallic.

Nonpolar bond – a covalent bond in which electron density is symmetrically distributed.

Nuclear – of or about the atomic nucleus.

Nuclear binding energy – the energy equivalent of the mass deficiency; the energy released in forming an atom from the subatomic particles.

Nuclear fission – when a heavy nucleus splits into nuclei of intermediate masses and protons are emitted.

Nuclear magnetic resonance spectroscopy – a technique that exploits the magnetic properties of specific nuclei; helpful in identifying unknown compounds.

Nuclear reaction – involves a change in the composition of a nucleus and can emit or absorb a tremendous amount of energy.

Nuclear reactor – a system in which controlled nuclear fission reactions generate heat energy on a large scale which is subsequently converted into electrical energy.

Nucleons – particles comprising the nucleus; protons, and neutrons.

Nucleus – the very small and dense, positively charged center of an atom containing protons and neutrons, as well as other subatomic particles; the net charge is positive.

Nuclides – refers to different atomic forms of elements; in contrast to isotopes, which refer only to different atomic forms of a single element.

Nuclide symbol – designation for an atom A/Z E, in which E is the symbol of an element, Z is its atomic number, and A is its mass number.

Number density – a measure of the concentration of countable objects (e.g., atoms, molecules) in space; the number per volume.

O

Octahedral – molecules and polyatomic ions with one atom in the center and six atoms at the corners of an octahedron.

Octane number – a number that indicates how smoothly a gasoline burns.

Octet rule – during bonding, atoms tend to reach an electron arrangement with eight electrons in the outermost shell. Many representative elements attain at least a share of eight electrons in their valence shells when they form molecular or ionic compounds; there are some limitations.

Open sextet – species with only six electrons in the highest energy level of the central element (many Lewis acids).

Orbital – may refer to an atomic orbital or a molecular orbital.

Organic chemistry – the chemistry of substances that contain carbon-hydrogen bonds.

Organic compound – substances that contain carbon.

Osmosis – when solvent molecules pass through a semi-permeable membrane from a dilute solution into a more concentrated solution.

Osmotic pressure – the hydrostatic pressure produced on the surface of a semi-permeable membrane.

Outer orbital complex – valence bond designation for a complex in which the metal ion utilizes d orbitals in the outermost (occupied) shell in hybridization.

Overlap – the interaction of orbitals on different atoms in the same region of space.

Oxidation – the addition of oxygen or the loss of electrons. An algebraic increase in the oxidation number; may correspond to a loss of electrons.

Oxidation numbers – quantitative values used as mechanical aids in writing formulas and balancing equations; for single-atom ions, they correspond to the charge on the ion; more electronegative atoms are assigned negative oxidation numbers; known as *oxidation states*.

Oxidation-reduction reactions – reactions in which oxidation and reduction occur; known as *redox reactions*.

Oxide – a binary compound of oxygen.

Oxidizing agent – the substance that oxidizes another substance and is reduced.

P

Pairing – a favorable interaction of two electrons with opposite m values in the same orbital.

Pairing energy – the energy required to pair two electrons in the same orbital.

Paramagnetism – attraction toward a magnetic field, stronger than diamagnetism but still weak compared to ferromagnetism.

Partial pressure – the force exerted by one gas in a mixture of gases.

Particulate matter – fine, divided solid particles suspended in polluted air.

Pauli exclusion principle – no electrons in the same atom may have identical sets of four quantum numbers.

Percentage ionization – the percentage of the weak electrolyte that will ionize in a solution of given concentration.

Percent by mass – 100% times the actual yield divided by the theoretical yield.

Percent composition – the mass percent of each element in a compound.

Percent purity – the percent of a specified compound or element in an impure sample.

Period – the elements in a horizontal row of the periodic table.

Periodicity – regular periodic variations of properties of elements with their atomic number (and position in the periodic table).

Periodic Law – the properties of the elements are periodic functions of their atomic numbers.

Periodic table – an arrangement of elements by increasing atomic numbers, emphasizes periodicity.

Peroxide – a compound with oxygen in –1 oxidation state; metal peroxides contain the peroxide ion, O_2^{2-}.

pH – the measure of acidity (or basicity) of a solution; negative logarithm of the concentration (mol/L) of the H_3O^+ [H^+] ion; scale is commonly used over a range 0 to 14.

Phase diagram – shows the equilibrium temperature-pressure relationships for different phases of a substance.

pH scale – a scale from 0 to 14. If the pH of a solution is 7 it is neutral; if the pH of a solution is less than 7 it is acidic; if the pH of a solution is greater than 7 it is basic.

Permanent hardness – hardness (relative to lathering soap) in water that cannot be removed by boiling; caused by calcium sulfate.

Photoelectric effect – emission of an electron from the surface of a metal caused by impinging electromagnetic radiation of specific minimum energy; the current increases with increasing radiation intensity.

Photon – a carrier of electromagnetic radiation of all wavelengths, such as gamma rays and radio waves; known as *quantum of light*.

Physical change – when a substance changes from one physical state to another, but no substances with different composition are formed; physical change may involve a phase change (e.g., melting, freezing) or another physical change such as crushing a crystal or separating one volume of liquid into different containers; does not produce a new substance.

Plasma – a physical state of matter that exists at extremely high temperatures in which molecules are dissociated, and most atoms are ionized.

Polar bond – a covalent bond with an unsymmetrical distribution of electron density.

Polarimeter – a device used to measure optical activity.

Polarization – the buildup of a product of oxidation or reduction of an electrode, preventing further reaction.

Polydentate – refers to ligands with more than one donor atom.

Polyene – a compound that contains more than one double bond per molecule.

Polymerization – the combination of many small molecules to form large molecules.

Polymer – a large molecule consisting of chains or rings of linked monomer units, usually characterized by high melting and boiling points.

Polymorphous – refers to substances that can crystallize in more than one crystalline arrangement.

Polyprotic acid – forms two or more hydronium ions per molecule; often, at least one ionization step is weak.

Positron – a nuclear particle with the mass of an electron but opposite charge (positive).

Potential difference (or *voltage*) – the force which moves the electrons around the circuit; the unit is Volt (V).

Potential energy (*PE*) – energy stored in a body or a system due to its position in a force field or configuration.

Power – the rate at which energy is converted from one form to another; the unit is Watts (W). Power = voltage × current (P = VI).

Precipitate – an insoluble solid formed by mixing in solution the constituent ions of a slightly soluble solution.

Precision – how close the results of multiple experimental trials are; see *accuracy*.

Pressure – force per unit area; unit is Pascal (Pa).

Primary standard – a known high degree of purity substance that undergoes one invariable reaction with the other reactant of interest.

Primary voltaic cells – voltaic cells that cannot be recharged; no further chemical reaction is possible once the reactants are consumed.

Products – chemicals produced (from reactants) in a chemical reaction.

Proton – a subatomic particle having a mass of 1.0073 amu and a charge of +1, found in the nuclei of atoms.

Protonation – the addition of a proton (H^+) to an atom, molecule or ion.

Pseudobinaryionic compounds – contain more than two elements but are named like binary compounds.

Q

Quanta – the minimum amount of energy emitted by radiation.

Quantum mechanics – the study of how atoms, molecules, subatomic particles, etc. behave and are structured; a mathematical method of treating particles based on quantum theory, which assumes that energy (of small particles) is not infinitely divisible.

Quantum numbers – numbers that describe the energies of electrons in atoms; derived from quantum mechanical treatment.

Quarks – elementary particles and a fundamental constituent of matter, combining to form hadrons (i.e., protons and neutrons).

R

Radiation – 1) heat transfer through invisible rays, which travel outwards from the hot object without a medium. 2) high-energy particles or rays emitted during the nuclear decay processes.

Radical – an atom or group of atoms that contains one or more unpaired electrons; usually a very reactive species.

Radioactive dating – method of dating ancient objects by determining the ratio of amounts of mother and daughter nuclides present in an object and relating the ratio to the object's age via half-life calculations.

Radioactive tracer – a small amount of radioisotope replacing a nonradioactive isotope of the element in a compound whose path (e.g., in the body) or whose decomposition products are monitored by detection of radioactivity; known as a "radioactive label."

Radioactivity – the spontaneous disintegration of atomic nuclei.

Raoult's Law – the vapor pressure of a solvent in an ideal solution decreases as its mole fraction decreases.

Rate-determining step – the slowest step in a mechanism; the step that determines the overall reaction rate.

Rate-law expression – equation relating the reaction rate to the concentrations of the reactants and the specific rate of the constant.

Rate of reaction – the change in the concentration of a reactant or product per unit time.

Reactants – substances consumed in a chemical reaction; react together in a chemical reaction.

Reaction quotient – the mass action expression under any set of conditions (not necessarily equilibrium); its magnitude relative to K determines the direction in which the reaction must occur to establish equilibrium.

Reaction ratio – the relative amounts of reactants and products involved in a reaction; may be the ratio of moles, millimoles, or masses.

Reaction stoichiometry – describes the quantitative relationships among substances participating in chemical reactions.

Reactivity series (or activity series) – an empirical, calculated, and structurally analytical progression of a series of metals, arranged by "reactivity" from highest to lowest; used to summarize information about the reactions of metals with acids and water, double displacement reactions and the extraction of metals from ores.

Reagent – a substance (or compound) added to a system to cause a chemical reaction or to visualize if a reaction occurs; the terms reactant and reagent are often used interchangeably; however, a *reactant* is more specifically a substance consumed during a chemical reaction.

Reducing agent – a substance that reduces another substance and is itself oxidized.

Reduction – the removal of oxygen or the gaining of electrons.

Resonance – the concept in which two or more equivalent dot formulas for the same arrangement of atoms (resonance structures) are necessary to describe the bonding in a molecule or ion.

Reverse osmosis – forcing solvent molecules to flow through a semi-permeable membrane from a concentrated solution into a dilute solution by applying greater hydrostatic pressure on the concentrated side than the osmotic pressure opposing it.

Reversible reaction – proceses that do not go to completion and occur in the forward and reverse direction.

S

Saline solution – a general term for NaCl (i.e., sodium chloride) in water.

Salt – when a metal replaces the hydrogen of an acid.

Salts – ionic compounds composed of anions and cations.

Salt bridge – a U-shaped tube containing an electrolyte, connects the two half-cells of a voltaic cell.

Saturated solution –no more solute will dissolve at that temperature.

s-block elements – group 1 and 2 elements (alkali and alkaline metals), including hydrogen and helium.

Schrödinger equation – quantum state equation representing the behavior of an electron around an atom; describes the wave function of a physical system evolving.

Second Law of Thermodynamics – the universe tends toward a state of greater disorder in spontaneous processes.

Secondary standard – a solution that has been titrated against a primary standard; a standard solution.

Secondary voltaic cells – voltaic cells that can be recharged; original reactants can be regenerated by reversing the direction of the current flow.

Semiconductor – a substance that does not conduct electricity at low temperatures but will do so at higher temperatures.

Semi-permeable membrane – a thin partition between two solutions through which specific molecules can pass but others cannot.

Shielding effect – electrons in filled sets of *s*, *p* orbitals between the nucleus and outer shell electrons shield the outer shell electrons somewhat from the effect of protons in the nucleus; known as the "screening effect."

Sigma (σ) bonds – bonds resulting from the head-on overlap of atomic orbitals. The region of electron sharing is along and (cylindrically) symmetrical to the imaginary line connecting the bonded atoms.

Sigma orbital – molecular orbital resulting from the head-on overlap of two atomic orbitals.

Single bond – covalent bond resulting from the sharing of two electrons (one pair) between two atoms.

Sol – a suspension of solid particles in a liquid; artificial examples include sol-gels.

Solid – one of the states of matter, where the molecules are packed close, resistance to movement/deformation and volume change.

Solubility product constant – equilibrium constant that applies to the dissolution of a slightly soluble compound.

Solubility product principle – the solubility product constant expression for a slightly soluble compound is the product of the concentrations of the constituent ions, each raised to the power that corresponds to the number of ions in one formula unit.

Solute – the dispersed (i.e., dissolved) phase of a solution; the solution is mixed into the solvent (e.g., NaCl in saline water).

Solution – a homogeneous mixture of multiple substances; comprised of solutes and solvents; a mixture of a solute (usually a solid) and a solvent (usually a liquid).

Solvation – the process by which solvent molecules surround and interact with solute ions or molecules.

Solvent – the dispersing medium of a solution (e.g., H_2O in saline water).

Solvolysis – the reaction of a substance with the solvent in which it is dissolved.

s-orbital – a spherically symmetrical atomic orbital; one per energy level.

Specific gravity – the ratio of the density of a substance to the density of water.

Specific heat – the amount of heat required to raise the temperature of one gram of substance one degree Celsius.

Specific rate constant – an experimentally determined (proportionality) constant, which is different for different reactions, and which changes only with temperature; *k* in the rate-law expression: Rate = *k* [A] × [B].

Spectator ions – ions in a solution that do not participate in a chemical reaction.

Spectral line – any of several lines corresponding to definite wavelengths of an atomic emission or absorption spectrum; marks the energy difference between two energy levels.

Spectrochemical series – arrangement of ligands in order of increasing ligand field strength.

Spectroscopy – the study of radiation and matter, such as X-ray absorption and emission spectroscopy.

Spectrum – display of component wavelengths (colors) of electromagnetic radiation.

Speed of light – the speed at which radiation travels through a vacuum (299,792,458 m/sec).

Square planar – describes molecules and polyatomic ions with one atom in the center and four atoms at the corners of a square.

Square planar complex – relationship with metal in the center of a square plane, with ligand donor atoms at each of the four corners.

Standard conditions for temperature and pressure (STP) – a standardization used to compare experimental results (25 °C and 100.000 kPa).

Standard electrodes – half-cells in which the oxidized and reduced forms of a species are present at the unit activity (1.0 M solutions of dissolved ions, 1.0 atm partial pressure of gases, pure solids, and liquids).

Standard electrode potential – by convention, the potential (Eo) of a half-reaction as a reduction relative to the standard hydrogen electrode when species are present at unit activity.

Standard entropy – the absolute entropy of a substance in its standard state at 298 K.

Standard molar enthalpy of formation – the amount of heat absorbed in forming one mole of a substance in a specified state from its elements in their standard states.

Standard molar volume – the space occupied by one mole of an ideal gas under standard conditions; 22.4 liters.

Standard reaction – a process where the numbers of moles of reactants in the balanced equation, in their standard states, are entirely converted to the numbers of moles of products in the balanced equation, at their standard state.

State of matter – a homogeneous, macroscopic phase (e.g., gas, plasma, liquid, solid) in increasing concentration.

Stoichiometry – the quantitative relationships among elements and compounds undergoing chemical changes.

Strong electrolyte – a substance that conducts electricity well in a dilute aqueous solution.

Strong field ligand – ligand that exerts a strong crystal or ligand electrical field and generally forms low spin complexes with metal ions when possible.

Structural isomers – compounds that contain the same number and kinds of atoms with different geometry.

Subatomic particles – comprise an atom (e.g., protons, neutrons, electrons).

Sublimation – the direct vaporization of a solid by heating without passing through the liquid state; a phase transition from solid to limewater fuel or gas.

Substance – any matter, specimens with the same chemical composition and physical properties.

Substitution reaction – a reaction in which another atom or group of atoms replaces an atom or a group of atoms.

Supercooled liquids – liquids that, when cooled, apparently solidify but continue to flow very slowly under the influence of gravity.

Supercritical fluid – a substance at a temperature above its critical temperature.

Supersaturated solution – contains a higher than saturation concentration of solute; slight disturbance or seeding causes crystallization of excess solute.

Suspension – a heterogeneous mixture in which solute-like particles settle out of the solvent-like phase sometime after their introduction. A mixture of a liquid and a finely divided insoluble solid.

T

Talc – a mineral representing the Mohs Scale composed of hydrated magnesium silicate with the chemical formula $H_2Mg_3(SiO_3)_4$ or $Mg_3Si_4O_{10}(OH)_2$.

Temperature – a measure of heat intensity (i.e., the hotness or coldness of a sample); a measure of the kinetic energy of an object. Unit is degrees, and scales are Celsius, Fahrenheit, and Kelvin.

Temporary hardness – hardness in water that can be removed by boiling; caused by calcium hydrogen carbonate.

Ternary acid – a ternary compound containing H, O and another element, often a nonmetal.

Ternary compound – a compound consisting of three elements; may be ionic or covalent.

Tetrahedral – a term used to describe molecules and polyatomic ions with one atom in the center and four atoms at the corners of a tetrahedron.

Theoretical yield – the maximum amount of a specified product that could be obtained from specified amounts of reactants, assuming complete consumption of the limiting reactant according to only one reaction and complete recovery of the product; see *actual yield*.

Theory – a model describing the nature of a phenomenon.

Thermal conductivity – a property of a material to conduct heat (often noted as k).

Thermal cracking – decomposition by heating a substance in the presence of a catalyst and the absence of air.

Thermochemistry – the study of absorption/release of heat within a chemical reaction; studies heat energy associated with chemical reactions and physical transformations.

Thermodynamics – studying the effects of changing temperature, volume, or pressure (or work, heat, and energy) on a macroscopic scale.

Thermodynamic stability – when a system is in its lowest energy state with its environment (equilibrium).

Thermometer – a device that measures the average energy of a system.

Thermonuclear energy – energy from nuclear fusion reactions.

Third Law of Thermodynamics – entropy of a pure crystalline substance at absolute zero temperature is zero.

Titration – a procedure in which one solution is added to another solution until the chemical reaction between the two solutes is complete; the concentration of one solution is known, and that of the other is unknown. The process of adding one solution to a measured amount of another to find out exactly how much of each is required to react.

Torr – a unit to measure pressure; 1 Torr is equivalent to 133.322 Pa or 1.3158×10^{-3} atm.

Total ionic equation – the expression for a chemical reaction written to show the predominant form of species in aqueous solution or contact with water.

Transition elements (metals) – B Group elements except IIB in the periodic table; sometimes called transition elements, elements with incomplete *d* sub-shells; the *d*-block elements.

Transition state theory –reactants pass through high-energy transition states before forming products.

Transuranic element – an atomic number greater than 92; none of the transuranic elements are stable.

Triple bond – the sharing of three pairs of electrons within a covalent bond (e.g., N_2).

Triple point – where the temperature and pressure of three phases are the same; water has a unique phase diagram.

Tyndall effect – results from light scattering by colloidal particles (a mixture where one substance is dispersed evenly throughout another) or by suspended particles.

U

Uncertainty – any measurement that involves estimating any amount that cannot be precisely reproducible.

Uncertainty principle – knowing the location of a particle makes the momentum uncertain, while knowing the momentum of a particle makes the location uncertain.

Unit cell – the smallest repeating unit of a lattice.

Unit factor – statements used in converting between units.

Universal (or ideal) gas constant – proportionality constant in the ideal gas law (0.08206 L·atm/(K·mol)).

UN number – a four-digit code used to note hazardous and flammable substances.

Unsaturated hydrocarbons – hydrocarbons that contain double or triple carbon-carbon bonds.

V

Valence bond theory – proposes that covalent bonds are formed when atomic orbitals on different atoms overlap and the electrons are shared.

Valence electrons – outermost electrons of atoms; usually those involved in bonding.

Valence shell electron pair repulsion theory (VSEPR) – assumes electron pairs are arranged around the central element of a molecule or polyatomic ion with maximum separation (and minimum repulsion) among regions of high electron density.

Valency – the number of electrons an atom wants to gain, lose, or share to have a full outer shell.

Van der Waals' equation – a quantitative relationship of a state extending the ideal gas law to real gases by including two empirically determined parameters, which are specific for different gases.

Van der Waals force – one of the forces (attraction/repulsion) between molecules.

Van't Hoff factor – the ratio of moles of particles in solution to moles of solute dissolved.

Vapor – when a substance is below the critical temperature in the gas phase.

Vaporization – the phase change from liquid to gas.

Vapor pressure – the particle pressure of vapor at the surface of its parent liquid.

Viscosity – the resistance of a liquid to flow (e.g., oil has a higher viscosity than water).

Volt – one joule of work per coulomb; the unit of electrical potential transferred.

Voltage – the potential difference between two electrodes; a measure of the chemical potential for a redox reaction.

Voltaic cells – electrochemical cells in which spontaneous chemical reactions produce electricity; known as *galvanic cells*.

Voltmeter – an instrument that measures the cell potential.

Volumetric analysis – measuring the volume of a solution (of known concentration) to determine the substance's concentration within the solution; see *titration*.

W

Water equivalent – the amount of water absorbing the same heat as the calorimeter per degree of temperature increases.

Weak electrolyte – a substance that conducts electricity poorly in a dilute aqueous solution.

Weak field ligand – a ligand exerting a weak crystal or ligand field and generally forms high spin complexes with metals.

X

X-ray – electromagnetic radiation between gamma and UV rays.

X-ray diffraction – a method for establishing structures of crystalline solids using single wavelength X-rays and studying the diffraction pattern.

X-ray photoelectron spectroscopy – a spectroscopic technique used to measure the composition of a material.

Y

Yield – the amount of product produced during a chemical reaction.

Z

Zone melting – remove impurities from an element by melting and slowly traveling it down an ingot (cast).

Zone refining – a method of purifying a metal bar by passing it through an induction heater; this causes impurities to move along a melted portion.

Zwitterion (formerly called a dipolar ion) – a neutral molecule with a positive and negative electrical charge; multiple positive and negative charges can be present, distinct from dipoles at different locations within that molecule; known as *inner salts*.

Customer Satisfaction Guarantee

Your feedback is important because we strive to provide the highest quality prep materials. Email us comments or suggestions.

info@sterling–prep.com

We reply to emails – check your spam folder

Highest quality guarantee

Be the first to report a content error for a $10 reward

or a grammatical mistake to receive a $5 reward.

College Chemistry Review provides comprehensive coverage of inorganic chemistry topics taught in college chemistry. The content covers foundational principles and theories necessary to understand the material and answer test questions.

Visit our Amazon store

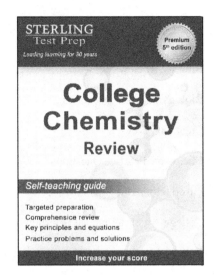

College study aids by Sterling Test Prep

Cell and Molecular Biology Review

Organismal Biology Review

Cell and Molecular Biology Practice Questions

Organismal Biology Practice Questions

Physics Review

Physics Practice Questions

Organic Chemistry Practice Questions

United States History 101

American Government and Politics 101

Environmental Science 101

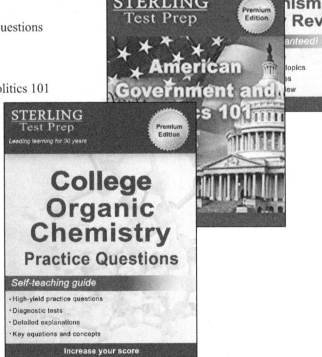

College Level Examination Program (CLEP)

Visit our Amazon store

Made in the USA
Monee, IL
22 December 2023